Coding for MIMO Communication Systems

Coding for MIMO Communication Systems

Tolga M. Duman

Arizona State University, USA

Ali Ghrayeb

Concordia University, Canada

BICENTENNIAL 1807 WILEY 2007 BICENTENNIAL

John Wiley & Sons, Ltd

Telephone (+44) 1243 779777

Email (for orders and customer service enquiries): cs-books@wiley.co.uk
Visit our Home Page on www.wileyeurope.com or www.wiley.com

Other Wiley Editorial Offices

John Wiley & Sons Inc., 111 River Street, Hoboken, NJ 07030, USA

Jossey-Bass, 989 Market Street, San Francisco, CA 94103-1741, USA

Wiley-VCH Verlag GmbH, Boschstr. 12, D-69469 Weinheim, Germany

John Wiley & Sons Australia Ltd, 42 McDougall Street, Milton, Queensland 4064, Australia

John Wiley & Sons (Asia) Pte Ltd, 2 Clementi Loop #02-01, Jin Xing Distripark, Singapore 129809

John Wiley & Sons Canada Ltd, 6045 Freemont Blvd, Mississauga, Ontario, L5R 4J3, Canada

Wiley also publishes its books in a variety of electronic formats. Some content that appears in print may not be available in electronic books.

Anniversary Logo Design: Richard J. Pacifico

Library of Congress Cataloging-in-Publication Data

Duman, Tolga M.
Coding for MIMO communication systems / Tolga M. Duman, Ali Ghrayeb.
p. cm.
ISBN 978-0-470-02809-4 (cloth)
1. Space time codes. 2. MIMO systems. 3. Wireless communication systems. I. Ghrayeb, Ali. II. Title.
TK5103.4877.D86 2007
621.3840285′572 – dc22

2007025115

British Library Cataloguing in Publication Data

A catalogue record for this book is available from the British Library

ISBN 978-0-470-02809-4 (HB)

Typeset in 10/12pt Times by Laserwords Private Limited, Chennai, India

MIX
Paper from responsible sources
FSC® C013604

Contents

About the Authors

Tolga M. Duman

Tolga M. Duman received the B.S. degree from Bilkent University, Ankara, Turkey, in 1993, M.S. and Ph.D. degrees from Northeastern University, Boston, in 1995 and 1998, respectively, all in electrical engineering. Since August 1998, he has been with the Electrical Engineering Department of Arizona State University, first as an Assistant Professor (1998–2004), and currently as an Associate Professor. He spent the 2004–05 academic year as a visiting associate professor at Bilkent University in Turkey. Dr. Duman's current research interests are in digital communications, wireless and mobile communications, MIMO systems, channel coding, underwater acoustic communications, and applications of coding to wireless and recording channels.

Dr. Duman is a recipient of the National Science Foundation CAREER Award and IEEE Third Millennium medal. He is a senior member of IEEE, and an editor for *IEEE Transactions on Wireless Communications* and *IEEE Transactions on Communications*.

Ali Ghrayeb

Ali Ghrayeb received the Ph.D. degree in electrical engineering from the University of Arizona, Tucson, AZ, in May 2000. He is currently an Associate Professor in the Department of Electrical and Computer Engineering, Concordia University, Montreal, Canada. He holds a Concordia Research Chair in High-Speed Wireless Communications. His research interests are in wireless and mobile communications, wireless networks, and coding and signal processing for data transmission and storage. He has co-instructed technical tutorials and short courses on Coding for MIMO Systems and on Synchronization for WCDMA Systems at several major IEEE conferences. He serves as an Associate Editor for *IEEE Transactions on Vehicular Technology* and *Wiley Wireless Communications and Mobile Computing Journal*.

Preface

Employing multiple transmit and receive antennas, namely using multi-input multi-output (MIMO) systems, has proven to be a major breakthrough in providing reliable wireless communication links. Since their invention in the mid-1990s, transmit diversity, achieved through space-time coding, and spatial multiplexing schemes have been the focus of much research in the area of wireless communications. Although many significant advancements have been made recently in MIMO communications, there is still much ongoing research in this area. Parallel to that, communication companies have already started looking into integrating MIMO systems in their current and future wireless communication systems. In fact, several standards for future wireless communication applications have already adopted MIMO systems as an option.

This book is intended to provide a comprehensive coverage of coding techniques for MIMO communication systems. The contents of this book have evolved over the past several years as a result of our own research in MIMO communications, and the tutorials and short courses we have given at several conferences (including IEEE International Conference on Communications (ICC), Global Telecommunications Conference (GLOBECOM), Vehicular Technology Conference (VTC), and Wireless Communications and Networking Conference (WCNC)). The feedback we have received motivated us to write this book in order to address the fundamentals of MIMO communications in an accessible manner.

At this time, several books have been published on MIMO systems. However, there are a number of factors that differentiate this book from the existing ones. First, we try to stay away from including very complicated derivations, mathematical expressions, and very specific systems. Instead, we focus more on the fundamental issues pertaining to MIMO systems. We use language that is easy to comprehend for a wide audience interested in this topic, including starting graduate or senior undergraduate students majoring in electrical engineering with some limited training in digital communications and probability theory. For certain topics, we present more details with some derivations in an effort to accommodate the needs of a more specific group of researchers or advanced graduate students. However, the book is organized in such a way that these subjects are easy to spot, and thus, these should not overwhelm the rest of the audience. Another major factor that differentiates this book from other books is the breadth of coverage of topics. For instance, in addition to our coverage of basic MIMO communication algorithms, such as space-time block codes, space-time trellis codes, unitary and differential signaling and spatial multiplexing schemes, we include a detailed coverage of turbo codes and iterative decoding for MIMO systems, antenna selection algorithms, practical issues such as spatial correlation and channel estimation, as well as MIMO systems for frequency selective fading channels. Finally, we provide numerous examples – some elementary, some more advanced – on various topics covered, and a large number of references on MIMO communications at the end of each chapter.

Audience

The primary audience of this book is senior undergraduate students, graduate students, practitioners and researchers who are interested in learning more about MIMO systems, or perhaps would like to get into this area of research. For the audience to get the full benefits of the book, it is recommended that they have some background in digital communications, linear algebra and probability theory.

Although this book is intended primarily for researchers and practitioners, it can also be adopted as a textbook for a graduate level, or an advanced undergraduate level, course on "Wireless MIMO Communications." The language, organization, and flow of the material should make this easy. The material could be covered in a one-semester course. In order to facilitate its use as a textbook, the book is also complemented with a set of problems at the end of each chapter which serve the purpose of making the main topics covered in each chapter more clear, and shedding some light on certain aspects that are not provided in detail in the text.

Acknowledgments

We thank the National Science Foundation of the United States and the Natural Sciences and Engineering Research Council of Canada for providing us with research funding in the area of MIMO communications over the past several years which enabled our collaboration on the subject, and made this project possible. Furthermore, we have received help from many individuals in completing this work. In particular, we appreciate the help we received from our former and current students in generating many of the figures throughout the book, and numerous suggestions they have provided. Tolga M. Duman wishes to thank Jun Hu, Subhadeep Roy, Mustafa N. Kaynak, Israfil Bahceci, Andrej Stefanov, Zheng Zhang, Vinod Kandasamy, Yunus Emre, Tansal Gucluoglu, and Renato Machado. Ali Ghrayeb would like to thank Xian Nian Zeng, Abdollah Sanei, Chuan Xiu Huang, Hao Shen, May Gomaa, Jeyadeepan Jeganathan and Ghaleb Al Habian. In addition, we would like to express our gratitude to John G. Proakis, Masoud Salehi, William E. Ryan, Cihan Tepedepenlioglu, Junshan Zhang and Walaa Hamouda for their feedback on various drafts of the book.

Finally, Tolga M. Duman would like to thank his wife, Dilek, for her understanding, love and support. Ali Ghrayeb wishes to express his gratitude to his wife, Rola, and his sons Adam and Mohamed for their continuous support, encouragement, patience and love throughout the course of writing this book.

Tolga M. Duman, Arizona State University

Ali Ghrayeb, Concordia University

List of Figures

List of Tables

Notation

$\approx$	approximately equal to
$\triangleq$	defined as equal to
$\gg$	much greater than
$\ll$	much less than
$\cdot$	multiplication operator
$\arg\max_x [f(x)]$	the value of x that maximizes the function $f(x)$
$\arg\min_x [f(x)]$	the value of x that minimizes the function $f(x)$
$\exp(x)$	exponential of x (i.e., e^x)
$\mathrm{Im}\{x\}$	the imaginary part of x
$\mathrm{Re}\{x\}$	the real part of x
$Q(x)$	Gaussian Q-function $\left(\frac{1}{\sqrt{2\pi}}\int_x^{\infty} e^{-t^2/2}dt\right)$
$\mathbb{R}$	the field of all real numbers
$X \sim p_X(x)$	the random variable X has p.d.f. $p_X(x)$
$E[X]$	the expected value of random variable X
$H(X)$	the entropy of random variable X
$H(Y\|X)$	the conditional entropy of random variable Y given random variable X
$I(X;Y)$	the mutual information between random variables X and Y
$\|x\|$	the absolute value of the complex number x
$\angle x$	the angle of the complex number x
x^*	the conjugate of a scalar or vector quantity
$\boldsymbol{x}$	the vector $\boldsymbol{x}$
$\\|\boldsymbol{x}\\|$	the norm of vector $\boldsymbol{x}$
$\boldsymbol{x}^T$	the transpose of vector $\boldsymbol{x}$
$\boldsymbol{x}^H$	the Hermitian (conjugate transpose) of vector $\boldsymbol{x}$
$\boldsymbol{A}$	the matrix $\boldsymbol{A}$
$\boldsymbol{A}^T$	the transpose of matrix $\boldsymbol{A}$
$\boldsymbol{A}^H$	the Hermitian (conjugate transpose) of matrix $\boldsymbol{A}$
$\boldsymbol{A}^*$	the conjugate of matrix $\boldsymbol{A}$
$\boldsymbol{A}^{-1}$	the inverse of matrix $\boldsymbol{A}$
$\\|\boldsymbol{A}\\|$	the Frobenius norm of the matrix $\boldsymbol{A}$ (i.e., sum of absolute value squares of all the entries of $\boldsymbol{A}$)
$\det(\boldsymbol{A})$	the determinant of matrix $\boldsymbol{A}$
$\mathrm{trace}(\boldsymbol{A})$	the trace of matrix $\boldsymbol{A}$

$\boldsymbol{I}_N$	the $N \times N$ identity matrix
$\mathbf{0}_N$	the $N \times N$ all zero matrix
$\mathbf{0}_{M\times N}$	the $M \times N$ all zero matrix
$diag\{a_1, a_2, \ldots, a_N\}$	the diagonal matrix with elements $a_1, a_2, \ldots, a_N$ on the main diagonal
N_t	number of transmit antennas
N_r	number of receive antennas
$h_{i,j}$	channel coefficient between the ith transmit and jth receive antennas
$h^{(l)}(k)$	ISI channel coefficient for the lth tap at time k
$h^{(l)}_{i,j}(k)$	channel coefficient from the ith antenna to the jth antenna at time k for the lth channel tap
$\boldsymbol{H}$	MIMO channel matrix
$\boldsymbol{X}$	transmitted signal
$\boldsymbol{Y}$	received signal
$\boldsymbol{N}$	AWGN noise
ρ	average signal-to-noise ratio at each receive antenna
L	number of intersymbol interference taps
L_r	number of selected antennas at the receiver side
L_t	number of selected antennas at the transmitter side
R_c	code rate
P_b	bit error probability
P_e	probability of error
T	coherence time in number of symbols
N	frame length at each transmit antenna
$\log_x \det[\boldsymbol{A}]$	the log, base x, of the determinant of matrix $\boldsymbol{A}$
$\mathrm{sinc}(x)$	the sinc function $(\sin(\pi x)/\pi x)$
$X \sim \mathcal{CN}(0, 1)$	the random variable X is circularly symmetric complex Gaussian with zero mean and variance $1/2$ in each dimension
W	bandwidth of a signal
$C(f;t)$	time-varying frequency response of a wireless channel
$c(\tau;t)$	impulse response of a wireless channel
T_m	multipath spread
B_D	Doppler spread
B_C	coherence bandwidth
$(\Delta t)_c$	coherence time (in seconds)
$S(\tau;\lambda)$	scattering function

Abbreviations

APP	a posteriori probability
AWGN	additive white Gaussian noise
BP	belief propagation
BICM	bit interleaved coded modulation
BLAST	Bell Laboratories layered space-time
BPSK	binary phase shift keying
BSC	binary symmetric channel
c.d.f.	cumulative distribution function
CSI	channel state information
DBLAST	diagonal Bell Laboratories layered space-time
DFE	decision feedback equalization
DFT	discrete Fourier transform
DPSK	differential phase shift keying
DSTC	differential space-time code
EGC	equal gain combining
EM	expectation maximization
FFT	fast Fourier transform
FS	frequency selective
FSK	frequency shift keying
HBLAST	horizontal Bell Laboratories layered space-time
HDD	hard decision decoding
IFFT	inverse fast Fourier transform
IIR	infinite impulse response
ISI	intersymbol interference
LAPP	log a posteriori probability
LDPC	low density parity check
LLR	log likelihood ratio
LOS	line of sight
LS	least squares
LSTC	layered space-time code
MAP	maximum a posteriori
MAPP	modified a posteriori probability
MIMO	multiple-input multiple-output
MISO	multiple-output single-input
ML	maximum likelihood
MLSD	maximum likelihood sequence detector

MLSTC	multilayered space-time code
MMSE	minimum mean-squared error
MMSE-IC	minimum mean-squared error with interference cancellation
M-PSK	M-ary phase shift keying
MRC	maximum ratio combining
MSOVA	modified soft output Viterbi algorithm
OFDM	orthogonal frequency division multiplexing
OFDMA	orthogonal frequency division multiple access
PAM	pulse amplitude modulation
PCCC	parallel concatenated convolutional code
PEP	pairwise error probability
p.d.f.	probability density function
PSK	phase shift keying
QAM	quadrature amplitude modulation
RF	radio frequency
RSC	recursive systematic convolutional
SC	selection combining
SCBLAST	single code Bell Laboratories layered space-time
SCCC	serial concatenated convolutional code
SDD	soft decision decoding
SISO	soft-input soft-output
SOVA	soft-output Viterbi algorithm
SSC	switch and stay combining
STBC	space-time block code
STC	space-time code
STCM	space-time coded modulation
STTC	space-time trellis code
SVD	singular value decomposition
TC-DSTC	turbo coded differential space-time code
TC-USTC	turbo-coded unitary space-time code
TCM	trellis-coded modulation
TDMA	time-division multiple access
TSTC	threaded space-time code
TuCM	turbo-coded modulation
USTC	unitary space-time code
VA	Viterbi algorithm
VBLAST	vertical Bell Laboratories layered space-time
ZF	zero forcing
ZF-IC	zero forcing with interference cancelation

1

Overview

1.1 Need for MIMO Systems

We have witnessed over the past decade or so an unprecedented growth in the demand for providing reliable high-speed wireless communication links in order to support a wide range of applications, including voice, video, e-mail, web browsing, to name a few. Providing such reliable links is challenging due to the fact that, in a wireless environment, unlike many other channels, transmitted signals are received through multiple paths which usually add destructively resulting in serious performance degradations. This phenomenon is normally referred to as *fading*. Furthermore, the medium is normally shared by many different users/applications, thus there is the possibility of significant interference as well. Other challenges for high-speed wireless applications include the scarcity of available bandwidth, highly constrained transmit powers, as well as hardware complexity and cost requirements.

With the limited frequency spectrum and the steady increase in the number of new wireless applications or expansion of existing ones, there is a clear problem in being able to accommodate all of them. A simple approach that naturally comes to mind is to use higher order modulation schemes in an effort to improve the bandwidth efficiency. However, one drawback of the naive application of this proposed solution is the poor reliability associated with it. That is, for the same level of transmit power, higher order modulation schemes yield performance that is inferior to that of the lower order modulation schemes. In fact, even for small signal constellations, i.e., low-order modulation schemes (e.g. binary), the reliability of uncoded communications over wireless links is very poor in general.

The single most effective technique to accomplish reliable communication over a wireless channel is *diversity* which attempts to provide the receiver with independently faded copies of the transmitted signal with the hope that at least one of these replicas will be received correctly. Diversity may be realized in different ways, including frequency diversity, time diversity, antenna diversity, modulation diversity, etc. We encounter the use of diversity in practical wireless communication scenarios all the time. For instance, most handsets in mobile communications are capable of polarization diversity.

Coding for MIMO Communication Systems Tolga M. Duman and Ali Ghrayeb

Channel coding may also be used to provide (a form of time) diversity for immunization against the impairments of the wireless channel. Examples of practical channel codes include convolutional and block codes, trellis-coded modulation, multi-level coding, bit-interleaved coded modulation, as well as the recently discovered capacity-approaching coding schemes such as turbo and low-density parity check codes, and coded modulation techniques. In the context of wireless communications, channel coding schemes are usually combined with interleaving to achieve time diversity in an efficient manner.

As another method, transmit and/or receive antenna diversity, also referred to as *spatial* diversity, represents a powerful means of combatting the deleterious effects of fading. Systems with multiple antennas are also referred to as multiple-input multiple-output (MIMO) systems. One of the major advantages of MIMO systems is the substantial increase in the channel capacity, which immediately translates to higher data throughputs. Another advantage of MIMO systems is the significant improvement in data transmission reliability, i.e., very low bit error rates. These advantages are achievable without any expansion in the required bandwidth or increase in the transmit power.

Various diversity techniques can be combined to further improve the system performance in a wireless environment. For example, one can combine spatial diversity obtained by transmit and receive antennas with the idea of channel coding. We refer to the resulting scheme as space-time coding, and to the resulting system as a coded MIMO system. Research over the last ten years or so has proven that the combination of coding with spatial diversity opens up new dimensions in wireless communications, and can offer effective solutions to the challenges faced in realizing reliable high-speed wireless communication links. Therefore, coding over MIMO systems is fundamental to the realization of the promises offered by MIMO systems in terms of reliability and achievable transmission rates.

There are different ways of exploiting multiple antennas at both ends of the MIMO communication link. For example, to achieve the best transmission reliability possible, the transmit antennas should be used such that transmit diversity is achieved. The transmission rate achieved in this case is normally comparable to that achieved in single-input single-output systems. That is, all the degrees of freedom of the MIMO channel are used for improving the transmission reliability and not the transmission rate. An alternative is to use the transmit antennas to maximize the transmission rate. In this case, independent signals are transmitted from the different transmit antennas, i.e., there is no correlation among the transmitted signals from different antennas. While this approach increases the transmission rate, the corresponding reliability is poor. A combination of these two approaches is also possible, that is, one could trade rate for reliability or vice versa. Examples of coding schemes that achieve the best spatial diversity are space-time trellis codes and space-time block codes. Whereas among the coding schemes that maximize the transmission rates are the Bell Laboratories layered space-time coding schemes. These two families of space-time codes represent two extremes in the sense that one achieves the best reliability and the other achieves the maximum transmission rate. Other space-time coding schemes that provide a trade-off between diversity and rate also exist.

We envision that the use of MIMO systems and associated signaling approaches are critical for the future of high data rate, extremely reliable wireless communications. Therefore, our main objective is to provide a comprehensive coverage of various coding schemes for MIMO systems.

1.2 MIMO Communications in Wireless Standards

Owing to their great promises, MIMO systems have found their way into several standards for future wireless communication systems, especially wireless local area networks and cellular networks. Examples of these standards include IEEE 802.11, 802.16 and the 3rd Group Partnership Project (3GPP). The IEEE 802.11 standard has been developed for *Wi-Fi* (wireless fidelity). Wi-Fi is normally used for public wireless access (known as hotspots) such as the ones at coffee shops, airports, shopping malls, etc. An earlier version of this standard (IEEE 802.11a) supports data rates up to 11 Mega bits per second (Mbps), whereas the IEEE 802.11b version supports data rates up to 54 Mbps. The newest version (IEEE 802.11n) is the one that incorporates MIMO communications. This standard has not been finalized yet, but it is expected to support data rates of 100 Mbps or higher. Several MIMO configurations are being considered for these applications, including 2×2 and 4×4 with space-time block coding schemes (IEEE P802.11n/D1.0 (2006)).

The IEEE 802.16 standard has been developed for the so-called WiMAX (short for World Interoperability for Microwave Access), which is intended to deliver high data rates over long distances. MIMO communications has been incorporated as an option in the IEEE 802.16e version of this standard, where 2×1 and 4×4 MIMO configurations are considered (IEEE 802.16e Part 16 (2004); IEEE 802.16e/D12: Part 16 (2006)). In some cases, the multiple antennas are used to deliver high data rates to the customers, and in others, particularly for cellular networks, the multiple antennas are used for beamforming to improve the overall network capacity, i.e., number of supported users.

The 3GPP technology, also known as wideband code division multiple access (W-CDMA) is an extension of the CDMA technology, which is used for the third generation (3G) cellular networks. The motivation behind the introduction of W-CDMA has been to achieve additional diversity through multipath fading where higher data rates are transmitted as compared to earlier versions of CDMA. MIMO has been incorporated in this standard, particularly in Releases 7 and 8. The latter release is referred to as LTE for long time evolution. In Release 7, 2×1 and 4×2 MIMO configurations employing space-time block coding are used, whereas 2×2 and 4×4 MIMO configurations are used for Release 8 (TSG-R1(04)0336 (2004)).

MIMO is also being considered, but not finalized, for the IEEE 802.20 and 802.22 standards. The former standard aims to put in place specifications for mobile broadband wireless access, which is a packet-based air interface designed for internet protocol (IP)-based services. The latter standard aims at constructing wireless regional area networks utilizing channels that are not used within the already allocated television frequency spectrum.

1.3 Organization of the Book

We discuss in this book a wide range of topics related to MIMO systems. In Chapter 2, we give an introduction to fading channels and diversity techniques. Several fading channels are considered, including fast, block and quasi-static fading. We briefly discuss the differences between flat fading and frequency selective fading. We describe various approaches to signal detection at the receiver, especially over frequency flat fading channels. Various diversity techniques, including the use of channel coding to achieve temporal diversity are also discussed. We conclude this chapter by motivating the need for employing multiple antennas at the transmitter and/or at the receiver, and by differentiating different uses for multiple antennas in various wireless environments.

In Chapter 3, we analyze the capacity of MIMO systems. We consider both ergodic and non-ergodic capacity. We first consider the basics of channel capacity, specifically for additive white Gaussian noise channels, and single-input single-output wireless channels. We consider both ergodic and non-ergodic channel models, and associated with them ergodic channel capacity and outage capacity. We extend the channel capacity results to wireless MIMO systems for various configurations, and present examples. Furthermore, we consider the information rates achieved by MIMO systems when specific input constellations are employed. These results basically demonstrate the tremendous increase in the channel capacity with the use of MIMO systems, and thus motivate the rest of the book which is mainly devoted to the exploration of various specific coding/signaling approaches for multiple antennas.

Space-time block codes (STBCs), a class of space-time codes used to achieve spatial diversity, are discussed in detail in Chapter 4. We first present the Alamouti scheme, which is a very simple example of an STBC. We then discuss generalized orthogonal and quasi-orthogonal STBCs. We give details of these codes in terms of encoding, decoding and performance analysis. Several numerical examples are given along the way to illustrate the performance of these codes. Furthermore, we briefly review linear dispersion codes.

Space-time trellis codes (STTCs), another class of space-time codes, are presented in detail in Chapter 5. These codes are comparable to STBCs in terms of their spatial diversity where both classes of codes can achieve the full spatial diversity for a given number of transmit and receive antennas. Unlike STBCs, the design of STTCs is based on trellises, and thus, in addition to achieving spatial diversity, they achieve coding gains. Since their encoder is represented by a trellis, STTCs are decoded using the Viterbi algorithm, which is more complex than the decoder used for STBCs. We give general code design principles, and review a number of good space-time trellis code designs. We also provide detailed performance analysis with appropriate modifications to the union bound.

In Chapter 6, we consider layered space-time codes (LSTCs). We discuss the various versions of the BLAST schemes. These schemes are suited for applications with high data rate transmission requirements, but they lack the spatial diversity provided by STTCs and STBCs. Other LSTC schemes such as the multilayered and threaded space-time coding are also discussed. These schemes provide some trade-off between spatial diversity and spatial multiplexing, which is needed in certain applications. We also consider various detection approaches for these layering schemes, including those detectors based on the zero forcing (ZF) and minimum mean-squared error (MMSE) criteria.

Concatenated codes and iterative decoding are treated in detail in Chapter 7. We consider serial and parallel concatenated codes for additive white Gaussian noise (AWGN) channels. We give details of the iterative decoding algorithms used to decode these concatenated codes, including the a posteriori probability (APP) algorithm and the soft-output Viterbi algorithm (SOVA). We also consider in detail several code concatenation schemes developed for MIMO channels. We show how channel coding and space-time coding can be combined to achieve further performance improvements.

Unitary and differential space-time codes are discussed in Chapter 8. These codes are used in applications where estimating the channel state information at the receiver is not possible due to the rapid variations of the channel. Thus, noncoherent detection is employed instead. Such channels are sometimes referred to as *noncoherent channels*. We begin the chapter by discussing the capacity of noncoherent channels and compare that with the capacity of coherent channels. We use this to motivate the use of unitary space-time signals, where we discuss their design, encoding, decoding and performance analysis. Differential space-time codes are discussed as well, which are another class of codes used for noncoherent channels. These two classes of codes have been shown to achieve capacity. We also consider various code concatenation schemes for noncoherent channels, which combine channel coding with unitary or differential space-time coding.

Chapter 9 deals with space-time coding for frequency selective fading channels. In particular, we discuss the capacity and information rates of MIMO frequency selective channels. Various coding approaches suitable for such channels are discussed, including space-time trellis coding, concatenated coding and spatial multiplexing. Several detection alternatives are presented as well, including linear and nonlinear equalization, and other reduced complexity detection schemes. Multi-carrier modulation in the form of orthogonal frequency division multiplexing (OFDM) is also considered.

MIMO communication systems with nonidealities are considered in Chapter 10. Among the nonidealities that every practical MIMO system is challenged with are the estimation of the channel state information at the receiver and the presence of spatial and time correlation. Estimation of the channel state information based on pilot symbols is discussed where different estimation methods are compared, including maximum likelihood (ML), maximum a posteriori (MAP), least squares and MMSE estimation. Measurement and modeling of the spatial correlation is briefly discussed and several relevant references are pointed out as well. The capacity of correlated MIMO channels is assessed analytically and through numerical examples.

In Chapter 11, we address the subject of antenna selection whereby the transmitter and/or receiver select a subset of the available antennas. The motivation behind antenna selection is to reduce the complexity associated with the deployment of MIMO systems while reaping the advantages of using multiple antennas. Among the selection criteria that we consider are the one that maximizes the channel capacity and the one that maximizes the instantaneous signal-to-noise ratio at the receiver. Transmit and receive antenna selection is considered for various fading channels and space-time coding schemes. The impact of several channel imperfections, including channel state information estimation error and spatial correlation, is assessed in conjunction with antenna selection.

Each chapter is concluded with a brief summary, a list of references for further reading and a number of problems on the respective subject.

1.4 Other Topics in MIMO Systems

As outlined in the previous section, the breadth of topics covered in this book is relatively wide. However, there are certain topics related to MIMO systems that we do not plan to delve into. For example, for the topics we cover throughout the book, as mentioned before, it is assumed that the channel state information is estimated at the receiver. The only exception to this is the material in Chapter 8 where unitary signaling and differential schemes for MIMO systems are studied. A very important complementary consideration is the availability of the statistical, or instantaneous knowledge of the MIMO channel at the transmitter. For instance, if the channel covariance matrix is available, it is possible to exploit this information to do some kind of waterfilling to improve the capacity. As another example, if the instantaneous channel knowledge is fed back to the transmitter from the receiver, one can use adaptive coding/modulation, as well as beamforming approaches to get the full benefits of MIMO systems. Clearly, schemes where limited, noisy or delayed feedback are also of great practical significance. Although we acknowledge that these are very important topics, in our exposition, we do not go into their details.

There are also many connections between multi-user communications and point-to-point MIMO communications. For example, *cooperative communications* has been receiving a lot of attention recently. In the context of cooperative communications, many recent papers consider capacity aspects, design of suitable cooperation protocols, and specific distributed coding schemes. Such schemes in a broad sense can also be considered as MIMO communication systems where different users basically refer to different transmit and/or receive elements in our context. Some of the studies on ad-hoc and sensor networks are also inherently related to MIMO communications. Although these all are very important subjects, and represent extremely important recent research directions, they are beyond the scope of this book. On a related note, we also would like to point out that multiple access with MIMO communications is also a recent area of study. For instance, there is a large volume of literature on MIMO CDMA and MIMO orthogonal frequency division multiple access (OFDMA) that fall beyond our scope. Basically, we limit our attention to point-to-point MIMO communications.

2

Fading Channels and Diversity Techniques

It is more and more desirable to communicate reliably at high data rates over wireless communication links in order to support various applications including multimedia transmission, sensing, etc. However, this is a very challenging task due to the many impairments including various forms of electromagnetic wave propagation, e.g., reflection, refraction and scattering, time variations in the medium, interference caused by other wireless applications, bandwidth and power limitations.

In this chapter, we review the basics of wireless communications. We study the channel characteristics, including path loss, shadowing and small-scale fading. We concentrate on the effects of small-scale fading, and classify wireless channels with respect to the length of multipath and rapidity of time variations. We consider the error rates and outage probabilities over commonly used fading channel models, and demonstrate that the performance of a digital communication system is severely degraded due to fading. We then focus on remedies to this problem, namely the use of diversity techniques and possible combining algorithms, and review their performance. We specifically talk about channel coding as a means of time diversity, and the use of multiple antennas to provide spatial diversity. Finally, we concentrate on multiple antenna communication techniques in wireless systems. We then briefly explain what is meant by space-time coding, and provide a number of references for further reading.

2.1 Wireless Channels

There are fundamental differences between wireless channels and usual wireline channels. For example, over wireless channels, signals are received through multiple paths. Also there are significant time variations caused by the relative motion of the transmitter and the receiver as well as changes in the environment. Furthermore, different signals being transmitted over a wireless medium might interfere with each other.

Coding for MIMO Communication Systems Tolga M. Duman and Ali Ghrayeb

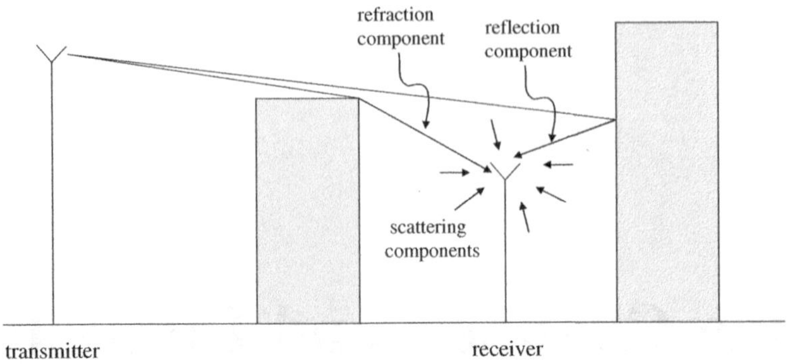

Figure 2.1 Illustration of the wireless propagation mechanisms.

Electromagnetic waves travel through three different mechanisms: reflection, refraction and scattering. When waves impinge upon a surface with surface variations significantly larger than their wavelength, part of the signal reflects as illustrated in Figure 2.1. This phenomenon is referred as reflection. The second mechanism, refraction, is basically the process of secondary wave production by "knife edge" as also illustrated in the figure. Finally, scattering usually arises due to surface variations of nearby objects that are smaller than the signal wavelength, e.g., due to the roughness of surfaces, due to foliage, etc. As a result of these different forms of electromagnetic wave propagation, the signals transmitted over a wireless channel are received via multiple secondary paths, in addition to the possible line of sight.

Different multipath propagation mechanisms, and also the fact that, depending on the materials involved and frequency of the operation, penetration of the electromagnetic waves through walls, allow reception of signals even when there is no direct line of sight between the transmitter and the receiver. However, the effects of multipath propagation are usually deleterious. This is due to the fact that different paths have different lengths, resulting in different attenuation factors and phase differences for replicas of the transmitted signals. Since frequencies employed in wireless communications are usually very high, the corresponding wavelengths are very small (compared with the differences in lengths of paths that different signal components travel). For instance, for cellular telephony, transmission bands are usually in the range of 800 MHz–2 GHz, thus wavelengths of the transmitted signals are in the range of 15–37.5 cm, whereas the path length differences can be in the order of tens of meters, or hundreds of meters. As a result, the phase differences between the different replicas of the received signal will be significant. The amplitude variations will also be significant since, for example, the components due to scattering, or due to refraction or reflection will not all have similar signal strengths.

The relative motion of the transmitter and the receiver, and the changes in the environment cause the multipath channel, through which the signals are being transmitted, to be time varying as well. This is also in sharp contrast with the wireline channels which are typically constant. For a clear understanding of wireless channels, their multipath and time variations should be characterized properly and suitable channel models should be developed.

2.1.1 Path Loss, Shadowing and Small-Scale Fading

The effects of multipath propagation and time variations in a wireless channel can be observed in the received signal power levels. A first order characterization of the average received signal power level is obtained by using path loss models. If we consider the average power of the received signal over a long period of time, denoted by $\bar{P}_r$, we can write

$$\bar{P}_r = \frac{c}{d^n} P_t, \tag{2.1}$$

where c is a constant, P_t is the transmit signal power level, d is the separation between the transmitter and the receiver, and n is the path loss exponent. The path loss exponent is typically between two and six, depending on the specific environment. For instance, for free space (when only a direct line of sight exists, no scatterers), $n = 2$, whereas for an indoor transmission in an office building with multiple floors, it could be as high as six. The constant c depends on a variety of factors including transmit and receive antenna gains, and frequency of operation.

Path loss is a useful characterization of the received signal strength, but it alone does not provide a satisfactory understanding of a wireless link. Instead of averaging over a very long period of time, if we use a smaller window, say in the order of a few seconds or minutes, we will find that the average received signal strength is a random variable. Let us denote it by P_r. We can model the received power at this scale as

$$P_r \text{ dBm} = \bar{P}_r \text{ dBm} + X_\sigma \text{ dB}, \tag{2.2}$$

where powers are expressed in dBm,[1] and X_σ is a zero mean random variable. For the common model of log-normal shadowing, X_σ is taken as a zero mean Gaussian random variable with standard deviation σ. In other words, as a more refined characterization of the wireless link, we see that the received power level, when expressed in dBm, is given by a Gaussian random variable. The shadowing standard deviation σ is heavily dependent on the environment, and could be in the range of three to 12 dBs. We further note that the shadowing component of the received signal has spatial dependence, that is, two different receiver locations separated by a few meters or even tens of meters may experience highly dependent shadow fades. We note that shadowing is also referred to as large-scale fading. The effects of the path loss and shadowing on the received signal powers are illustrated in Figure 2.2.

Path loss and log-normal shadowing are average quantities. In fact, the actual received signal power in a wireless channel is a much more rapidly varying random quantity (as illustrated in Figure 2.3) which needs to be characterized using statistical models. This is explained in detail in the following section.

For the rest of the book, with the understanding that the path loss and shadowing determine the average operating signal-to-noise ratios, we will be concerned with the small-scale variations of wireless channels. This point is easily justified by considering that the packet sizes in typical digital communication systems are very small compared with changes in path loss and shadowing, and thus path loss and shadowing components of the received signal do not change over transmission of many frames of data.

[1] The power unit dBm denotes the ratio of the amount of power to 1 mW in dBs.

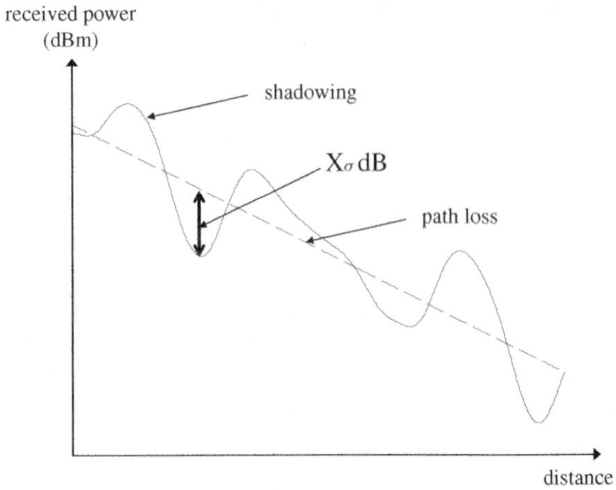

Figure 2.2 Effect of path loss and shadowing on the received signal power.

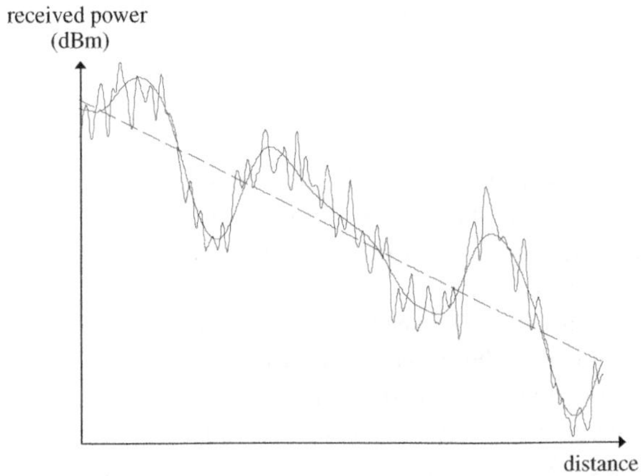

Figure 2.3 Received power when small-scale fading is also taken into account.

2.1.2 Fading Channel Models

In this section, we briefly present a statistical characterization of fading channels, and describe several relevant wireless channel models. These models will be used throughout the book.

Fading channels are modeled as linear time-varying systems where the time variations are random. Since most wireless transmission systems employ bandpass transmission centered around a carrier frequency f_c, more specifically, we model these channels as linear time-varying bandpass systems. Therefore, we can characterize them using a time-varying impulse response, i.e., the response of the channel at time t to an impulse applied at time

$t - \tau$ in the time domain, or using a time-varying frequency response in the frequency domain.

Let us denote the transmitted bandpass signal by $x(t)$, and its low-pass equivalent by $x_l(t)$. That is, $x(t) = \text{Re}\{x_l(t)e^{j2\pi f_c t}\}$. If the low-pass equivalent channel impulse response is denoted by $c(\tau; t)$, the received signal in the absence of noise can be written as

$$y_l(t) = \int_{-\infty}^{\infty} c(\tau; t)x_l(t - \tau)d\tau. \tag{2.3}$$

If we define the time-varying channel transfer function as the Fourier transform of the channel impulse response with respect to the τ variable, we obtain

$$C(f; t) = \int_{-\infty}^{\infty} c(\tau; t)e^{-2\pi f\tau}d\tau. \tag{2.4}$$

Clearly, we can also write

$$y_l(t) = \int_{-\infty}^{\infty} C(f; t)X_l(f)e^{j2\pi ft}df. \tag{2.5}$$

where $X_l(f)$ is the Fourier transform of the low-pass equivalent of the transmitted signal.

Let us now consider the transmission of a sinusoidal signal at a particular frequency, say $f_c + \Delta f$. The received signal through the multipath fading channel is the superposition of many different replicas of the sinusoid received through reflection, refraction and scattering components. There are typically a very large number of such replicas with time-varying amplitudes and phases. Therefore, it is very difficult to deterministically characterize these fluctuations in signal strength. In fact, the effective channel, as far as the user is concerned, is a random attenuation applied to the transmitted signal. Invoking the central limit theorem, we can argue that the channel coefficient observed due to a sinusoidal transmission is a complex Gaussian random process. It is zero mean if there is no dominant line of sight component, or specular component, otherwise the mean may be non-zero.

Going back to Equation (2.3), we see that for a sinusoidal input at a frequency $f_c + \Delta f$, i.e., $x_l(t) = e^{j2\pi\Delta ft}$, the received signal is given by

$$y_{l,sinusoid}(t) = \int_{-\infty}^{\infty} e^{j2\pi\Delta f(t-\tau)}c(\tau; t)d\tau, \tag{2.6}$$

$$= C(\Delta f; t)e^{j2\pi\Delta ft}, \tag{2.7}$$

which basically means that the channel coefficient that multiplies the sinusoidal signal is in fact the time-varying transfer function of the channel evaluated at the frequency of the transmission. Combining this with the previous result that this coefficient is well modeled as a complex Gaussian random process, we arrive at the conclusion that the time-varying channel frequency response $C(f; t)$ should be complex Gaussian. Since the channel impulse response $c(\tau; t)$ is the inverse Fourier transform of this quantity, i.e., it is a linear combination of jointly Gaussian random processes, we can conclude that $c(\tau; t)$ is also a complex Gaussian random process (in the t variable).

Statistical characterization of the complex Gaussian random processes $C(f; t)$ and $c(\tau; t)$ is done by using autocorrelation functions. For a detailed discussion, the reader is referred to standard textbooks such as Proakis (2001). Here, we only give a brief overview for completeness of our exposition.

The multipath structure can be characterized in the time domain using the "multipath intensity profile" of wireless channels, which basically shows relative powers of the received signal through different delays (we note that there are multiple paths corresponding to each delay). The extent of the non-negligible values of this function is called the multipath spread of the channel, T_m. The multipath spread is basically the time difference between the shortest and the longest paths that the transmitted signal goes through. Typical values could be in the order of nanoseconds for indoor applications, and up to several tens of microseconds for outdoor applications (depending on the exact propagation environment). The Fourier transform of the multipath intensity profile gives a frequency domain characterization of the wireless channel's multipath structure. The coherence bandwidth of the channel, B_C, is the range of non-negligible values of this Fourier transform. Roughly speaking, $B_C \sim \frac{1}{T_m}$. Two frequencies separated by less than the coherence bandwidth of the channel are affected in almost the same way by the channel. On the other hand, frequencies separated by more than the coherence bandwidth undergo different channel fades.

Time variations of fading channels are characterized by a correlation function (in the t variable), and its Fourier transforms. The Doppler power spectrum shows the amount of frequency shift and spectral broadening in the received signal. The frequency range for which the Doppler power spectrum is non-zero is defined as the Doppler spread, B_D, whose inverse is the coherence time of the channel $(\Delta t)_c$. The interpretation is that for $(\Delta t)_c$ seconds the channel remains roughly the same.

The scattering function, $S(\tau; \lambda)$, which shows the Doppler power spectrum for paths with different delays τ (where λ is the Doppler frequency), is a complete characterization of the second order statistics of wireless channels. A typical scattering function is shown in Figure 2.4.

Frequency Flat versus Frequency Selective Fading Channels

Consider digital modulation over a wireless channel, and assume that the bandwidth of the signal used in the transmission is W. Let us now characterize the type of fading that will be experienced using the multipath structure of the channel. If the signal bandwidth W is significantly smaller than the coherence bandwidth of the channel B_c, clearly, all the frequency components of the transmitted signal see the same effective channel, i.e., $C(f; t)$ for the frequency range of interest will be independent of f. This is illustrated on the left-hand side of Figure 2.5. For this case, we can write

$$y_l(t) = C(0; t) \int_{-\infty}^{\infty} X_l(f) e^{j2\pi f t} df, \tag{2.8}$$

$$= C(0; t) x_l(t), \tag{2.9}$$

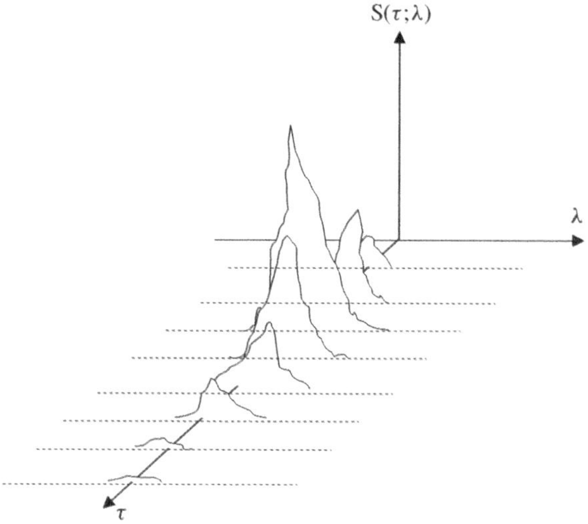

Figure 2.4 A typical scattering function.

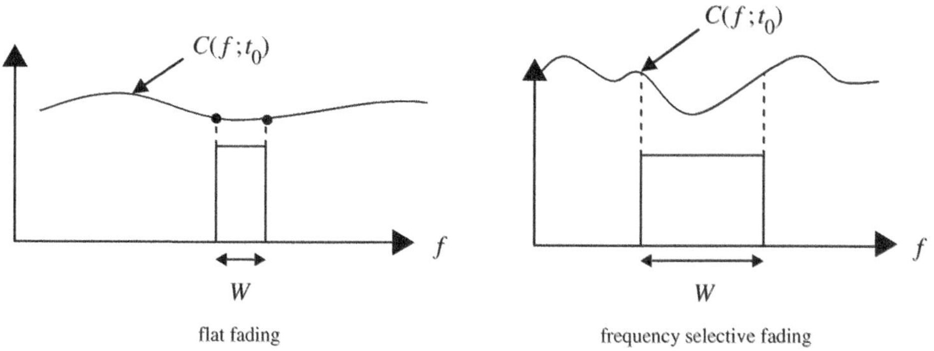

Figure 2.5 Frequency flat versus frequency selective fading (in the frequency domain).

in the absence of channel noise. This expression basically means that the effect of the wireless channel is simply multiplicative if the bandwidth of the signal transmitted is significantly smaller than the coherence bandwidth of the channel, where the multiplicative term is a complex Gaussian random process. We note that the condition $W \ll B_c$ is equivalent to saying that the multipath spread of the channel is significantly smaller than the signal duration in time, and therefore, there is no intersymbol interference between consecutive transmitted symbols.

If the condition $W \ll B_c$, or equivalently, $T_m \ll T_s$ (where T_s is the symbol duration), is not satisfied, then different frequency components of the signal undergo different channel fades as illustrated on the right-hand side of Figure 2.5. In such a case, the channel is said

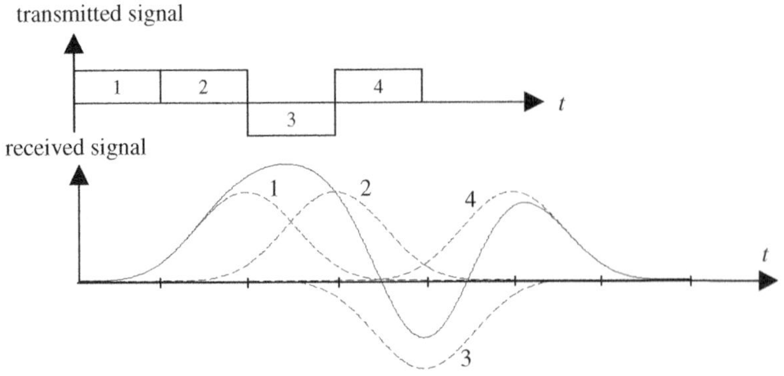

Figure 2.6 Illustration of frequency selective fading in the time domain.

to be frequency selective, and it exhibits intersymbol interference. We illustrate the effect of a frequency selective fading channel on the transmitted signal in the time domain in Figure 2.6.

Although we have a simple multiplicative description for the frequency flat fading scenario, exact characterization of a frequency selective fading channel is more difficult. An appropriate simplified characterization is the symbol spaced tapped delay line model as will be described a little later.

Slow versus Fast Fading Channels

In most wireless applications, the rate of transmission is several orders of magnitude larger than the Doppler spread of the channel. In other words, the symbol duration is significantly smaller than the coherence time of the channel, meaning that the channel remains the same over an entire symbol period. As an example, consider the case of a vehicle moving with a constant speed of 100 miles per hour, and carrier frequency 1 GHz. In this case, the maximum Doppler shift can be calculated to be approximately 150 Hz, which is a very high value in practical scenarios. Then the coherence time of the channel is in the order of 6.7 ms, meaning that the wireless channel will effectively remain constant over an interval of 6.7 ms. For practical wireless applications, such a duration typically corresponds to a large number of symbols. For instance, if the symbol rate is 100 kilo symbols per second, then the channel will not change for about 670 symbols. Therefore, for flat fading channels, we can then say that the multiplicative factor $C(0; t) \approx C(0, nT)$ is approximately constant over an entire symbol duration (for $(nT, (n+1)T)$). That is, the channel gain is simply a (Gaussian) random variable. For the case of frequency selective fading channels, if the symbol spaced tapped delay line is used as a model, then different tap coefficients will remain the same over an entire symbol duration.

Time variations of wireless channels can be described using several channel models. For quasi-static fading channels, the channel remains the same for an entire frame of data, and it changes from one frame to the next. In such a case, there is no means of "averaging" over the fading statistics by increasing the frame length, i.e., the channel is non-ergodic. Block fading channels are used as proper models if channel fading is constant for a block of symbols, and changes from one block to another. However, the frame may consist of many differently faded blocks. Block fading models could be encountered if, for example, frequency hopping is employed and different parts of the frame are transmitted in different frequency bands. Independent and identically distributed (i.i.d.) fading (or, fully interleaved fading) is the proper model if the fading coefficients change from one symbol to the next independently. Such models are obtained by employing interleavers to decorrelate the channel coefficients adjacent symbols observe.

Rayleigh, Rician and Nakagami Fading Channels

We have just argued that the effect of the wireless channel on the transmitted signals is multiplicative (for frequency flat fading channels), where the multiplicative term is a complex Gaussian random variable. If the channel coefficient has zero mean, then such a channel is considered as Rayleigh fading since the absolute value of the channel gain is a Rayleigh random variable. If the channel gain has a non-zero mean, then its absolute value is Rician distributed, and the channel is said to be Rician fading.

Let us denote the channel gain by a complex random variable h, where $|h|$ is the absolute value of the channel gain, and $\phi = \angle h$ is the channel phase. For Rayleigh fading, the probability density function (p.d.f.) of $|h|$ is given by

$$f_{|h|}(u) = \frac{2u}{P_r} \exp\left(-\frac{u^2}{P_r}\right), \quad \text{if } u > 0, \tag{2.10}$$

where P_r is the average channel power (due to path loss and shadowing), and the phase term ϕ is a uniform random variable on $(0, 2\pi)$.

For Rician fading channels, the p.d.f. of the channel envelope is given by

$$f_{|h|}(u) = \frac{2u(K+1)}{P_r} \exp\left(-K - \frac{(K+1)u^2}{P_r}\right) I_0\left(2u\sqrt{\frac{K(K+1)}{P_r}}\right) \quad \text{for } u > 0, \tag{2.11}$$

where $I_0(\cdot)$ is the zeroth order modified Bessel function of the first kind, and K is the Rician factor, which denotes the ratio of the power in the line of sight (LOS) (or the specular) component of the channel to the power in the diffuse component. Clearly, if $K = 0$, we have Rayleigh fading, i.e., there is no LOS component, and if $K = \infty$, we obtain a simple AWGN channel (with a fixed channel gain) as there is no diffuse component. The phase of the channel gain is not uniformly distributed in this case.

There is another popular fading channel model called Nakagami fading which has its basis on experimental observations as opposed to theoretical models used to develop Rayleigh and Rician models. For Nakagami fading, the p.d.f. of the channel envelope $|h|$ is given by

$$f_{|h|}(u) = \frac{2m^m u^{2m-1}}{\Gamma(m) P_r^m} \exp\left(-\frac{mu^2}{P_r}\right), \quad \text{if } u > 0, \tag{2.12}$$

where $m \geq 1/2$ is the Nakagami fading parameter, and $\Gamma(\cdot)$ is the Gamma function.

Although it is not developed using a theoretical basis, Nakagami fading is useful to characterize fading in certain environments. Also, it generalizes Rayleigh and Rician fading channel models, that is, if $m = 1$ is chosen, we obtain Rayleigh fading, and if we select $m = \frac{(K+1)^2}{2K+1}$, we can approximate Rician fading. We further note that the system performance analysis with the Nakagami fading model is usually more tractable than Rician fading.

For most of this book, we will be concerned with Rayleigh fading channels. Only occasionally, we will refer to the Rician and Nakagami fading channel models.

Summary of Fading Channel Models

Let us now summarize the wireless channel models that will be used extensively throughout this book. We only describe the single-input single-output case here. We will consider multi-antenna generalizations later in the book. We consider frequency flat fading and frequency selective fading models separately.

Flat Fading Channels:

Mathematically, the received signal at time k, $y(k)$, for a flat fading channel is given by

$$y(k) = \sqrt{\rho} h(k) x(k) + n(k), \tag{2.13}$$

where $x(k)$ is the kth transmitted symbol, $h(k)$ is the fading channel coefficient corresponding to this symbol, and $n(k)$ is the additive zero mean white complex Gaussian noise term. We assume that $E[|h(k)|^2] = 1$, the average signal power is normalized to unity, and the noise term has zero mean and $1/2$ variance per dimension. Thus, ρ can be interpreted as the average signal-to-noise ratio at the receiver. The distribution of the random channel gain depends on the exact channel model. For example, if the channel is Rayleigh fading, then $|h(k)|$ is Rayleigh distributed, and $\phi(k) = \angle h(k)$ is uniform on $(0, 2\pi)$. Dependence of the channel coefficient on the time index k is determined by the rapidity of the fading. For instance, if we have a fully interleaved fading channel, all the coefficients are independent of each other, whereas for a quasi-static fading channel, they are identical for an entire frame of data.

Frequency Selective Fading Channels:

As mentioned before, a popular model for a frequency selective fading channel is the symbol spaced tapped delay line model, which basically states that the received signal at time k is given by

$$y(k) = \sqrt{\rho} \sum_{l=0}^{L-1} h^{(l)}(k)\; x(k-l) + n(k), \tag{2.14}$$

where L denotes the number of intersymbol interference (ISI) terms, and $h^{(l)}(k)$ is the complex channel coefficient for the lth ISI tap at time k. We normalize the channel coefficients such that $\sum_{l=0}^{L-1} E[|h^{(l)}(k)|^2] = 1$, the signal power and noise distribution are the same as the flat fading case, thus ρ refers to the signal-to-noise ratio. For Rayleigh fading channels, the channel tap coefficients are modeled as complex Gaussian random variables. The coefficients corresponding to different channel paths are usually assumed to be independent. Finally, the time correlations (with respect to k) depend on the rapidity of the fading.

The average channel gains for different paths are determined using the power delay profile (i.e., the multipath intensity profile) of the wireless channel. For example, for a uniform power delay profile, all the channel gains (for different paths) have equal average channel power. As another example, the exponential power delay profile refers to the case where the average powers of different ISI paths decay exponentially.

2.2 Error/Outage Probabilities over Fading Channels

As we know from basic digital communications, error rates for digital modulation techniques such as phase shift keying (PSK), quadrature amplitude modulation (QAM) and frequency shift keying (FSK) all reduce exponentially with the signal-to-noise ratio for additive white Gaussian noise (AWGN) channels. However, for fading channels, the situation is quite different, and if not mitigated properly, the effects of channel fading are deleterious on the digital communication system performance. We illustrate this point in detail in this section.

2.2.1 Outage Probability for Rayleigh Fading Channels

We first consider a scenario where fading is extremely slow, and we observe only a single state of the channel over an entire frame of data, i.e., quasi-static fading. Assume that a certain frame error rate or bit error rate is tolerable for a particular application, however if the error rate goes above this level, the resulting performance is simply unacceptable. For instance, if we consider speech transmission, frame error rates of 5% may be tolerable, but beyond this value, we cannot communicate reliably. Assuming that this specified error rate corresponds to a certain minimum acceptable signal-to-noise ratio, if the instantaneous signal-to-noise ratio over the channel is below this level, the system is in outage. This is because, for this channel model, we are unable to "average" over different fading coefficients. Therefore, the channel outage probability can be used to evaluate the system performance.

Let us compute the outage probability over a flat Rayleigh fading channel for illustration. With the notation introduced in the previous section, the instantaneous signal-to-noise ratio over this channel is given by $\rho|h|^2$ where ρ is the average signal-to-noise ratio. Since h is zero mean complex Gaussian with variance $1/2$ per dimension, the instantaneous signal-to-noise ratio is an exponential random variable with mean ρ. Therefore, denoting the minimum acceptable signal-to-noise ratio level by ρ_{min}, the outage probability is given by

$$P_{out} = \int_0^{\rho_{min}} \frac{1}{\rho} e^{-x/\rho} dx, \tag{2.15}$$

$$= 1 - e^{-\frac{\rho_{min}}{\rho}}, \tag{2.16}$$

which can be approximated at large signal-to-noise ratios by

$$P_{out} \approx \frac{\rho_{min}}{\rho}. \tag{2.17}$$

This result clearly shows that the communication failure probability is inversely proportional to the average signal-to-noise ratio. For many applications, this is simply unacceptable, because, to make the outage probability low, excessive transmission powers will be required.

2.2.2 Average Error Probabilities over Rayleigh Fading Channels

Let us deviate from the slow fading channel assumption to illustrate what happens if we are able to see different states of the channel, thus we can average over the channel statistics. As an example, consider binary PSK (BPSK) transmission over frequency flat Rayleigh fading channels. Assuming that the channel phase can be estimated and tracked perfectly, it can be compensated for at the receiver, and the optimal detector becomes the same as the one for an AWGN channel. Using results from basic digital communications, the conditional bit error rate for a given channel coefficient h can be written as

$$P_b(h) = Q\left(\sqrt{2|h|^2\rho}\right), \tag{2.18}$$

since the effective (instantaneous) signal-to-noise ratio is scaled by $|h|^2$. Averaging this conditional error rate over Rayleigh fading channel statistics, we obtain the bit error rate for BPSK modulation as

$$P_b = E_h[P_b(h)], \tag{2.19}$$

$$= \int_0^\infty Q\left(\sqrt{2u\rho}\right) e^{-u} du, \tag{2.20}$$

where the second equality follows since $|h|^2$ is exponential with parameter 1. Using integration by parts, it is easy to see that

$$P_b = \frac{1}{2}\left(1 - \sqrt{\frac{\rho}{1+\rho}}\right), \tag{2.21}$$

which can be approximated for large signal-to-noise ratios, $\rho \gg 1$, as

$$P_b \approx \frac{1}{4}\frac{1}{1+\rho}, \tag{2.22}$$

$$\approx \frac{1}{4\rho}. \tag{2.23}$$

This expression clearly shows that the error probability for a Rayleigh fading channel is only inversely proportional to the signal-to-noise ratio. This is in sharp contrast with error rates over AWGN channels as they decay exponentially with ρ. For instance, for BPSK modulation, over an AWGN channel, $P_e = Q(\sqrt{2\rho}) \leq \frac{1}{2}e^{-\rho}$.

For other modulation schemes, following similar steps, it can also be shown that error rates only decay inversely with the signal-to-noise ratio for Rayleigh fading channels. This is a very slow decrease, demonstrating the deleterious effects of Rayleigh fading on the system performance. Clearly, increasing the transmission power alone to reduce the error rates is not acceptable. This is because, for example, to reduce the error rates by an order of magnitude, the transmission power has to be increased by the same factor, which is very inefficient.

We need ways of mitigating the adverse effects of channel fading for reliable operation of digital communication systems. Such methods are collectively known as diversity techniques, and they will be described a little later in the chapter.

2.2.3 Extensions to Other Fading Channels

Similar results can be produced for other fading channel models as well, such as Nakagami fading or Rician fading. Since the Rayleigh fading channel is a realistic model with a sound theoretical foundation, and it is basically the "worst case" scenario with no line of sight or specular component, we mostly deal with this model in detail. This will be the case throughout the book.

Due to the existence of the specular component, the outage probability expressions and average error rates for the case of Rician fading will be better than those of Rayleigh fading. Depending on the value of the K parameter, the resulting error rates could be close to those of AWGN or Rayleigh fading channels. As an example, Figure 2.7 illustrates the average bit error rates for BPSK modulation for several values of the Rician factor.

2.2.4 Performance over Frequency Selective Fading Channels

Similar calculations for the outage and average error probabilities can be performed for frequency selective fading channels as well. However, the expressions will not be as simple

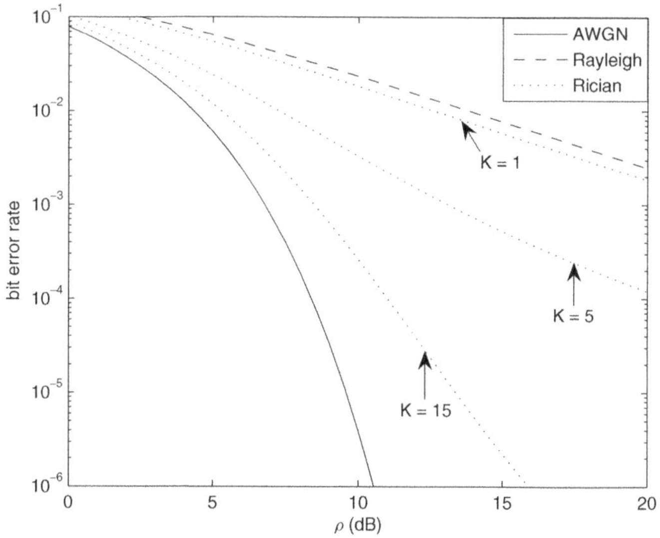

Figure 2.7 Error rates of BPSK modulation over Rayleigh and Rician fading channels.

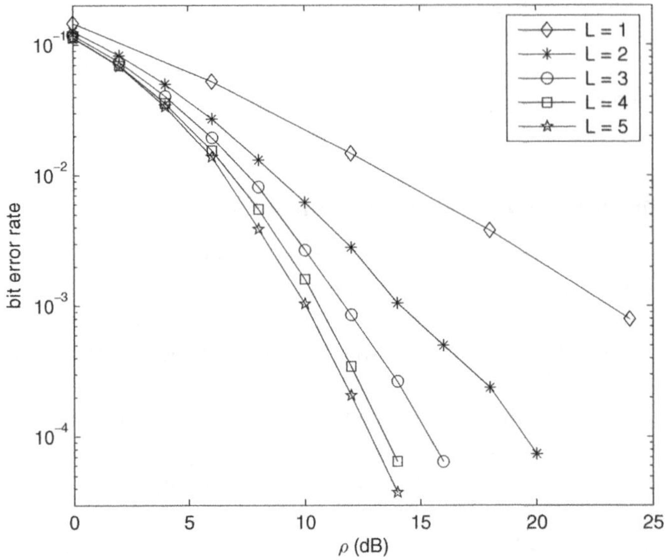

Figure 2.8 Error rates of BPSK modulation over several frequency selective fading channels.

to derive. Here, we do not go into the details of these derivations, instead we only note the following: Although the existence of ISI may seem like a degrading factor, it effectively provides frequency diversity (or, multipath diversity), that is, effectively each signal gets transmitted through several differently faded channels. Therefore, error rates may improve with frequency selectivity compared with flat fading channels.

In Figure 2.8, we show the bit error rates of BPSK modulation over several frequency selective fading channels with different numbers of taps. The power delay profile is assumed to be uniform. The fading coefficients are assumed to be known at the receiver. For the case of $L > 1$, a maximum likelihood sequence detector (MLSD), implemented using the Viterbi algorithm, is employed. We observe that the error rate performance improves with increasing the number of taps, as expected.

2.3 Diversity Techniques

We have just illustrated some of the adverse effects of channel fading on the error rate performance or the outage probability of digital communication systems. Diversity techniques collectively refer to methods of improving this performance by effectively transmitting the same information multiple times, where each replica sees a different (ideally, independent) channel. The hope is that at least one of the replicas will be received correctly, or at least one of the diversity branches will have a large enough instantaneous signal-to-noise ratio, so that the transmitted signal is received correctly, and the system is not in outage.

Assume that we have L diversity branches available, and each of these branches fades independently. If the error probability for one of the branches is p, then the error probability for L parallel transmissions (assuming that error will occur only if all the replicas are received incorrectly) is p^L. Therefore, the performance of the system is improved.

In the context of Rayleigh fading channels, if there are effectively L independent sub-channels, the error rate will decay inversely with the Lth power of the average signal-to-noise ratio. Such a system is said to provide a diversity of order L. The diversity order is one, i.e., $L = 1$, if there is only one fading channel over which information is being transmitted.

2.3.1 Types of Diversity

There are many methods by which diversity can be achieved. Examples include time diversity, frequency diversity, space (spatial) diversity, channel coding (as an efficient means of time diversity) among others.

Time diversity can be realized by transmitting the same signal several times using different time intervals. Obviously, the separation of these time intervals has to be sufficiently large, i.e., it should be more than the coherence time of the channel, so that the fading channel coefficients change, and different channel gains are observed as illustrated in Figure 2.9. Considering idealized channel models discussed before, it is clear that we can achieve time diversity over fully interleaved (i.e., i.i.d.) fading channels, or block fading channels. However, over the quasi-static fading channel, fading is so slow that, regardless of how long we wait for transmitting the replicas, the channel encountered remains the same. Therefore, no time diversity is available.

Frequency diversity is obtained by transmitting different replicas of the signal over different frequency bands. To make sure that the channels seen are different (ideally, independent), separation of these frequency bands has to be more than the coherence bandwidth of the channel. As an example, assuming that the multipath spread of the channel is 200 μs, this means that a minimum separation of 5 KHz would be needed. This diversity scheme is also illustrated in Figure 2.9.

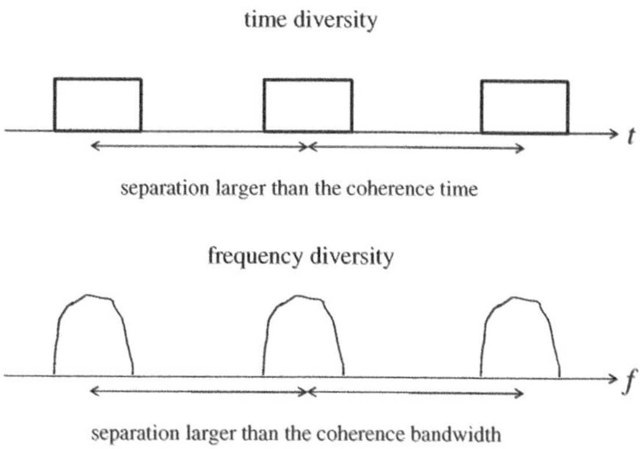

Figure 2.9 Illustration of time and frequency diversity techniques.

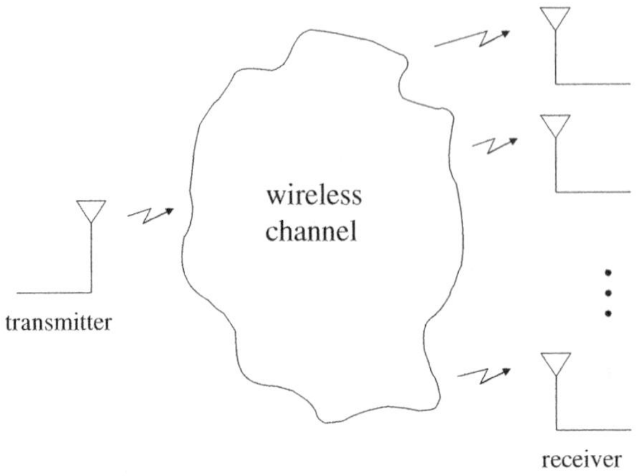

Figure 2.10 Spatial diversity scheme.

The Lth order time diversity can be considered as repetition coding with a code rate of $1/L$. By employing channel coding techniques, we can also obtain diversity over fading channels that can be considered as a more sophisticated form of time diversity. In this case, a channel code of rate R_c is used to encode a sequence of length k, to a coded sequence of length $n = k/R_c$, and the coded bits are transmitted over the channel. The channel is assumed to change over the duration of an entire codeword (clearly, a quasi-static fading channel will not be suitable), thereby providing diversity. We will elaborate more on the use of channel coding for communicating over fading channels later in this chapter.

"Space" can also be used as a resource that can efficiently provide diversity. Assume that the receiver is equipped with multiple antennas. Then, different replicas of the transmitted signal will be picked up by each of these antennas as illustrated in Figure 2.10. If the separation of the receive antennas is sufficient (as a rule of thumb, more than half a wavelength in a uniform scattering environment), then the received signals will undergo different channel fades, thereby providing "spatial" diversity.

Signals obtained through different diversity branches have to be combined at the receiver to detect the transmitted symbols. There are different methods to accomplish this which include selection combining, maximal ratio combining, equal-gain combining, etc. Let us describe the system set up and explain these combining schemes in a little more detail.

2.3.2 System Model for Lth Order Diversity

Consider an M-ary digital modulation scheme with constellation points $x_1, x_2, \ldots, x_M$. Assume that a certain constellation point, denoted by x, is transmitted over L diversity branches. The channel gain for the lth branch is given by h_l, and the received signals are corrupted by independent Gaussian noise terms, denoted by n_l. Assume that the average

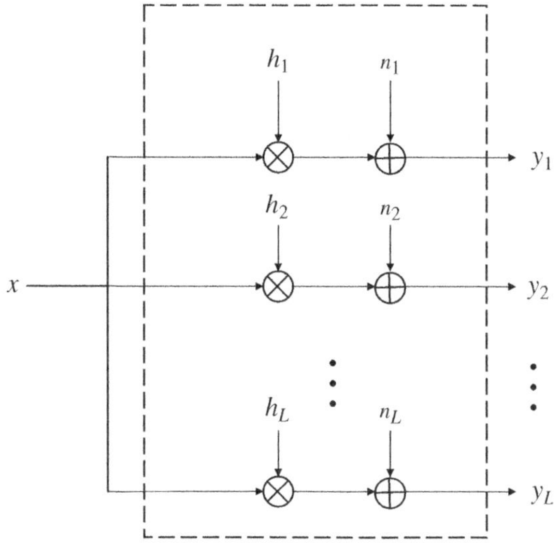

Figure 2.11 Channel model for an Lth order diversity scheme.

signal energy is normalized to unity, the channel powers are normalized such that $E[|h_l|^2] = 1$, and the noise variance per complex dimension is 1/2. Also assume that the average signal-to-noise ratio per branch is ρ. This model for an L branch diversity scheme is illustrated in Figure 2.11.

Mathematically, the set of received signals is given as

$$\begin{aligned} y_1 &= \sqrt{\rho} h_1 x + n_1, \\ y_2 &= \sqrt{\rho} h_2 x + n_2, \\ &\vdots \\ y_L &= \sqrt{\rho} h_L x + n_L. \end{aligned}$$

The problem is to demodulate the transmitted signal from the set of received signals. There are different ways to accomplish this. In the following subsections we consider several approaches.

2.3.3 Maximal Ratio Combining (MRC)

Let us assume that the channel coefficients are perfectly available at the receiver. Then, the maximum likelihood decision rule on the symbol transmitted is given by

$$\hat{x} = \underset{j=1,2,\ldots,M}{\arg\max}\; p(y_1, y_2, \ldots, y_L | h_1, h_2, \ldots, h_L, x_j), \tag{2.24}$$

where $p(\cdot|\cdot)$ denotes conditional (joint) p.d.f. Since, conditioned on the channel gains and the transmitted signal, the received signals are independent, the above expression can be written as

$$\hat{x} = \underset{j=1,2,\ldots,M}{\arg\max} \prod_{l=1}^{L} p(y_l|h_l, x_j), \tag{2.25}$$

$$= \underset{j=1,2,\ldots,M}{\arg\max} \prod_{l=1}^{L} \frac{1}{\pi} \exp\left(-|y_l - \sqrt{\rho} h_l x|^2\right). \tag{2.26}$$

This can be simplified by ignoring the multiplicative constant, taking the logarithm and ignoring the common terms as

$$\hat{x} = \underset{j=1,2,\ldots,M}{\arg\min} \ \mathrm{Re}\left\{\left(\sum_{l=1}^{L} h_l^* y_l\right) x_j^*\right\} - \frac{1}{2}\sqrt{\rho}\left(\sum_{l=1}^{L} |h_l|^2\right)|x_j|^2. \tag{2.27}$$

Furthermore, if we assume that the constellation points have equal energies, e.g., as in PSK, the expression is further simplified to

$$\hat{x} = \underset{j=1,2,\ldots,M}{\arg\min} \ \mathrm{Re}\left\{\left(\sum_{l=1}^{L} h_l^* y_l\right) x_j^*\right\}. \tag{2.28}$$

Therefore, the optimal decision rule linearly combines the received signals through different diversity branches after co-phasing and weighting them with their respective channel gains. This result illustrates that the branches that have better channel gains, i.e., larger instantaneous signal-to-noise ratios, will be emphasized more than others. This is intuitive since the received signals through better channels are more reliable, and thus provide us with more accurate information. After the received signals are combined, the decision rule is the same as the case with no diversity. The resulting combining scheme is named *maximal ratio combining*.

We note that we have not made any assumptions on the statistics of the channel gains or on their independence. The maximal ratio combining rule above is general and will apply as long as the noise terms have identical variances, and they are independent. If their variances are not the same, it is straightforward to scale each observation appropriately, and apply the same rule.

Equivalent Channel Model with MRC

Let us now consider the performance of a digital communication system when L diversity branches are available. The effective channel model with MRC is given by

$$y = \left(\sum_{l=1}^{L} |h_l|^2\right) x + n', \tag{2.29}$$

where the noise term is still Gaussian, but its variance per complex dimension is $(\sum_{l=1}^{L} |h_l|^2)/2$. Therefore, the effective instantaneous signal-to-noise ratio with MRC is $(\sum_{l=1}^{L} |h_l|^2)\rho$. This clearly shows that the signal-to-noise ratio of L branch diversity with MRC is the sum of the instantaneous signal-to-noise ratios for each branch. Thus, the outage and the error probabilities will be reduced.

Outage Probability with MRC

Let us compute the outage probability for the case of independent Rayleigh fading branches when MRC is employed. The effective signal-to-noise ratio after combining is chi-square distributed with $2L$ degrees of freedom. That is, its p.d.f. is given by

$$p_{\rho_{eff}}(u) = \frac{u^{L-1} e^{-\frac{u}{\rho}}}{\rho^L (L-1)!}, \qquad u > 0. \tag{2.30}$$

The outage probability is then given by

$$P_{out} = P\left(\rho_{eff} < \rho_{min}\right), \tag{2.31}$$

$$= \int_0^{\rho_{min}} \frac{u^{L-1} e^{-\frac{u}{\rho}}}{\rho^L (L-1)!} du, \tag{2.32}$$

$$= 1 - e^{-\frac{\rho_{min}}{\rho}} \left(\sum_{l=1}^{L} \frac{(\rho_{min}/\rho)^{l-1}}{(l-1)!} \right). \tag{2.33}$$

It is easy to see that this expression decays with $1/\rho^L$ at high signal-to-noise ratios. This is in contrast to the $P_{out} \sim \frac{1}{\rho}$ behavior for the system with no diversity. Therefore, the system performance is improved considerably.

Average Error Rates with MRC

A similar computation can be carried out for the average error rates as well. For instance, if BPSK is employed, the average error probability over Rayleigh fading with MRC is expressed as

$$P_{b,MRC} = \int_0^{\infty} Q\left(\sqrt{2u}\right) p_{\rho_{eff}}(u) du, \tag{2.34}$$

which, after some straightforward algebra, can be shown to be

$$P_{b,MRC} = \left(\frac{1-\Gamma}{2}\right)^L \sum_{l=0}^{L-1} \binom{L-1+l}{l} \left(\frac{1}{2}(1+\Gamma)\right)^l, \tag{2.35}$$

with $\Gamma = \sqrt{\frac{\rho}{1+\rho}}$.

This is the exact error rate expression for BPSK with MRC over Rayleigh fading. Although it is somewhat difficult to see the diversity order obtained in this relatively complicated form, we note that the high signal-to-noise ratio behavior of this expression is still $P_{b,MRC} \sim \frac{1}{\rho^L}$, demonstrating that the average error rates decay dramatically faster than the case with no diversity. To see the diversity order in an easier way, we can use the upper bound on the Q-function, $Q(u) \leq \frac{1}{2}e^{-\frac{u^2}{2}}$, and obtain

$$P_{b,MRC} \leq \int_0^\infty \frac{1}{2}e^{-u}\frac{u^{L-1}e^{-\frac{u}{\rho}}}{\rho^L(L-1)!}du, \tag{2.36}$$

$$= \frac{1}{2}\frac{1}{(1+\rho)^L}, \tag{2.37}$$

$$\leq \frac{1}{2}\frac{1}{\rho^L}, \tag{2.38}$$

which clearly shows the diversity order achieved as L.

Before concluding this section, we stress that the average error rate performance is meaningful when the channel fading is relatively fast so that for an entire frame of transmission we can observe different channel states.

2.3.4 Suboptimal Combining Algorithms

MRC is not the only way of combining signals observed through different diversity branches. There are many other alternatives including selection combining (SC), equal-gain combining (EGC), switch-and-stay combining (SSC), etc.

In selection combining, the main idea is to work with the branch that sees the best channel conditions for any given transmission. Among all the L transmissions, the branch with the highest signal-to-noise ratio is picked, and the decision is made based on this link alone. In equal-gain combining, the signals of all branches are co-phased (multiplied by $e^{-j\angle h_l}$ using baseband notation) and summed together to form the equivalent channel output. In switch-and-stay combining, we use a particular branch for demodulation until the signal-to-noise ratio of the branch falls below a certain threshold. When it falls below the threshold, we switch to another branch (for instance, the one with the largest instantaneous signal-to-noise ratio), and stay with it until its signal-to-noise ratio falls below the given threshold. This combining algorithm is also called threshold combining due to the way it is implemented.

Different combining algorithms have their advantages and disadvantages. For instance, MRC is optimal, however its implementation requires exact knowledge of the instantaneous channel gains, and thus is more complicated. On the other hand, for selection combining the performance is inferior, however the implementation is simpler as exact knowledge of the channel gains is not needed in practice. We can simply measure the received signal power for different branches (perhaps this can be done at the radio frequency stage), and make the selection based on the received signal power. This may result in a performance degradation, but doing this will allow us to work with the channel estimate of a single branch, i.e., the best one at a given time. We can even go one step further and eliminate the channel estimation requirement altogether by using differential modulation schemes. Furthermore, we can trade-off the performance with the complexity even further by employing switch-and-stay combining as this technique will avoid frequent changes in the branch used for demodulation.

Different diversity techniques are covered in many digital communications and wireless communications texts (see the end of the chapter for a list). Therefore, we will not go into too much depth. However, we will consider selection combining in the following in more detail as it will become important for one of the later topics treated in this book, i.e., antenna selection for MIMO systems.

2.3.5 Selection Combining

As summarized in the previous section, in selection combining, at any given transmission interval, the branch with the largest signal-to-noise ratio is used in demodulation. Therefore, mathematically, we have the following input–output relationship between the transmitted and received signals

$$y = \left(\max_{j=1,2,\ldots,L} |h_j| \right) x + n', \tag{2.39}$$

where n' is a complex Gaussian random variable with variance $1/2$ per dimension. Thus the effective instantaneous signal-to-noise ratio after combining is given by

$$\rho_{SC,eff} = \left(\max_{j=1,2,\ldots,L} |h_j|^2 \right) \rho, \tag{2.40}$$

whose p.d.f. is given by (using order statistics)

$$p_{\rho_{SC,eff}}(u) = \frac{L}{\rho} e^{-\frac{u}{\rho}} \left(1 - e^{-\frac{u}{\rho}} \right)^{L-1}, \qquad u > 0. \tag{2.41}$$

Using this intermediate result we can find the outage and average error probabilities of interest.

Outage Probability with SC

Using a similar line of thought with the case of MRC, the outage probability with SC is obtained as

$$\begin{aligned} P_{out,SC} &= \int_0^{\rho_{min}} \frac{L}{\rho} e^{-\frac{u}{\rho}} \left(1 - e^{-\frac{u}{\rho}} \right)^{L-1} du, \\ &= \left(1 - e^{-\frac{\rho_{min}}{\rho}} \right)^L . \end{aligned} \tag{2.42}$$

For large signal-to-noise ratios, this can be approximated as

$$P_{out,sc} \approx \left(\frac{\rho_{min}}{\rho} \right)^L , \tag{2.43}$$

which clearly shows that the outage probability is greatly reduced due to the use of Lth order diversity just like the case of MRC.

Average Error Rates with SC

Let us illustrate the average error rate performance of SC with a specific modulation scheme. An interesting modulation scheme to be used in conjunction with the SC is differential PSK (DPSK). This is because no explicit channel estimation is needed for detection, and in practice, selection can be done using the received power (which may degrade the detector performance slightly). For binary DPSK, the error rate in the presence of Gaussian noise conditioned on the channel signal-to-noise ratio is given by $P_b(\rho) = \frac{1}{2}e^{-\rho}$. Therefore, the average error rate using SC is found to be

$$\begin{aligned} P_{b,bin\ DPSK} &= \int_0^\infty \frac{1}{2}e^{-u}\frac{L}{\rho}e^{-\frac{u}{\rho}}\left(1-e^{-\frac{u}{\rho}}\right)^{L-1} du, \\ &= \frac{L}{2}\sum_{l=0}^{L-1}\binom{L-1}{l}(-1)^l\frac{1}{1+l+\rho}, \end{aligned} \tag{2.44}$$

which also behaves like $\sim \frac{1}{\rho^L}$ for $\rho \gg 1$, and thus a diversity of order L is achieved.

2.3.6 Examples

In this section, we give several examples of system performance with and without diversity. Figure 2.12 shows the outage probabilities over Rayleigh fading (as functions of average signal-to-noise ratio per branch) when two, three, four and five branches are employed with $\rho_{min} = 0$ dB for MRC and SC. Also shown is the outage performance with no diversity. Clearly, the use of diversity improves the outage probabilities significantly. We further observe that the marginal gains with increased number of diversity branches are reduced, and that MRC outperforms SC.

In Figure 2.13, we illustrate the average error rates with binary DPSK with differential detection for several cases. We again observe the great benefits of using diversity, superiority of MRC over SC (which is more noticeable at lower signal-to-noise ratios), and the fact that gains become marginal with increased number of diversity branches (if the total signal-to-noise ratio (ρL) is kept constant).

2.4 Channel Coding as a Means of Time Diversity

Time diversity can be considered as repetition coding where different parts of the codeword corresponding to a particular symbol are transmitted using different time intervals which are separated by at least the coherence time of the wireless channel. This is clearly very

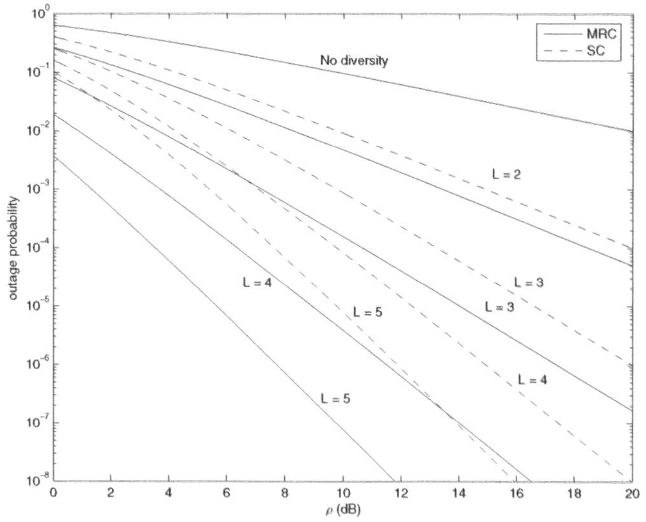

Figure 2.12 Outage probability of MRC and SC over a Rayleigh fading channel (assuming minimum acceptable signal-to-noise ratio is 0 dB).

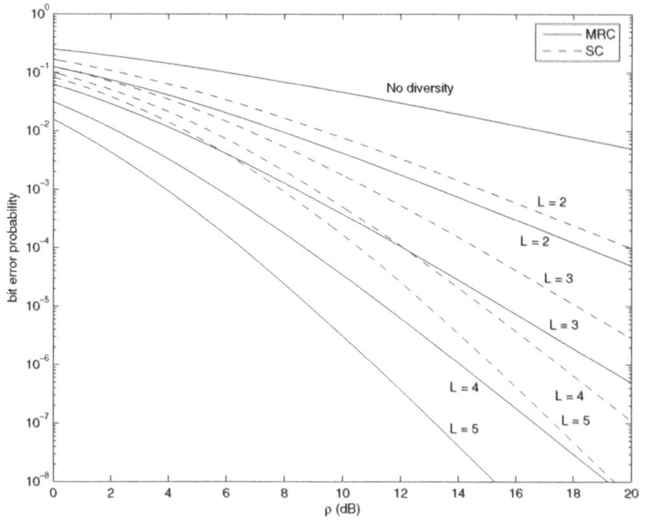

Figure 2.13 Average error probability of binary DPSK with MRC and SC over a Rayleigh fading channel.

inefficient. Instead, one can employ a more sophisticated channel code, as we will discuss in this section, to obtain the same diversity performance with a smaller overhead.

Block coding or convolutional coding techniques can be employed to obtain time diversity (see, for example, Biglieri (2005); Lin and Costello (2004); Wicker (1995)). As for the specific modulation scheme, for simplicity, one may employ BPSK. However, if the bandwidth efficiency is a concern, higher order modulation techniques can also be used in almost the same manner, i.e., one can design a coded modulation scheme (based on an underlying block code) or a trellis coded modulation scheme (see Caire et al. (1998); Divsalar and Simon (1987, 1988a,b); Jamali and Le-Ngoc (1994); Schlegel and Costello (1989); Ungerboeck (1982); Zehavi (1992)). Here since our objective is to only briefly describe diversity that can be obtained with channel coding, we will not explicitly consider the use of higher order modulation schemes.

As described earlier, fading channels exhibit correlation in time, that is, in a typical wireless channel, consecutive bits normally see very similar channel fades, depending on the coherence time of the channel. However, in order to obtain diversity using a channel code, different parts of the codeword should undergo different channel fades. Therefore, the use of an interleaver, whose role is to scramble the coded bits (or, symbols) before transmission, is necessary. Clearly, a de-interleaver whose function is exactly the opposite is needed at the receiver before channel decoding can be performed. See Figure 2.14 for a generic block diagram of the system.

The exact design of the interleaver is not essential. Its function is simply to decorrelate the "effective" channel that consecutively transmitted bits see, thereby providing a means of recovery of information bits even when some of the components of the corresponding codeword are deeply faded. A popular choice which is simple to implement is the "block interleaver" which basically writes the bits to be transmitted into the rows of a matrix, and reads them columnwise. The corresponding de-interleaver needs to write incoming bits to the columns of a matrix (of the same size as the one used for interleaving), and then read them row-wise.

Let us now compute the diversity order obtained with channel coding using an ideal (or, infinite depth) interleaver for which the channel effectively becomes i.i.d. fading, i.e., fully interleaved fading.

2.4.1 Block Coding over a Fully Interleaved Channel

Assume that a linear block code with minimum distance d_{min} is used over a fully interleaved fading channel. Let us compute an upper bound on the codeword error probability. Since the

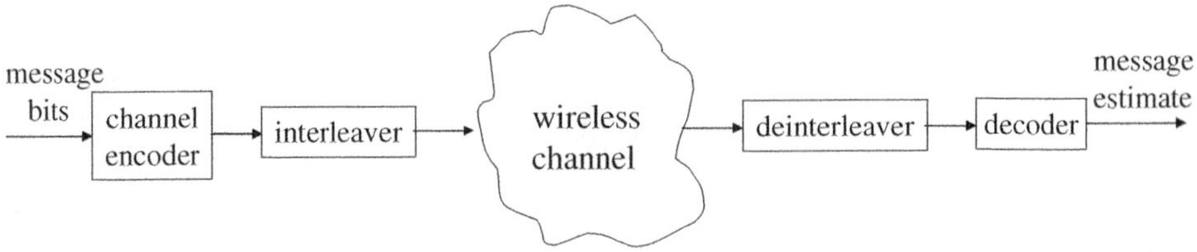

Figure 2.14 Coding over a wireless channel.

code is assumed to be linear, we can assume that the all-zero codeword ($\mathbf{0}$) is transmitted without any loss of generality. That is, denoting the probability of error by P_e, we have

$$P_e = P(\text{error}|\mathbf{0} \text{ is transmitted}). \tag{2.45}$$

This is very difficult to compute in general, therefore using the union bound, we upper bound this expression by a simpler one given by

$$P_e \le \sum_{c \neq \mathbf{0}} P(\mathbf{0} \to \boldsymbol{c}), \tag{2.46}$$

where $P(\mathbf{0} \to \boldsymbol{c})$ is the pairwise error probability (PEP) of receiving a signal "closer" to the incorrect codeword $\boldsymbol{c}$ when the all-zero codeword is transmitted.

At this point we note that the value of the PEP, hence the bound on the error probability, depends on the decoding method employed. We consider two different cases: hard-decision decoding and soft-decision decoding. In each case, we use the corresponding maximum likelihood decoder and compute an upper bound on the error probability above.

Hard-Decision Decoding (HDD)

In hard-decision decoding, tentative hard decisions are made first (based on matched filter or correlator outputs) without using the code constraints. The received sequence is then nothing but a binary sequence of bits. Therefore, the equivalent channel is a binary symmetric channel (BSC) with a certain cross-over probability, say p. Without loss of generality, assume that $p \le \frac{1}{2}$. With the assumption that the channel is fully interleaved, the BSC is memoryless, and errors that occur in bit transmissions are independent of each other.

In HDD, the maximum likelihood decoding rule picks the codeword closest to the received sequence in the Hamming distance sense as the decoder output. For the computation of the PEP, we note that if the received sequence (denoted by $\boldsymbol{y}$) is closer to $\boldsymbol{c}$ in the Hamming distance sense compared with the all zero codeword $\mathbf{0}$, then a "pairwise" error event occurs. This will happen if there are more than half of the bits in error (among the d positions for which the component of $\boldsymbol{c}$ is "1") where d is the weight of the codeword $\boldsymbol{c}$. In case of exactly $d/2$ errors among these positions, a fair coin is tossed, i.e., there is an error with probability $1/2$. In other words, we have the following PEP

$$P(\mathbf{0} \to \boldsymbol{c}) = \sum_{l=\frac{d+1}{2}}^{d} \binom{d}{l} p^l (1-p)^{d-l} \quad \text{if } d \text{ is odd}, \tag{2.47}$$

or,

$$P(\mathbf{0} \to \boldsymbol{c}) = \frac{1}{2}\binom{d}{\frac{d}{2}} p^{\frac{d}{2}} (1-p)^{\frac{d}{2}} + \sum_{l=\frac{d}{2}+1}^{d} \binom{d}{l} p^l (1-p)^{d-l} \quad \text{if } d \text{ is even}. \tag{2.48}$$

In either case, we can upper bound the PEP as

$$P(\mathbf{0} \to \boldsymbol{c}) \le (4p(1-p))^{\frac{d}{2}}. \tag{2.49}$$

Substituting this upper bound in the expression (2.46), we obtain an upper bound on the error probability of the overall code as

$$P_e \leq \sum_{\mathbf{c}\neq\mathbf{0}} (4p(1-p))^{\frac{d}{2}}. \tag{2.50}$$

Noting that the argument of the sum is the largest when d has its smallest possible value (i.e., $d = d_{min}$), and that there are exactly $2^k - 1$ non-zero codewords, we obtain

$$P_e \leq (2^k - 1)(4p(1-p))^{\frac{d_{min}}{2}}. \tag{2.51}$$

Let us now specialize the result to the case of Rayleigh flat fading with BPSK modulation. For this case, the cross-over probability of the equivalent BSC is given by

$$p \approx \frac{1}{4\rho_c}, \tag{2.52}$$

$$= \frac{1}{4R_c\rho}, \tag{2.53}$$

where R_c is the code rate, ρ_c is the average signal-to-noise ratio per coded bit and ρ is the average signal-to-noise ratio per information bit transmitted. Using this result, and assuming $\rho \gg 1$, we find that

$$P_e \leq (\text{constant}) \frac{1}{\rho^{\frac{d_{min}}{2}}}, \tag{2.54}$$

which shows that the diversity order obtained with HDD is simply $d_{min}/2$.

To be precise, we also need to find a lower bound on the codeword error probability that decays like $\sim \frac{1}{\rho^{d_{min}/2}}$. However, this is immediate, since the error probability will be larger than the PEP between the actual transmitted codeword and its nearest neighbor in the Hamming distance sense.

Soft-Decision Decoding (SDD)

Although HDD is relatively simple to implement, it does not give the best possible performance. Instead, using soft-decision decoding may be a better idea if the receiver complexity is not a major issue. In this case, the channel decoder directly works with the matched filter outputs to perform maximum likelihood decoding. Let us now illustrate the diversity order that can be achieved with SDD.

Let us consider BPSK modulation where coded bits are transmitted (equivalently) as -1 (for "0") and as 1 (for "1"), i.e., energy per coded bit is normalized to unity. Over a Rayleigh flat fading channel, the received signal is given by the vector $\boldsymbol{y} = (y(1), y(2), \ldots, y(n))$, where $y(l) = \sqrt{\rho_c} h(l)u(l) + n(l)$, where $h(l)$ are the zero mean complex Gaussian channel gains with variance $1/2$ per dimension, $n(l)$ are the additive Gaussian noise terms with variance $1/2$ per dimension, $u(l) = \pm 1$, and ρ_c is the signal-to-noise ratio per coded bit. The fading coefficients over the entire codeword are assumed to be independent of each other due to the fully interleaved fading assumption.

Assuming that the channel coefficients are perfectly available at the receiver, the optimal ML decoder declares the codeword whose corresponding clean (noiseless) signal is closest to the received signal in the Euclidean distance sense. That is, the ML decoder minimizes the Euclidean distance between $\boldsymbol{y}$ and the vector $(\sqrt{\rho_c}h(1)u(1), \sqrt{\rho_c}h(2)u(2), \ldots, \sqrt{\rho_c}$ $h(n)u(n))$ over all the codewords $\boldsymbol{u}$ (where $\boldsymbol{u} = (u(1), u(2), \ldots, u(n))$).

Let us now turn to the computation of the PEP. Conditioned on the fading channel coefficients, the PEP is only a function of the squared Euclidean distance between the all-zero codeword and the codeword $\boldsymbol{c}$, which is given by $4\rho_c(|h(j_1)|^2 + |h(j_2)|^2 + \cdots + |h(j_d)|^2)$ where d is the weight of $\boldsymbol{c}$ and $j_1, j_2, \ldots, j_d$ are the places with the component of $\boldsymbol{c}$ being "1". We can write

$$P(\mathbf{0} \to \boldsymbol{c}) = E\left[Q\left(\sqrt{2\rho_c(|h(j_1)|^2 + |h(j_2)|^2 + \cdots + |h(j_d)|^2)}\right)\right], \tag{2.55}$$

where the expectation is over the fading coefficients $h(j_1), h(j_2), \ldots, h(j_d)$. Using the p.d.f. of the instantaneous signal-to-noise ratios ($\gamma_l = \rho_c|h(j_l)|^2$) with Rayleigh fading, this can be written explicitly as

$$P(\mathbf{0} \to \boldsymbol{c}) = \int\int \cdots \int Q\left(\sqrt{2(\gamma_1 + \gamma_2 + \cdots + \gamma_d)}\right) \prod_{l=1}^{d} \frac{1}{\rho_c} e^{-\frac{\gamma_l}{\rho_c}} d\gamma_1 \ldots d\gamma_d. \tag{2.56}$$

By upper bounding the Q-function using $Q(x) \le \frac{1}{2}e^{-\frac{x^2}{2}}$, we obtain

$$P(\mathbf{0} \to \boldsymbol{c}) \le \prod_{l=1}^{d} \left(\frac{1}{2\rho_c} \int_0^\infty e^{-\gamma_l} e^{-\frac{\gamma_l}{\rho_c}} d\gamma_l\right), \tag{2.57}$$

$$= \frac{1}{(2(1 + \rho_c))^d}, \tag{2.58}$$

$$\le \frac{1}{(2\rho_c)^d}. \tag{2.59}$$

By substituting this upper bound on the PEP expression in (2.46), we get an upper bound on the codeword error rate as

$$P_e \le \sum_{\boldsymbol{c} \neq \mathbf{0}} \frac{1}{(2\rho_c)^d}, \tag{2.60}$$

$$\le (2^k - 1)\frac{1}{(2\rho_c)^{d_{min}}}, \tag{2.61}$$

which clearly shows that $P_e \le (\text{constant})\frac{1}{\rho^{d_{min}}}$ (since $\rho_c = R_c\rho$). As in the case of HDD, it is easy to see that the error probability is lower bounded by the PEP of two closest codewords which decays with $\sim \frac{1}{\rho^{d_{min}}}$. Therefore a diversity order of d_{min} is achieved. This is clearly much better than the performance offered by the HDD.

2.4.2 Convolutional Coding

In addition to block coding techniques, one can also employ convolutional coding over fading channels to obtain diversity. For a convolutional code, long sequences of message bits are encoded using a finite state machine implemented with shift registers. The memory of the encoder (specified by the number of shift registers) determines the complexity of the code. If for each set of k information bits input to the encoder, there are n output bits produced, then the code rate is k/n. These codes are significantly different than the block codes since they are not obtained by mapping a fixed-length message to a fixed-length codeword. Instead, for convolutional codes, both the messages and the coded sequences are of infinite length. An example of a convolutional code with rate $1/2$ produced by $(21, 37)_{octal}$ generators is shown in Figure 2.15.

Optimal decoding of a convolutionally coded sequence (in AWGN) is achieved using the Viterbi algorithm (if the sequence error probability is to be minimized), or a maximum a-posteriori (MAP) type of an algorithm can be used (if the bit error rate is to be minimized). For the Viterbi algorithm, hard-decision decoding (minimizing the Hamming distance between the received sequence and the set of codewords), or soft-decision decoding can be employed. For flat fading channels (with channel state information at the receiver), soft-decision decoding minimizes the Euclidean distance between the received sequence and the set of "weighted" coded sequences. The decoding complexity is determined by the number of branches emanating from each state (in the trellis representation), and the number of states (determined by the memory of the code).

As in the case of block coding, if achieving a high spectral efficiency is a major concern due to limited bandwidth availability, convolutional codes can be combined with higher order modulation schemes efficiently. The resulting scheme is referred as "trellis coded modulation" (Ungerboeck (1982, 1987a,b)).

Since the codeword sequence is very long, one question that arises is the length of the interleaver needed to decorrelate the channel to obtain diversity in a practical system (see Forney (1971) and references therein). It turns out that most of the errors are due to short error events that diverge from the correct path at some state and merge with it again after a (relatively small) number of steps. Therefore, the interleaver can be used to effectively make sure that fading coefficients observed over the "short" error events are different. This can easily be accomplished by using a relatively short interleaver whose length can be

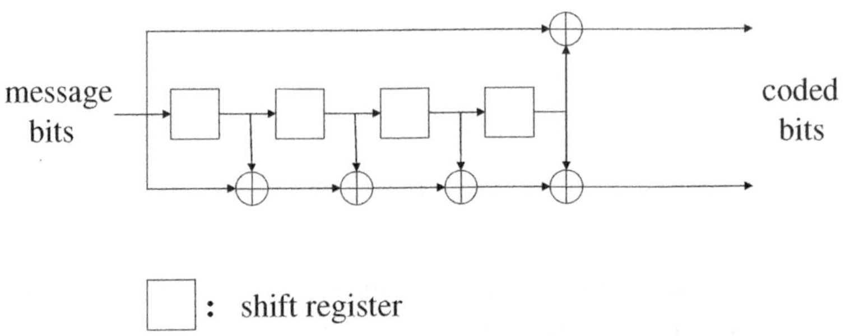

Figure 2.15 An example of a convolutional code with generators $(21, 37)_{octal}$.

determined using the length of dominant error events and the channel coherence time. The exact nature of the interleaver is not crucial, i.e., a block or a convolutional interleaver can be used efficiently.

To see the diversity order that will be provided by a convolutional code over a fully interleaved Rayleigh flat fading channel, we note that if the convolutional code is terminated by adding several bits at the end to make sure that the final state of the code is the same as the initial (all-zero) state, we obtain a linear block code with a minimum distance given by the free distance of the convolutional code d_{free}. Since this is true regardless of the block length selected, and for large block lengths, we expect the convolutional code to behave very similar to the block code obtained by its termination, and the derivation of the previous section applies. Therefore, we can conclude that the diversity order with HDD is $d_{free}/2$ and with SDD is d_{free}.

In fact, we can obtain a more accurate characterization of the code performance over fading channels using the transfer function of the code. That is, we can compute upper bounds on the bit error probability. However, we omit them here.

We finally note that the results and techniques developed for space-time trellis codes in Chapter 5 would be applicable to convolutional codes, or trellis codes as well. This point will become clear later.

2.5 Multiple Antennas in Wireless Communications

An important resource in achieving reliable communications over unreliable wireless links is "space", i.e., the use of multiple antennas. Different ways of using multiple antenna elements can be identified as described below.

2.5.1 Receive Diversity

We have already talked about the use of multiple receive antennas that can be employed to provide spatial diversity. This is a well-known technique, and it has been studied thoroughly in the literature.

For receive diversity, as shown in Figure 2.10, we simply have L replicas of the transmitted signal where each replica is effectively transmitted through a different (spatial) channel. If the receive antennas are placed sufficiently far apart, then the signals received by these antennas undergo (almost) independent fading. For instance, in a uniform scattering environment, half a wavelength separation is sufficient. For other environments, a larger separation may be needed. Using any of the diversity combining techniques, these signals can be used to make a decision on the transmitted symbol and an Lth order diversity can be achieved over a Rayleigh flat fading channel.

Multiple antennas can also be employed at the transmitter as discussed in the following two subsections.

2.5.2 Smart Antennas and Beamforming

Consider an environment that has a relatively small number of local scatterers, and hence the wireless communication is achieved mainly through line of sight, or a small number of specular components. For instance, this may be the scenario in an ad-hoc network

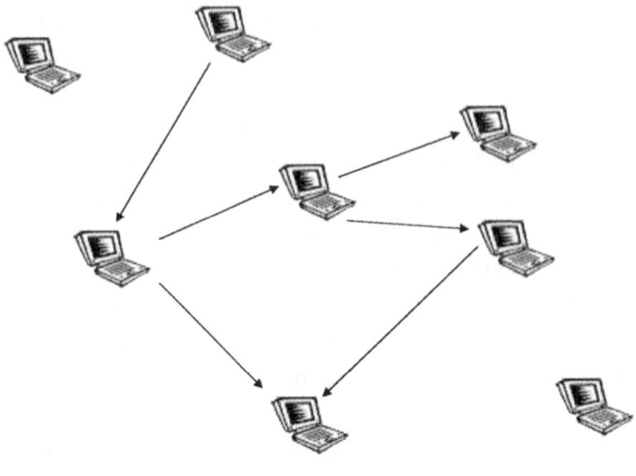

Figure 2.16 A simple ad-hoc network illustration.

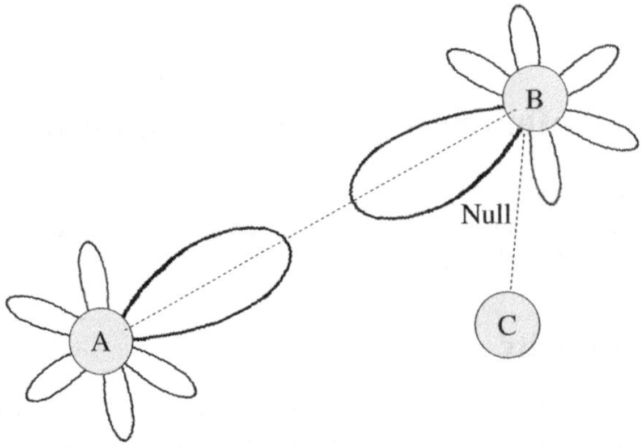

Figure 2.17 Illustration of beamforming being used for improving signal-to-noise ratio and reducing interference.

formed by mobile computers in a large conference room. A simple illustration is provided in Figure 2.16. In such an environment, multiple antenna elements can be employed at the transmitters in order to provide directionality for the electromagnetic waves, hence improving the effective signal-to-noise ratio of the channel. For instance, as shown in Figure 2.17, mobile A can "focus" on the intended receiver (mobile B), thus significantly increasing the received signal power at the intended destination, and dramatically reducing possible interference caused to other users in the system. In fact, multiple antennas can also be employed at the receiver. For instance, node B can effectively obtain a receive antenna pattern as shown in the figure to pick up the transmitted signal efficiently, and to reduce the amount of interference it sees.

The use of multiple antenna elements (antenna arrays, or smart antennas) in such an environment, either at the transmitter or at the receiver side, is referred to as "beamforming" for obvious reasons (see Liberti and Rappaport (1999); Litva and Lo (1996)). This technique has been known for over fifty years, and it is used successfully in many applications, including cellular systems to provide sectoring (that will give a better spatial reuse) and underwater acoustic communications.

A fundamental approach that makes beamforming work is simply the transmission of the same signal from each of the antenna elements, with a certain gain and a phase shift (or, equivalently, with a complex gain). This can be implemented either by analog or digital means, though using digital techniques offer much more flexibility. Different techniques can be used to find suitable coefficients for use. For instance, a direction-of-arrival algorithm can be used (which could be training based), and the coefficients can be computed to make the effective antenna pattern have a main beam in this direction. At the transmitter side, with some feedback from the receiver, position information can be obtained to accomplish the same. Interference suppression can also be accomplished where some of the interference being suppressed could be due to the different replicas of the desired signal (this is needed since they may be out of phase with the main arrival).

Although beamforming or the use of smart antennas is a very important subject, we do not go into its details in this book. Our main objective is to consider a completely different wireless scenario, and study the recent techniques of space-time communications (that provide transmit and receive diversity and/or increased transmission rates). This is introduced in the next subsection, and studied thoroughly in the rest of the book.

2.5.3 Space-Time Coding – Basic Ideas

Consider again a wireless communications scenario, but assume that unlike the set-up in the previous subsection, there are many local scatterers, and the channel between the transmitter and the receiver is not due to a line of sight, or a specular component, hence it is a "rich scattering" fading channel. In such a case, there is no specific direction of arrival for the desired signal (or the interference), instead we observe a large number of multipath components that are not resolvable, i.e., the proper channel model is that of fading.

In this setting, the use of spatial receive diversity is straightforward as discussed before. However, it turns out we can also develop transmit diversity schemes, and combine them with receive diversity. This is a recent approach that can be used to provide spatial diversity (transmit and receive) and/or to improve the transmission rates of a wireless communication system without requiring additional bandwidth or power. The rest of the book is dedicated to multiple-input multiple-output transmission techniques that facilitate this approach.

The idea in space-time coding is to encode information both spatially and temporally and transmit the encoded sequence over multiple antenna elements using the same bandwidth. Encoding in either dimension is in fact optional, and results in variations of space-time coding. For instance, if independent uncoded streams of symbols are transmitted over different transmit antenna elements, we simply obtain what is called a spatial multiplexing scheme.

It is worth emphasizing the fundamental difference between "beamforming" and "space-time coding". In a beamforming scenario, there is a certain desired direction of transmission and reception, and antenna arrays are used at the transmitter and the receiver to place the main beams of the antenna patterns in this direction, while suppressing possible interference. The transmitted signals are the same (scaled by certain coefficients), and the received signals

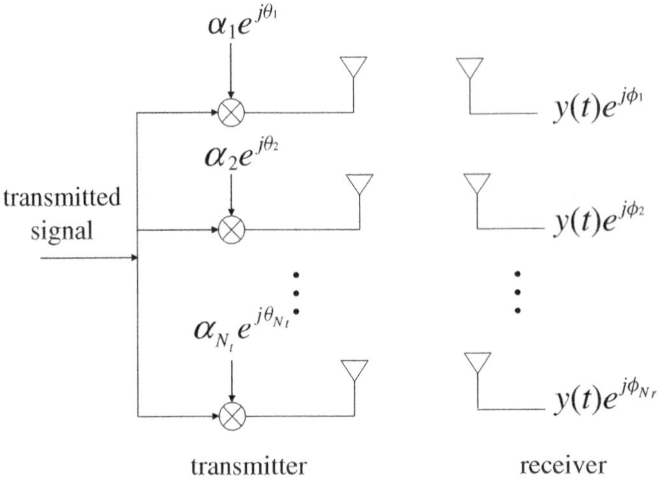

Figure 2.18 Multiple antennas being used for beamforming.

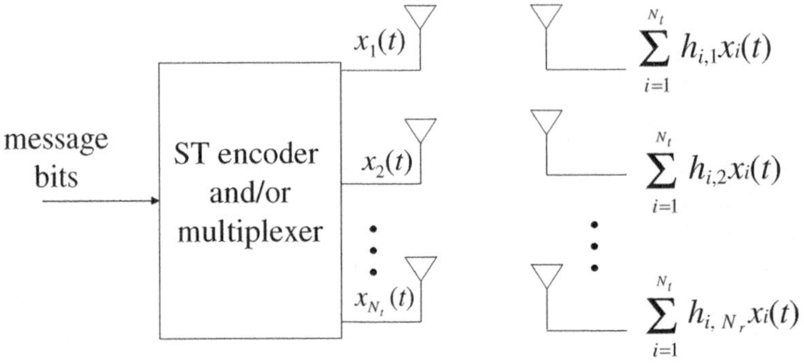

Figure 2.19 Multiple antennas being used for space-time coding or spatial multiplexing.

are simply phase-shifted versions of each other. The block diagram of the transmitter and the receiver is illustrated in Figure 2.18. The main objective is to increase the effective signal-to-noise (plus interference) ratio. In the space-time coding set-up, there is no specific direction that we would like to communicate. In general, the transmitted signals are completely different, i.e., they may be produced by the same encoder, so they may be correlated, or they may be completely independent information streams. The received signals are obtained via completely different fading channels. This is illustrated in Figure 2.19. In this case, the main objective is to obtain diversity and/or increased transmission rates.

2.6 Chapter Summary and Further Reading

In this chapter, we have reviewed some of the basics of wireless communications, developed suitable fading channel models and considered digital modulation techniques over these channels. In addition, we have considered the use of diversity techniques, including channel coding, as a means of improving the communication reliability over wireless links. Finally, we have reviewed the use of multiple antennas, and identified multiple-input multiple-output transmission over fading channels as an important topic. The latter is the subject of the rest of this book.

There is an extensive literature on wireless communications. Biglieri et al. (1998) provide an excellent survey of communications through fading channels. There are also many books available that provide thorough discussions of characteristics of such channels and various signaling techniques, for example, Goldsmith (2005); Molisch (2001); Rappaport (2001); Simon and Alouini (2004); Stuber (2001); Tse and Viswanath (2005). In addition, several digital communications textbooks, including Proakis (2001) contain detailed coverage of wireless channels in general. For detailed coverage of smart antenna systems, the reader may want to refer to Kaiser et al. (2004); Lehne and Pettersen (1999); Liberti and Rappaport (1999); Litva and Lo (1996); Winters (1998). Specifically, in the area of space-time coding, there are several recent books, including Biglieri et al. (2006); Gershman and Sidiropoulos (2005); Giannakis et al. (2007); Jafarkhani (2005); Jankiraman (2004); Larsson et al. (2003); Paulraj et al. (2003); Vucetic and Yuan (2003), and many survey type papers, such as Paulraj et al. (2004) and Boelcskei (2006).

Problems

2.1 A scattering function for a fading channel is given by $S(\tau, \lambda) = 1$ if $0 \leq \tau \leq 50$ μs and $|\lambda| < 5$ Hz, and it is zero otherwise.

a) What is the multipath intensity profile of the channel? What is its Doppler power spectrum?

b) What are the multipath delay spread, Doppler spread, coherence time and coherence bandwidth of the channel?

c) Can we design a digital communication system such that this channel can be viewed as a slow frequency flat fading channel? If so, what should the symbol period be selected as?

2.2 Consider BPSK modulation over a flat Rayleigh fading channel and assume that the low pass equivalent received signal is $y_l(t)$, and the pulse shaping function used is rectangular.

a) Describe the coherent maximum likelihood receiver (make sure to give a block diagram) if the channel state information is available at the receiver.

b) If a noisy version of the channel state information is available, how would the maximum likelihood receiver change? (Assume that the channel phase is known perfectly but not the amplitude).

c) What is the maximum likelihood receiver if no channel state information is available at the receiver? (Again assume that the channel phase is available).

2.3 Assume that two-branch diversity with BPSK modulation is used to transmit digital data. The received signals through the two diversity branches are given by

$$y_i = \sqrt{\rho} h_i x + n_i, \qquad i = 1, 2,$$

where y_i is the received signal, x is the transmitted signal (where $x = \pm 1$ with equal probability), and n_i is a zero mean (white) Gaussian noise with variance $1/2$.

Assume that the joint probability mass function of h_1 and h_2 is given by

$$p_{h_1,h_2}(h_1, h_2) = \begin{cases} 0.1 & \text{if } h_1 = h_2 = 1, \\ 0.1 & \text{if } h_1 = 1, \quad h_2 = 2, \\ 0.1 & \text{if } h_1 = 2, \quad h_2 = 1, \\ 0.7 & \text{if } h_1 = 2, \quad h_2 = 2. \end{cases}$$

a) What is the probability of bit error if maximal ratio combining is used at the receiver?

b) What is the probability of bit error if selection combining is used?

c) What are the marginal probability mass functions of h_1 and h_2? Are the two diversity branches independent? Why or why not?

d) How would your answers to parts (a) and (b) change if the two diversity branches were independent with the same marginal densities (as computed in part (c))? Comment on your result.

2.4 Assume that differential BPSK (with differential detection) is used to transmit digital data over the channel described by

$$y = \sqrt{\rho} h x + n,$$

where y is the received signal if x is the transmitted signal (± 1). The probability density function of $A = |h|^2$ is as shown in Figure 2.20, and n is a zero mean (complex) white Gaussian noise with variance $1/2$ per dimension.

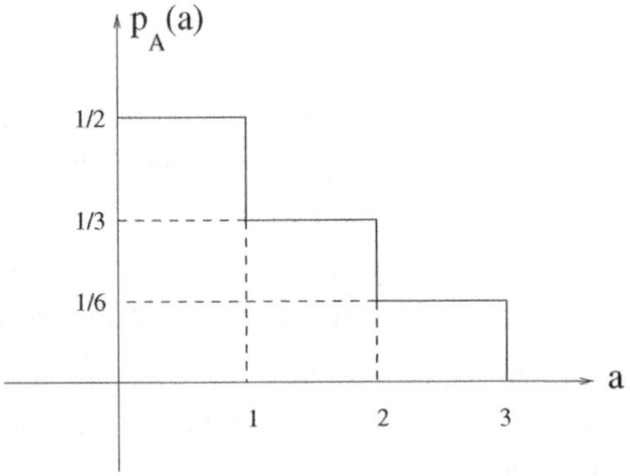

Figure 2.20 Figure for Problem 2.4.

a) Assume that the fading is so slow that the outage probability is important. Assuming that the minimum acceptable signal-to-noise ratio threshold is $\rho_{\min} = \frac{1}{10}\rho$, find the outage probability.

b) Assuming that the fading is relatively rapid (i.e., we can average over the channel fading), what is the average probability of error? How can it be approximated for large ρ?

2.5 Assume that BPSK modulation is used over a wireless channel and coherent demodulation is employed at the receiver. Also assume that the instantaneous signal-to-noise ratio at the receiver is given by $\gamma = \beta\rho$ where the probability density function of β is

$$p(\beta) = \sum_{i=0}^{\infty} a_i \beta^{t+i}, \qquad \beta \geq 0,$$

for some t.

a) What is the probability of error?

b) Give an approximate expression for the probability of error for large signal-to-noise ratios. What is the diversity order achieved?

c) Answer part (b) for the specific case of a Nakagami fading channel for which the probability density function of β is given by

$$p(\beta) = \frac{m^m \beta^{m-1}}{\Gamma(m)} \exp(-m\beta), \qquad \beta \geq 0,$$

where the Gamma function is defined as

$$\Gamma(z) = \int_0^{\infty} x^{z-1} e^{-x} dx.$$

2.6 Derive Equation (2.35).

2.7 Assume that a sinusoid with frequency f_c is transmitted to a mobile moving with a constant velocity v. The mobile receives the signal transmitted via five multipath components (with equal strengths). The multipath components have the angles (measured counter-clockwise with the direction of motion) $\theta_1 = 30^o$, $\theta_2 = 60^o$, $\theta_3 = 90^o$, $\theta_4 = 120^o$, and $\theta_5 = 150^o$. An isotrophic antenna is used at the mobile.

a) If the total power received is 1 mW, what is the power spectral density of the low pass equivalent received signal?

b) Assume further that we know the delays associated with the five multipath components, which are 0 ms, 0.1 ms (two of them), 0.2 ms and 0.25 ms. What is the multipath delay profile of the channel? Do you have sufficient information to find the scattering function?

2.8 Three independent diversity branches are used to transmit a BPSK modulated digital data over a channel described by

$$y = \sqrt{\rho} h x + n,$$

where y is the received signal, x is the transmitted signal (± 1 with equal probabilities). Assume that h is a random variable with probability density function $p_h(h) = 0.1\delta(h - 1) + 0.9\delta(h - 3)$, n is a zero mean (white) Gaussian noise with variance $1/2$. Assume that the channel state information is available at the receiver for each of the three branches.

a) What is the optimal decision rule at the receiver?

b) What is the probability of error with equal-gain combining?

c) What is the probability of error with maximal-ratio combining?

d) What is the probability of error with selection combining?

e) What is the probability of error if we first make a hard decision on each branch and then use majority-logic combining to make the final decision?

f) What is the probability of error if we first select the "best" two branches, and then use maximal-ratio combining with these two branches?

g) Order the schemes of (b) through (f) from best to worst in terms of performance for large signal-to-noise ratios.

2.9 Two independent diversity branches with average signal-to-noise ratios ρ and 2ρ are used over a flat Rayleigh fading channel. Assume that the modulation scheme is binary DPSK and the demodulator employs differential detection. Assume that the signals of the two branches are combined using selection combining.

a) What is the probability of error?

b) Which one is more effective to reduce the bit error probability (for binary differential PSK with differential detection with selection combining) – two branches with equal power or unequal power?

3

Capacity and Information Rates of MIMO Channels

One of the primary concerns in the study of noisy channels is the computation of channel capacity, which is defined as the maximum rate at which we can transmit information with an arbitrarily low probability of error. This establishes a fundamental limit on reliable communications. Therefore, the channel capacity is widely used for evaluating the performance of communication systems. Our main objective in this chapter is to study channel capacity issues pertaining to MIMO wireless channels to demonstrate that the use of MIMO systems will increase the transmission rates significantly without requiring additional power or bandwidth. This will establish our motivation for the rest of the book, i.e., detailed coverage of various coding techniques for MIMO communication systems.

Parallel to the capacity of MIMO channels, we will also consider achievable information rates for various communication scenarios. By achievable information rates, we refer to the transmission rates that can be supported with an arbitrarily low probability of error for a given type of input, such as inputs drawn independently from a certain signal constellation (e.g., BPSK, PAM, QAM). These results will be very useful in evaluating practical communication system performance, since in practice we are limited to these types of input signals anyway.

We begin the chapter with a brief discussion of capacity and achievable information rates, and then specialize the results to the case of AWGN channels. We then extend the discussion to fading channels. After establishing the channel capacity for single-input single-output systems, we move on to the study of wireless MIMO channels and demonstrate the tremendous capacity improvements promised by the use of multiple antennas.

3.1 Capacity and Information Rates of Noisy Channels

Consider the generic communication channel model depicted in Figure 3.1. The message W is encoded to a sequence of n channel inputs $X_1, X_2, \ldots, X_n$, and the sequence $Y_1, Y_2, \ldots, Y_n$ is observed at the channel output. The received sequence is then used to

Coding for MIMO Communication Systems Tolga M. Duman and Ali Ghrayeb

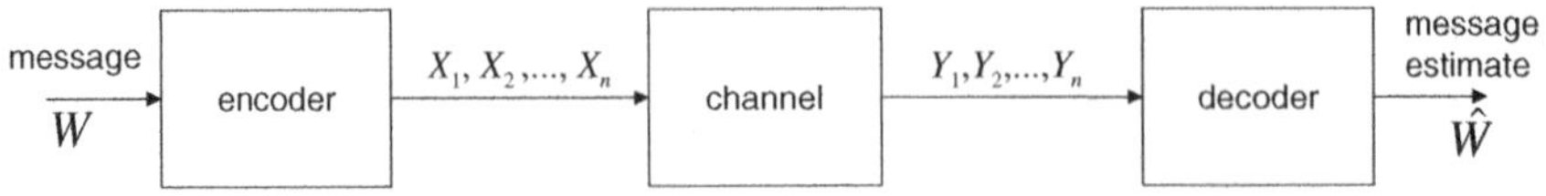

Figure 3.1 Generic block diagram for a channel coded communication system.

recover the transmitted message, producing the estimate $\hat{W}$. The channel input is characterized by a joint p.d.f. $p(x_1, x_2, \ldots, x_n)$, the channel output has p.d.f. $p(y_1, y_2, \ldots, y_n)$, and the channel statistics are described by the conditional p.d.f. $p(y_1, y_2, \ldots, y_n|x_1, x_2, \ldots, x_n)$.

Let us now characterize the limits of reliable communications over this generic channel model. In his pioneering work, Shannon (1948) has proved that over a noisy channel, the channel capacity which is defined as the highest transmission rate per channel use that can be supported with an arbitrarily low probability of error $P(\hat{W} \neq W)$ as the block length n goes to infinity, is given by the maximum mutual information between the channel input and the channel output (denoted by $I(\cdot;\cdot)$, where the maximum is taken over the joint distribution of the input sequence, i.e., $p(x_1, x_2, \ldots, x_n)$). To prove this result, Shannon employed a random coding argument. The idea is to pick codes with long block lengths at random, and to upper bound the resulting error probability averaged over all possible codes. He showed that if the transmission rate is below the channel capacity, then the average error probability goes to zero as the block length n goes to infinity. Therefore, there must be at least one code (with a long block length) for which the probability of error can be reduced to any desired level, i.e., reliable communication is possible.

Mathematically, the channel capacity C is given by

$$C = \lim_{n\to\infty} \frac{1}{n} \max_{p(x_1,x_2,\ldots,x_n)} I(X_1, X_2, \ldots, X_n; Y_1, Y_2, \ldots, Y_n). \tag{3.1}$$

For memoryless channels, i.e., if the input–output relationship can be written as

$$p(y_1, y_2, \ldots, y_n|x_1, x_2, \ldots, x_n) = \prod_{i=1}^{n} p(y_i|x_i), \tag{3.2}$$

then the channel capacity can be simplified to a single letter representation given by

$$C = \max_{p(x)} I(X; Y), \tag{3.3}$$

where the mutual information $I(X; Y)$ is defined as

$$I(X; Y) = E\left[\log \frac{p(X, Y)}{p(X)p(Y)}\right], \tag{3.4}$$

where the expectation is over the joint density of X and Y. If the logarithm is taken with respect to base 2, then the resulting channel capacity is given in bits per channel use. If instead, the natural logarithm is used, then the unit is nats per channel use.

In practical communication systems, channel inputs are taken from specific signal constellations. For instance, if we employ BPSK, the possible channel inputs are $+1$ and -1.

In this case, although maximization of the mutual information over a general input distribution gives the channel capacity, another important quantity is what is called the "achievable information rate" where maximization is carried out with specific constraints on the channel inputs. This achievable information rate is definitely smaller than or equal to the channel capacity, and it is a useful quantity that represents the maximum possible rate for reliable communications when a specific signal constellation is employed.

3.2 Capacity and Information Rates of AWGN and Fading Channels

We now consider widely encountered AWGN and fading channel models in more detail, and characterize their capacity and achievable information rates with constrained signaling.

3.2.1 AWGN Channels

As a basic but important channel model, let us consider the capacity of a widely encountered communication channel, namely, the discrete time AWGN channel. The input–output relationship is given by

$$y = \sqrt{\rho}x + n, \tag{3.5}$$

where x is the channel input, y is the channel output, and n represents the additive noise term. To keep the terminology consistent with the rest of the chapter (and the book), we assume that the input, output and noise terms are complex (two-dimensional). The noise is modeled as a complex Gaussian random variable with zero mean and variance $1/2$ per dimension. The noise terms at different uses of the channel are assumed to be independent. The channel input is assumed to satisfy the power constraint $E[|x|^2] \leq 1$, thus the constant ρ represents the signal-to-noise ratio at the receiver.

It can be shown that for this channel model the input distribution that maximizes the mutual information is zero mean complex Gaussian with variance $1/2$ per dimension (independent for different uses of the channel), and the resulting channel capacity is given by

$$C = \log(1+\rho). \tag{3.6}$$

In practice, using Gaussian channel inputs is not feasible. Therefore although the capacity expression given above is an upper limit for reliable transmission, it may not be possible to achieve. For example, assume that the modulation scheme used is BPSK, then with the given power constraint, the channel input x is either $+1$ or -1. With this constraint, the input that maximizes the mutual information expression uses these two possibilities with the same probability ($1/2$ each). Therefore, the achievable information rate for BPSK modulation over an AWGN channel is given by

$$\begin{aligned} I &= \frac{1}{2}\sum_{x'=\pm 1}\int p(y|x=x')\log\frac{2p(y|x=x')}{p(y|x=-1)+p(y|x=1)}\,dy, \\ &= 1-\sqrt{\frac{1}{\pi}}\int_{-\infty}^{\infty} e^{-(u-\sqrt{\rho})^2}\log(1+e^{-4\sqrt{\rho}u})du. \end{aligned} \tag{3.7}$$

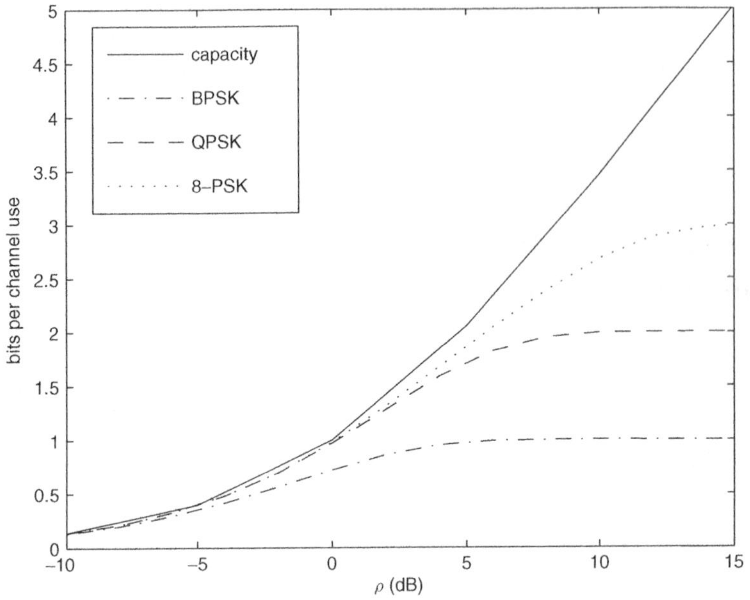

Figure 3.2 Capacity and information rates for several modulation schemes over AWGN channels.

In Figure 3.2, we illustrate the capacity of an AWGN channel as well as the achievable information rates for several modulation schemes (BPSK, QPSK and 8-PSK). It is clear that for an M-ary modulation scheme, the maximum achievable rate is limited by $\log_2 M$ bits per channel use. It is also observed that for low signal-to-noise ratios, the constrained information rate results are very close to the channel capacity obtained by using Gaussian inputs. This is explained as follows. The optimal input distribution is the one that induces Gaussian distribution at the channel output. However, if the signal-to-noise ratio is very low, the noise variance is high relative to the signal power, and regardless of the input selected, the channel output becomes approximately Gaussian.

3.2.2 Fading Channels

Let us now consider flat fading channels as extensions of AWGN channels studied in the previous section. Instead of a comprehensive coverage of capacity and information rate issues, we only highlight some of the important results that will be needed for extensions to multi-antenna wireless systems.

Consider the flat fading channel model, where the input–output relationship is given by

$$y = \sqrt{\rho} h x + n, \tag{3.8}$$

where h is the complex channel gain. The input x has the same power constraint as in the AWGN case, i.e., $E[|X|^2] \leq 1$, and the noise is complex Gaussian with zero mean and variance $1/2$ per dimension. For the specific case of Rayleigh fading, the channel gain is zero mean complex Gaussian, or equivalently, $|h|$ is a Rayleigh random variable and $\phi = \angle h$ is a uniform random variable on $(0, 2\pi)$. We normalize the channel gain such that $E[|h|^2] = 1$, thus ρ is the received signal-to-noise ratio. Regarding the availability of the channel state information, we consider only one case, that is, the channel fade coefficient is known precisely at the receiver, and its distribution is known both at the receiver and the transmitter.

For fading channels, there are two capacity definitions that are important: ergodic (Shannon type) capacity and outage capacity for non-ergodic channels. We will deal with these two cases separately in the following.

Ergodic Capacity and Information Rates

Assume that the fading channel is ergodic. In this case, the channel coefficients vary over time, and it is possible to average over their statistics by coding over large blocks of data. For this scenario, the channel capacity is obtained by taking the expected value of capacity of an AWGN channel with the signal-to-noise ratio $|h|^2\rho$ (over the statistics of the channel coefficient h). That is, defining $z = |h|$, we have

$$C_{fading} = \int_0^\infty \log\left(1 + z^2\rho\right) p(z)dz, \tag{3.9}$$

or, equivalently by averaging over the p.d.f. of the instantaneous signal-to-noise ratio ρ' of the channel as

$$C_{fading} = \int_0^\infty \log\left(1 + \rho'\right) p(\rho')d\rho', \tag{3.10}$$

where $\rho' = |h|^2\rho$. Clearly, for some cases, this integral can be evaluated analytically. However, in general a closed form solution may not exist, and one would need to resort to numerical integration techniques. For instance, we can use Monte Carlo techniques for evaluation, i.e., we can generate a large number of realizations for the instantaneous signal-to-noise ratio ρ' using the channel statistics, compute $\log(1 + \rho')$ corresponding to each, and average these quantities to estimate the capacity. This is guaranteed to converge using the law of large numbers.

In order to see why the channel capacity is obtained by the above expectation, we note the following. Since the additive noise is independent Gaussian, the capacity is achieved by transmitting a very long sequence of independent Gaussian inputs with zero mean, and a fixed variance (determined by the power constraint). Since the transmitter does not have the channel state information, all the symbols are transmitted with the same variance. At the receiver, the resulting channel outputs are classified with respect to their fading states. This can be done since the fading coefficients are known at the receiver. Therefore, equivalently, for each state of the channel, long blocks of independent identically distributed Gaussian inputs are employed, i.e., for a given signal-to-noise ratio, a transmission rate of $\log(1 + \rho')$ is obtained, and the ergodic capacity is simply their average.

As mentioned previously, another important quantity is the achievable information rates with constrained inputs. This can also be calculated for fading channels. For instance, consider a fading channel with Gaussian noise, for BPSK modulation, the information rate

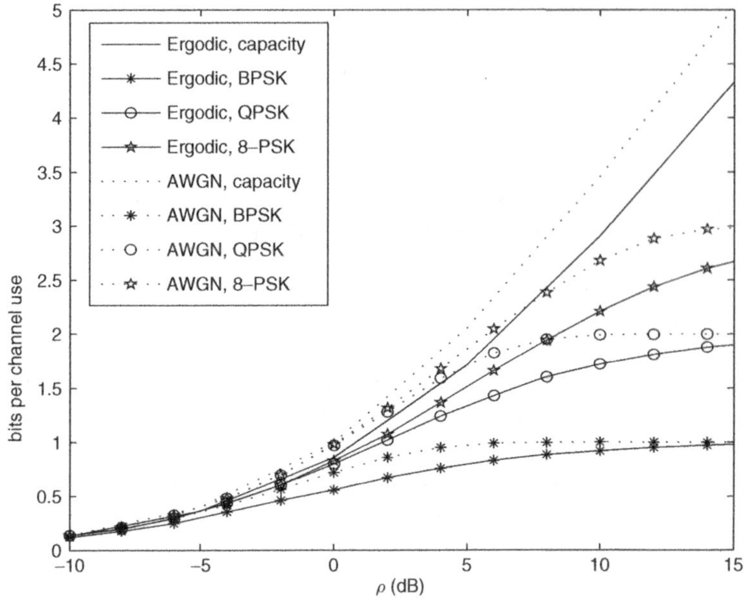

Figure 3.3 Capacity and information rates for several modulation schemes over ergodic Rayleigh fading channels.

can be computed by simply averaging the expression in (3.7) over the instantaneous signal-to-noise ratio, $\rho|h|^2$. Clearly, Monte Carlo techniques can also be employed as in the case of AWGN channels.

As an example, Figure 3.3 shows the capacity and achievable information rates with BPSK, QPSK and 8-PSK modulation schemes over ergodic Rayleigh flat fading channels as a function of the average signal-to-noise ratio. We observe that the channel fading deteriorates the capacity and information rates considerably compared with the case of AWGN channels, as expected. Also, the information rates with BPSK and QPSK inputs are very close to the channel capacity for low signal-to-noise ratios. However, they level off at one bit/channel use and two bits/channel use for high signal-to-noise ratios. We also observe from the figure a large difference between the information rates and capacity results for large signal-to-noise ratios.

Outage Capacity and Outage Information Rates

Let us now assume that fading is non-ergodic, that is, it is not possible to code across different states of the channel, and we have to live with a limited number of realizations for an entire frame of data. For instance, the quasi-static Rayleigh channel model introduced in the previous chapter falls in this category. In this case, the ergodic or Shannon type capacity of the channel is simply zero. This is because, regardless of the frame length, there is a non-zero probability that the instantaneous mutual information between the channel input and the channel output is below any given fixed rate. Therefore, another capacity definition is needed, namely, the outage capacity.

Consider a quasi-static fading channel, for a given channel realization, or instantaneous signal-to-noise ratio, the channel capacity is a random variable. For instance, for a quasi-static Rayleigh flat fading channel, the instantaneous channel capacity as a function of the signal-to-noise ratio is given by $C(\rho') = \log(1 + \rho')$, where ρ' is an exponential random variable. Therefore, for any given rate R, there is a probability that any coding scheme will not be supported reliably over this channel. That is, the outage probability is given by

$$P_{out} = P(C(\rho') < R). \tag{3.11}$$

To interpret this in another way, for any given outage probability level, there is an outage capacity associated with it, with the interpretation that when the system is not in outage (which occurs with probability $1 - P_{out}$), this particular transmission rate can be supported.

A similar definition holds for the outage information rates. That is, when a specific channel input is employed, the outage probability is given by

$$P_{out} = P(I(\rho') < R) \tag{3.12}$$

for a given rate R where the mutual information is computed for the given signal-to-noise ratio level using specific input constraints.

As a simple example, in Figure 3.4, we provide the $P_{out} = 10\%$ outage capacity and outage information rates with BPSK, QPSK and 8-PSK signals over a Rayleigh fading channel. Again for low signal-to-noise ratios, the capacity and information rates are similar, but for large signal-to-noise ratios, they differ significantly.

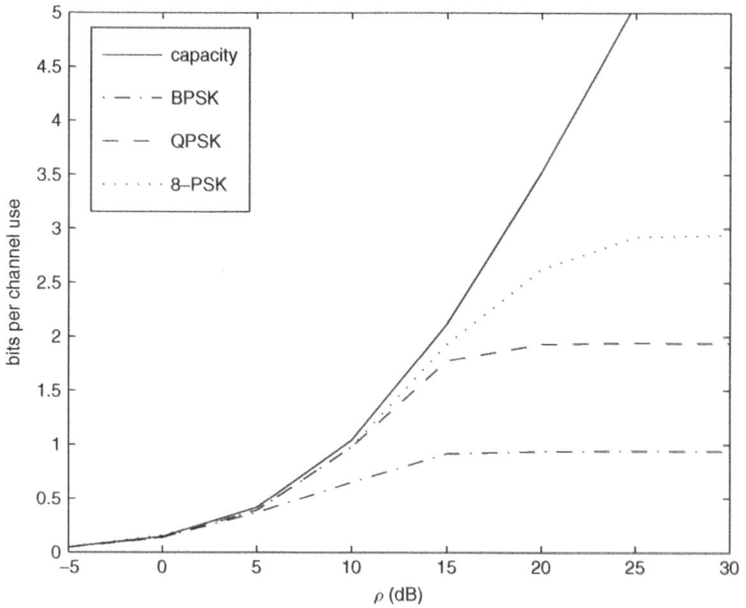

Figure 3.4 Outage capacity and information rates for quasi-static Rayleigh fading channels.

Other Possible Extensions

Admittedly, the above coverage of fading channel capacity is limited to the case of known channel state information at the receiver, and known input distribution both at the transmitter and the receiver. Other assumptions are certainly possible.

An important but difficult problem is the calculation of channel capacity when neither the transmitter nor the receiver knows the channel state information, but they have access to its distribution. In this case, the results shown in this section are not applicable. In fact, a full characterization is very difficult. Even for the simple case of independent identically distributed Rayleigh fading, the capacity is obtained using a discrete input with a finite number of mass points (Abou-Faycal et al. (2001)), and should be calculated numerically.

Another possible extension is the knowledge of the channel state information both at the transmitter and the receiver. In this case, since the transmitter knows when the channel fading is severe, it is possible to allocate the overall power in an optimal manner by performing waterfilling. The main idea is to allocate more power to more reliable transmissions, and less to the unreliable ones (deep fades) while keeping a total power constraint over a long block of symbols. We do not go into the details of waterfilling here. Instead, we will study the idea to some extent in the following section in the context of communications over deterministic MIMO channels.

3.3 Capacity of MIMO Channels

Let us now consider capacity and information rate issues for wireless MIMO systems. Our main objective is to illustrate that the channel capacity improves significantly when MIMO systems are used without requiring additional bandwidth or power, thereby motivating the rest of the book.

We assume that there are N_t transmit and N_r receive antennas as illustrated in Figure 3.5, and that there is no intersymbol interference (i.e., the sub-channels are flat fading). The input–output relationship of the MIMO channel is given by

$$\boldsymbol{y} = \sqrt{\rho}\boldsymbol{x}\boldsymbol{H} + \boldsymbol{n}, \tag{3.13}$$

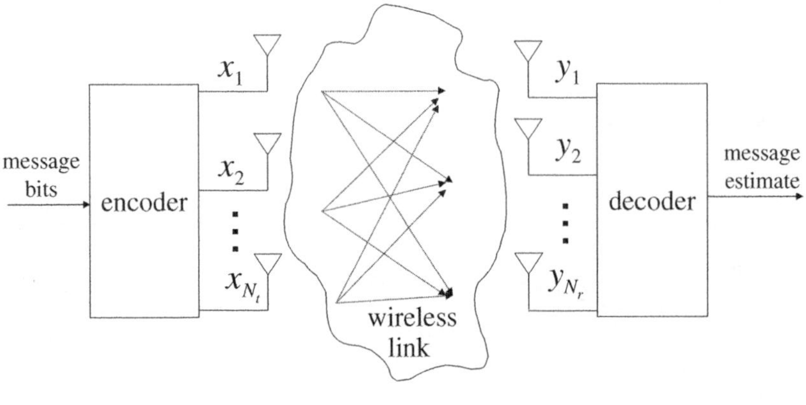

Figure 3.5 Generic block diagram for a channel coded MIMO communication system.

where $\boldsymbol{x}$ is the $1 \times N_t$ vector of transmitted signals (whose components are complex numbers), $\boldsymbol{H}$ is the $N_t \times N_r$ matrix denoting the channel gains for each transmit and receive antenna pair, $\boldsymbol{n}$ is the $1 \times N_r$ vector of independent complex Gaussian noise terms. The signal vector satisfies the power constraint $E[\boldsymbol{x}\boldsymbol{x}^H] \leq 1$. Each component of the noise vector has a variance of $1/2$ per dimension, and the noise terms at different receive antennas and for different uses of the channel are independent. If the channel does not introduce additional power, we can interpret the constant ρ as the signal-to-noise ratio at each receive antenna.

In the following, we will consider deterministic MIMO channels and random (fading) MIMO channels. For the case of fading channels, we will study the ergodic and non-ergodic cases separately.

3.3.1 Deterministic MIMO Channels

For deterministic MIMO channels, the assumption is that the channel gain matrix $\boldsymbol{H}$ is fixed. This may be the case for instance for fixed wireless links when the variations in the environment are negligible.

Using singular value decomposition (SVD), we can write

$$\boldsymbol{H} = \boldsymbol{U}\boldsymbol{\Sigma}\boldsymbol{V}^H, \tag{3.14}$$

where $\boldsymbol{U}$ and $\boldsymbol{V}$ are $N_t \times N_t$ and $N_r \times N_r$ unitary matrices (i.e., $\boldsymbol{U}^H\boldsymbol{U} = \boldsymbol{I}_{N_t}$, $\boldsymbol{V}^H\boldsymbol{V} = \boldsymbol{I}_{N_r}$), and $\boldsymbol{\Sigma}$ is an $N_t \times N_r$ non-negative diagonal matrix whose diagonal elements are the singular values of the matrix $\boldsymbol{H}$. Let $\sigma_1, \sigma_2, \ldots, \sigma_\nu$ denote the non-zero singular values, clearly, $\nu \leq \min\{N_t, N_r\}$. Without loss of generality, we assume that the singular values are sorted in a non-increasing order, i.e., $\sigma_1 \geq \sigma_2 \geq \cdots \geq \sigma_\nu$.

Using the SVD of $\boldsymbol{H}$, as an equivalent channel model, we can write (3.13) as

$$\tilde{\boldsymbol{y}} = \sqrt{\rho}\tilde{\boldsymbol{x}}\boldsymbol{\Sigma} + \tilde{\boldsymbol{n}}, \tag{3.15}$$

where $\tilde{\boldsymbol{x}} = \boldsymbol{x}\boldsymbol{U}$, $\tilde{\boldsymbol{y}} = \boldsymbol{y}\boldsymbol{V}$ and $\tilde{\boldsymbol{n}} = \boldsymbol{n}\boldsymbol{V}$. Since the matrices $\boldsymbol{U}$ and $\boldsymbol{V}$ are invertible, knowing $\tilde{\boldsymbol{x}}$ and $\tilde{\boldsymbol{y}}$ is the same as knowing $\boldsymbol{x}$ and $\boldsymbol{y}$. Also, the power constraint on the new input vector is the same as the one on the original input vector since $E[\tilde{\boldsymbol{x}}\tilde{\boldsymbol{x}}^H] = E[\boldsymbol{x}\boldsymbol{U}\boldsymbol{U}^H\boldsymbol{x}^H] = E[\boldsymbol{x}\boldsymbol{x}^H] \leq 1$. Furthermore, the new noise vector has the same statistics as the original one (since $\boldsymbol{V}$ is unitary and $\boldsymbol{n}$ has independent Gaussian elements). Therefore, equivalent to the original channel model, we have the following set of parallel channels

$$\tilde{y}_1 = \sqrt{\rho}\sigma_1\tilde{x}_1 + \tilde{n}_1, \tag{3.16}$$

$$\tilde{y}_2 = \sqrt{\rho}\sigma_2\tilde{x}_2 + \tilde{n}_2, \tag{3.17}$$

$$\vdots$$

$$\tilde{y}_\nu = \sqrt{\rho}\sigma_\nu\tilde{x}_\nu + \tilde{n}_\nu, \tag{3.18}$$

$$\tilde{y}_{\nu+1} = \tilde{n}_{\nu+1}, \tag{3.19}$$

$$\vdots$$

$$\tilde{y}_{N_r} = \tilde{n}_{N_r}, \tag{3.20}$$

where the power constraint on the input is $\sum_{i=1}^{N_t} E[|\tilde{x}_i|^2] \leq 1$ and the noise terms are independent complex Gaussian with variance 1/2 per dimension. Clearly, this is nothing but a set of usual parallel Gaussian channels with independent noise terms having identical variances.

Equal Transmit Power Allocation

Let us first assume that although the MIMO channel is deterministic (fixed $\boldsymbol{H}$), the transmitter does not have access to the channel matrix (but the receiver does). Therefore the transmitter cannot optimize its power allocation among its antennas. In this case, the capacity of the MIMO channel is given by

$$C = \max_{p(\boldsymbol{X})} I(\boldsymbol{X}; \boldsymbol{Y}), \tag{3.21}$$

$$= \max_{p(\boldsymbol{X})} H(\boldsymbol{Y}) - H(\boldsymbol{Y}|\boldsymbol{X}). \tag{3.22}$$

When the channel input is given, the remaining uncertainty in the output is simply the entropy of the noise. Thus we can write

$$C = \max_{p(\boldsymbol{X})} H(\boldsymbol{Y}) - H(\boldsymbol{N}). \tag{3.23}$$

As the noise components are independent complex Gaussian with variance 1/2 per dimension, the entropy of $\boldsymbol{N}$ is $H(\boldsymbol{N}) = N_r \log(\pi e)$. Therefore, the capacity is achieved if the entropy of the channel output is maximized. On the other hand, for the entropy of the channel output, we can write

$$H(\boldsymbol{Y}) \leq \log\left(\det(\pi e \boldsymbol{R}_y)\right),$$

where $\boldsymbol{R}_y$ denotes the covariance matrix of the output vector $\boldsymbol{y}$, and we have equality if $\boldsymbol{y}$ is complex Gaussian. Clearly the channel output vector $\boldsymbol{y}$ is complex Gaussian if the channel input vector $\boldsymbol{x}$ is complex Gaussian. Noting that $\boldsymbol{R}_y = E[\boldsymbol{y}^H \boldsymbol{y}] = \rho \boldsymbol{H}^H \boldsymbol{R}_x \boldsymbol{H} + \boldsymbol{I}_{N_r}$, we can write the channel capacity as

$$C = \max_{\boldsymbol{R}_x} \log\det(\rho \boldsymbol{H}^H \boldsymbol{R}_x \boldsymbol{H} + \boldsymbol{I}_{N_r}), \tag{3.24}$$

where the maximization is subject to the constraint $E[\boldsymbol{x}\boldsymbol{x}^H] = \text{trace}(\boldsymbol{R}_x) \leq 1$.

Since we assume that the transmitter does not have the channel state information, with the given trace constraint, it selects an input with covariance $\boldsymbol{R}_x = \frac{1}{N_t}\boldsymbol{I}_{N_t}$. Namely the capacity-achieving input vector is independent complex Gaussian with equal power on each of the antennas. Therefore, the channel capacity is given by

$$C = \log\det\left(\boldsymbol{I}_{N_r} + \frac{\rho}{N_t}\boldsymbol{H}^H \boldsymbol{H}\right). \tag{3.25}$$

Since the product $\boldsymbol{H}^H \boldsymbol{H}$ is positive semidefinite with positive eigenvalues ($\lambda_1, \lambda_2, \ldots, \lambda_\upsilon$) as the squares of the non-zero singular values of $\boldsymbol{H}$ (i.e., $\lambda_1 = \sigma_1^2, \lambda_2 = \sigma_2^2, \ldots, \lambda_\upsilon = \sigma_\upsilon^2$), it can be diagonalized using a unitary matrix $\boldsymbol{W}$ as $\boldsymbol{H}^H \boldsymbol{H} = \boldsymbol{W}\boldsymbol{\Lambda}\boldsymbol{W}^H$, where $\boldsymbol{\Lambda}$ is a diagonal matrix containing the eigenvalues $\lambda_1, \lambda_2, \ldots, \lambda_\nu$ (and possibly zero eigenvalues). Noting that

$$\log \det \left(\boldsymbol{I}_{N_r} + \frac{\rho}{N_t} \boldsymbol{H}^H \boldsymbol{H} \right) = \log \det \left(\boldsymbol{I}_{N_r} + \frac{\rho}{N_t} \boldsymbol{W} \boldsymbol{\Lambda} \boldsymbol{W}^H \right), \tag{3.26}$$

$$= \log \det \left(\boldsymbol{W} \left(\boldsymbol{I}_{N_r} + \frac{\rho}{N_t} \boldsymbol{\Lambda} \right) \boldsymbol{W}^H \right), \tag{3.27}$$

we obtain

$$C = \sum_{i=1}^{\nu} \log \left(1 + \frac{\rho}{N_t} \lambda_i \right), \tag{3.28}$$

as the MIMO channel capacity.

It is easy to interpret the final result using the equivalent set of parallel channels as described previously. Since the optimal input with equal power allocation among the transmit antennas is independent complex Gaussian, and the transformation from $\boldsymbol{x}$ to $\tilde{\boldsymbol{x}}$ is unitary, the optimal choice of $\tilde{\boldsymbol{x}}$ is also independent complex Gaussian. Therefore, for capacity-achieving signaling, independent Gaussian inputs over the set of N_r parallel channels are used (each with power $\frac{1}{N_t}$), and the overall capacity is simply the sum of the capacities of these parallel channels.

Beamforming

In the previous subsection, we have assumed that the transmitter does not have access to the channel state information, hence it cannot optimize its power allocation. Let us now assume that the transmitter knows the channel matrix $\boldsymbol{H}$. With the ordering of the equivalent parallel channels from good to bad (i.e., with $\lambda_1 \geq \lambda_2 \geq \cdots \geq \lambda_\nu$), the transmitter may choose to allocate all its power to the first channel and no power to the remaining links, resulting in a capacity of

$$C = \log(1 + \rho\lambda_1). \tag{3.29}$$

This is referred to as the beamforming scheme (Paulraj et al. (2003)). This is because, in order to accomplish the use of only the best channel, the transmitter emits the same signal from each of the antennas scaled by a certain coefficient. To see this, we again note that $\tilde{\boldsymbol{x}} = \boldsymbol{x}\boldsymbol{U}$, or $\boldsymbol{x} = \tilde{\boldsymbol{x}}\boldsymbol{U}^H$. Therefore, in order to use the best parallel channel, we need to select $\boldsymbol{x} = [\tilde{x}, 0, 0, \ldots, 0]\boldsymbol{U}^H$ where $\tilde{x}$ is picked to be complex Gaussian. But this is equivalent to saying that all the antennas transmit the same symbol (scaled by the conjugates of the entries in the first column of the matrix $\boldsymbol{U}$).

We note that although this is also referred to as beamforming, the idea is different than the one presented in the previous chapter in the context of smart antennas (Liberti and Rappaport (1999); Litva and Lo (1996)).

Waterfilling

When the transmitter has access to the channel state information, the use of equal power allocation or beamforming (transmitting the same signal from each of the antennas) is suboptimal in general. Instead, the optimal solution is obtained by applying what is known as the waterfilling principle as described next.

Denoting the power of the symbol transmitted on the ith parallel channel by P_i (with the nonzero gain), the capacity is given by

$$C = \max_{\boldsymbol{P}:\sum_{j=1}^{\nu} P_j \leq 1} \sum_{i=1}^{\nu} \log\left(1 + \rho\lambda_i P_i\right), \tag{3.30}$$

with $\boldsymbol{P} = [P_1, P_2, \ldots, P_\nu]$. Clearly, this capacity is achieved if the inputs are independent complex Gaussian.

This constrained optimization problem yields the following well-known solution

$$P_{i,opt} = \left(\mu - \frac{1}{\rho\lambda_i}\right)^+, \tag{3.31}$$

where $(z)^+ = \max\{z, 0\}$ and μ is the solution of

$$\sum_{i=1}^{\nu} \left(\mu - \frac{1}{\rho\lambda_i}\right)^+ = 1, \tag{3.32}$$

and the resulting capacity is

$$C = \sum_{i=1}^{\nu} \left(\log(\mu\rho\lambda_i)\right)^+. \tag{3.33}$$

This solution shows that the optimal scheme only uses some of the equivalent parallel channels depending on the signal-to-noise ratio. If the signal-to-noise ratio is very low, only the channel with the best gain will be excited making beamforming optimal. As the signal-to-noise ratio increases, more and more of the parallel channels will be used for transmission.

Examples

Let us now illustrate the capacity of a deterministic MIMO channel by an example. Assume that there are three transmit and four receive antennas, i.e., $N_t = 3$, $N_r = 4$, and the channel matrix is given by

$$\boldsymbol{H}_1 = \begin{bmatrix} 0.5+0.5j & j & 1 & -0.5+j \\ -0.7 & 0.3+0.5j & 1.2-0.5j & 0.8 \\ 0.6j & -0.8 & -0.2+0.9j & 0.4j \end{bmatrix}. \tag{3.34}$$

By SVD, it is easy to see that the nonzero singular values of $\boldsymbol{H}_1$ are $\sigma_1 = 2.53$, $\sigma_2 = 1.36$ and $\sigma_3 = 0.81$, and the eigenvalues of $\boldsymbol{H}_1^H \boldsymbol{H}_1$ are $\lambda_1 = 6.41$, $\lambda_2 = 1.86$ and $\lambda_3 = 0.65$. If beamforming is to be employed, the weights of the symbols transmitted from the three antennas are $(-0.67, -0.34 - 0.44j, -0.13 + 0.47j)$.

The capacity of this channel obtained by equal power allocation, beamforming and optimal waterfilling is shown in Figure 3.6. We observe that for low signal-to-noise ratios, i.e., when the noise power is very high, beamforming is optimal, this is because all the power should be spent on exciting only one of the parallel channels. Equal power allocation is not efficient for this range as the scarce power is distributed among different parallel links. On the other hand, for large signal-to-noise ratios, the opposite is true. That is,

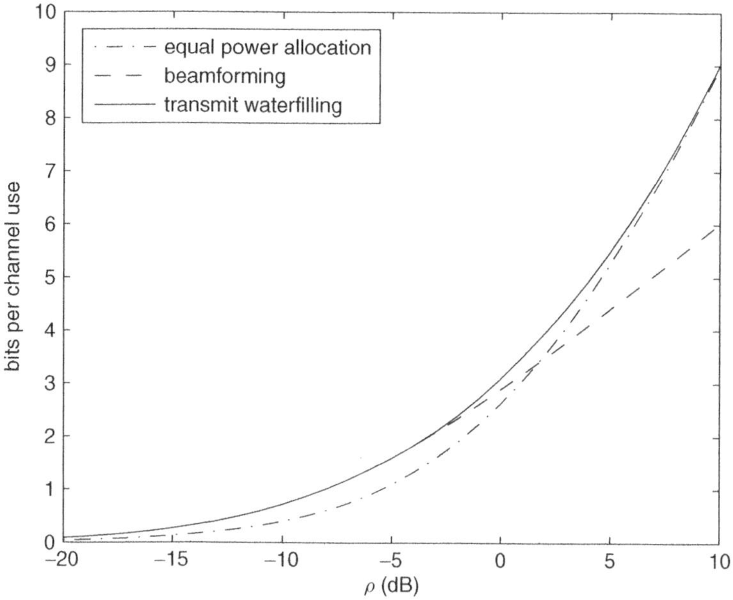

Figure 3.6 Capacity of the channel $\boldsymbol{H}_1$.

beamforming is significantly suboptimal, whereas equal power allocation is almost optimal since excitation of all three parallel channels is the right choice, and the power levels should be almost the same (as the noise variances are significantly small).

As another example, we consider a MIMO system with $N_t = 4$ and $N_r = 2$ whose channel matrix is given by

$$\boldsymbol{H}_2 = \begin{bmatrix} 0.5j & 0.5 + j \\ 1 - j & j \\ -0.7j & 0.8 \\ 0.3 + 0.2j & 1 - 0.5j \end{bmatrix}. \tag{3.35}$$

In this case, since there are only two receive antennas, effectively there are two parallel channels between the transmitter and the receiver. Using SVD, we see that the nonzero singular values of the channel matrix are $\sigma_1 = 2.16$ and $\sigma_2 = 1.52$, and the eigenvalues of $\boldsymbol{H}_2^H \boldsymbol{H}_2$ are $\lambda_1 = 4.69$ and $\lambda_2 = 2.32$. The resulting channel capacities for the three different cases (equal power allocation, beamforming and waterfilling) are shown in Figure 3.7.

The observations for the case of low signal-to-noise ratios are similar to the previous example. However, for large signal-to-noise ratios we see that there is a gap between equal power allocation and the optimal (waterfilling) solution. Equal power allocation divides the power equally among all four transmit antennas, and tries to excite all four parallel channels with identical power levels. However, in this example, there are only two parallel channels with nonzero channel gains, and the power allocated to noise-only channels is simply wasted. Since this is half of the total power, equal power allocation is 3 dB inferior to the optimal solution.

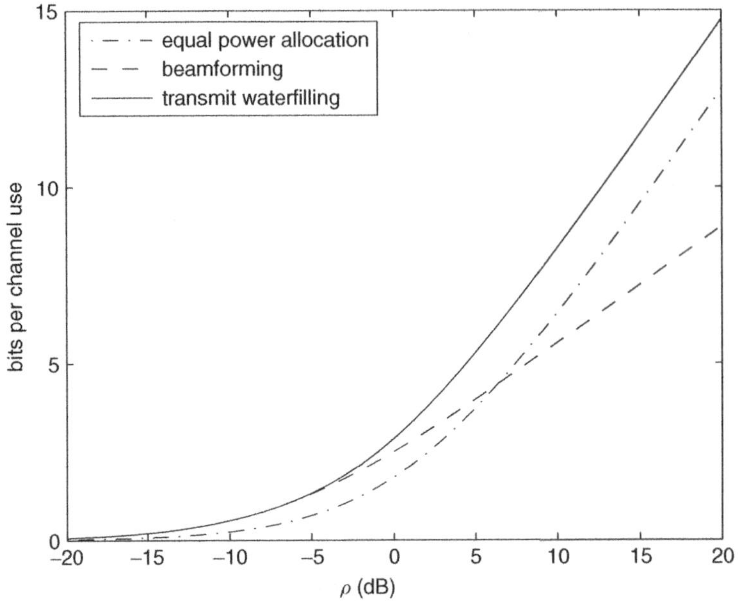

Figure 3.7 Capacity of the channel $\boldsymbol{H}_2$.

3.3.2 Ergodic MIMO Channels

Let us now consider MIMO channels where the channel matrix $\boldsymbol{H}$ is random, but ergodic. Assume that the channel matrix is known by the receiver precisely, but it is unknown to the transmitter. From the discussion in the previous section, we know that for this case the optimal signaling uses (spatially and temporally) independent complex Gaussian inputs with equal power. Therefore, taking an approach similar to the case of single-input single-output systems, we can write the channel capacity as

$$C = E\left[\log\det\left(\boldsymbol{I}_{N_r} + \frac{\rho}{N_t}\boldsymbol{H}^H\boldsymbol{H}\right)\right], \tag{3.36}$$

where the expectation is taken over the statistics of the random matrix $\boldsymbol{H}$. To see that this is in fact the case, we use the following argument: Since the channel is assumed to be ergodic, roughly speaking, by using a very long block length code, all the possible states of the channel are observed for long periods of time. The receiver (using its knowledge of the channel state information) classifies the observations corresponding to the specific realizations of the channel coefficients. The corresponding inputs are still independent Gaussian, hence for each state we can achieve the "log det" capacity. Then the overall capacity is simply the average of all these quantities, hence the expectation.

Assume that the channels are independent Rayleigh fading, then the entries of the channel matrix $\boldsymbol{H}$ $(h_{i,j})$ are independent zero mean complex Gaussian random variables. Assume that the channel power is normalized such that the variances of the real and imaginary parts of the channel coefficients are equal to $1/2$. Then, ρ can be interpreted as the average signal-to-noise ratio per receive antenna. The ergodic capacity is given by the integral (see Telatar (1999))

$$C = \int_0^\infty \log\left(1 + \frac{\rho\lambda}{N_t}\right) \sum_{k=0}^{m-1} \frac{k!}{(k+n-m)!} [L_k^{n-m}(\lambda)]^2 \lambda^{n-m} e^{-\lambda} d\lambda, \tag{3.37}$$

where $m = \min\{N_r, N_t\}$, $n = \max\{N_r, N_t\}$, and

$$L_k^{n-m}(x) = \frac{1}{k!} e^x x^{m-n} \frac{d^k}{dx^k}(e^{-x} x^{n-m+k}),$$

is the associated Laguerre polynomial of order k.

Another way to calculate the capacity is to use Monte Carlo simulation techniques. That is, we can simply generate many realizations of the channel matrix using its distribution, compute the channel capacity for each realization (i.e., $\log\det(\boldsymbol{I}_{N_r} + \frac{\rho}{N_t}\boldsymbol{H}^H\boldsymbol{H})$), and then take their average. This is guaranteed to converge to the actual channel capacity due to the law of large numbers. Note that this approach is not limited to independent Rayleigh fading channels only, and it can be used in general.

Several special cases deserve some more attention.

Reciprocity

Using the notation above, ρ denotes the signal-to-noise ratio for each receive antenna. If we define ρ_{total} as the total signal-to-noise ratio ($\rho_{total} = N_r\rho$), we can rewrite the capacity expression as

$$C = E\left[\log\det\left(\boldsymbol{I}_{N_r} + \frac{\rho_{total}}{N_t N_r}\boldsymbol{H}^H\boldsymbol{H}\right)\right].$$

Observing that the nonzero eigenvalues of $\boldsymbol{H}^H\boldsymbol{H}$ and $\boldsymbol{H}\boldsymbol{H}^H$ are identical, we can also write

$$C = E\left[\log\det\left(\boldsymbol{I}_{N_t} + \frac{\rho_{total}}{N_t N_r}\boldsymbol{H}\boldsymbol{H}^H\right)\right].$$

Therefore, we observe that, for Rayleigh fading, if the numbers of transmit and receive antennas are interchanged, the capacity remains the same (for a fixed total signal-to-noise ratio, ρ_{total}).

Single Transmit Antenna

For the case of a single transmit antenna, i.e., $N_t = 1$, we simply have receive antenna diversity. The channel gain matrix is a row vector of size $1 \times N_r$, denoted by $\boldsymbol{h}$. The capacity is given by

$$C = E[\log(1 + \rho\|\boldsymbol{h}\|^2)], \tag{3.38}$$

which, for the case of independent Rayleigh fading, can be evaluated to be

$$C = \frac{1}{(N_r - 1)!}\int_0^\infty \log(1 + \rho\lambda)\lambda^{N_r - 1} e^{-\lambda} d\lambda. \tag{3.39}$$

The capacity approaches $\log(1 + \rho N_r)$ as the number of receive antennas increases which shows that capacity increases only logarithmically with N_r.

Single Receive Antenna

For the case of a single receive antenna, the channel matrix is a column vector of size $N_t \times 1$, denoted by $\boldsymbol{h}$. The channel capacity is given by

$$C = E\left[\log\left(1 + \frac{\rho}{N_t}\|\boldsymbol{h}\|^2\right)\right], \tag{3.40}$$

and for independent Rayleigh fading, we have

$$C = \frac{1}{(N_t - 1)!}\int_0^\infty \log\left(1 + \frac{\rho\lambda}{N_t}\right)\lambda^{N_t-1}e^{-\lambda}d\lambda, \tag{3.41}$$

which approaches a constant, $\log(1+\rho)$, as the number of transmit antennas is increased. This seems to be contradictory to the reciprocity result. However, we note that this is due to the definition of signal-to-noise ratio used. For a fixed total received signal-to-noise ratio (ρ_{total}), the capacities for both single receive antenna and single transmit antenna cases approach a constant.

Equal Number of Transmit and Receive Antennas

Let us now consider the case of equal numbers of transmit and receive antennas, i.e., $N_r = N_t = n$. For independent Rayleigh fading, the capacity is given by

$$C = n\int_0^\infty \log(1+\rho u)\sum_{k=0}^{n-1} L_k(nu)^2 e^{-nu}du, \tag{3.42}$$

which, as n is increased, can be well approximated as (Telatar (1999))

$$C \approx n\int_0^4 \log(1+\rho u)\frac{1}{\pi}\sqrt{\frac{1}{u} - \frac{1}{4}}du. \tag{3.43}$$

This result is very important as it shows that, for a given transmit power level, the MIMO channel capacity scales linearly with the number of receive and transmit antennas used. This is a tremendous increase that motivates the search for good coding techniques for practical wireless MIMO communications.

Examples

Let us present several examples for the case of independent Rayleigh fading channels. In Figure 3.8, we show the capacity of independent Rayleigh fading channel for a single transmit antenna as a function of the number of receive antennas for several signal-to-noise ratios. Clearly, the capacity increases logarithmically with N_r as expected from our previous discussion.

For independent Rayleigh fading, the ergodic channel capacity for the case of a single receive antenna as a function of the number of transmit antennas is illustrated in Figure 3.9. In this case, the capacity approaches a constant as N_t is increased. We point out again that

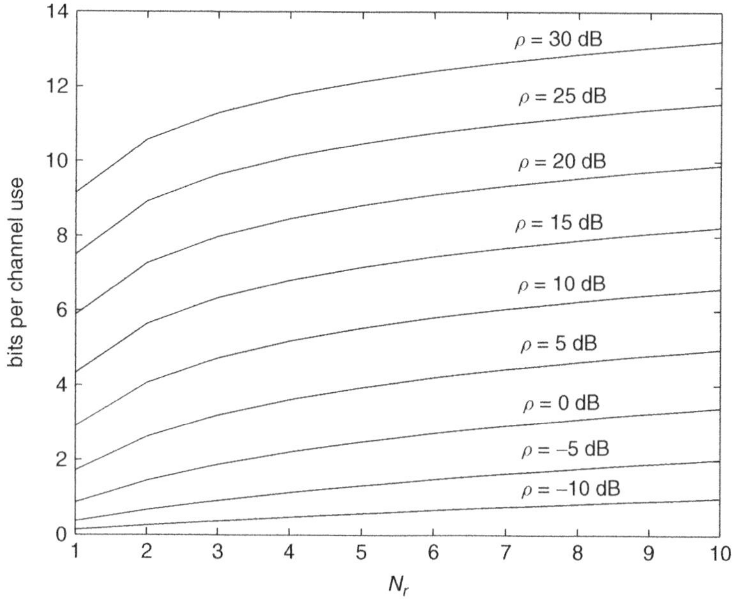

Figure 3.8 Ergodic capacity of MIMO Rayleigh fading channels with $N_t = 1$.

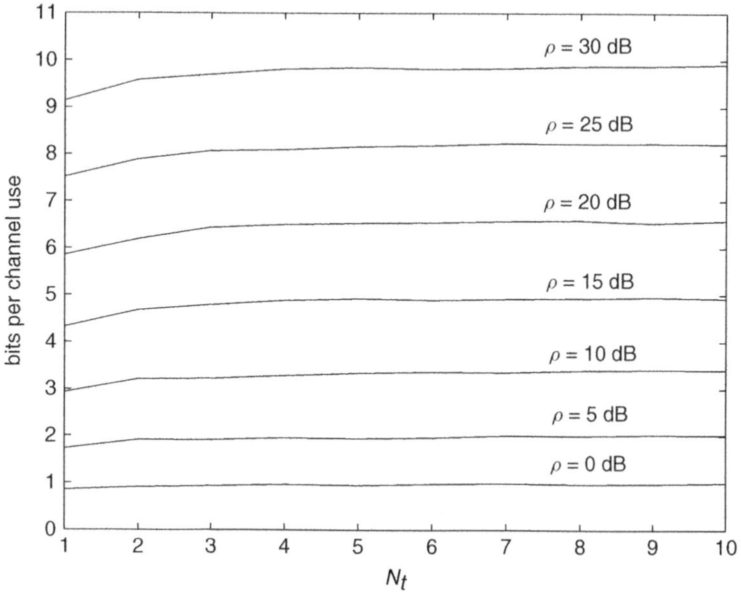

Figure 3.9 Ergodic capacity of MIMO Rayleigh fading channels with $N_r = 1$.

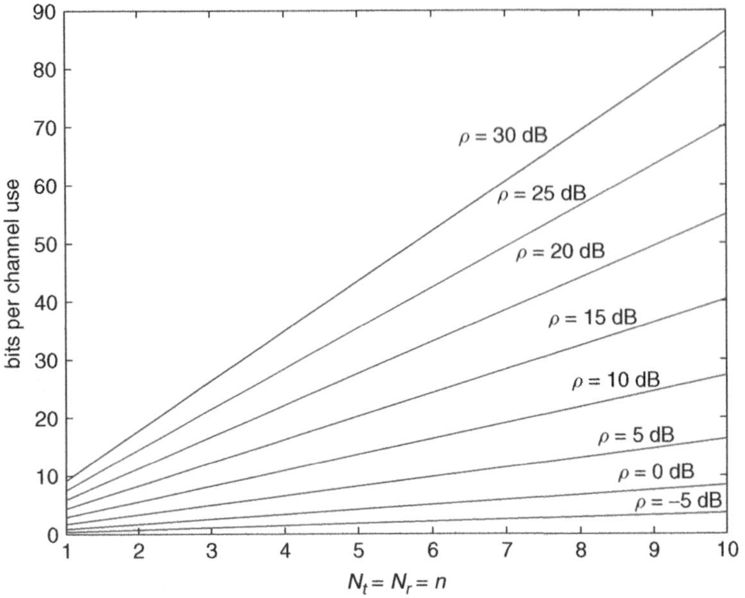

Figure 3.10 Ergodic capacity of MIMO Rayleigh fading channels with equal number of transmit and receive antennas.

this difference in behavior compared with the $N_t = 1$ case is only due to the signal-to-noise ratio definition, otherwise the two cases behave the same way.

Another example is shown in Figure 3.10 where the channel is again independent Rayleigh fading, but the numbers of transmit and receive antennas are equal. We observe that the channel capacity scales linearly with $N_r = N_t = n$. Clearly there is no increase in the total transmit power or the bandwidth with increasing n. From our earlier discussion, we already expected this result for large n, however, the example shows that it is valid for a lower number of antennas as well. This is a very encouraging result as it shows that the "spatial" dimension can be exploited as another resource to improve the capabilities of wireless communication systems considerably.

3.3.3 Non-Ergodic MIMO Channels and Outage Capacity

Let us now turn our attention to the case of non-ergodic fading channels. Specifically, let us assume that the channel is quasi-static, i.e., fading coefficients remain constant over an entire codeword. As discussed earlier in the case of single-input single-output systems, the Shannon type capacity is zero. This is because, regardless of the length of the code chosen, there is a certain probability that the channel cannot support any given fixed rate of transmission. Therefore, we talk about outage capacity.

In the discussion of single-input single-output systems, it was clear that the optimal scheme was to use independent Gaussian inputs with variances given by the power constraint. In the case of multiple antenna systems, for the outage capacity, it is not obvious what the input should be. Telatar (1999) conjectures that the optimal scheme still uses independent Gaussian inputs, but potentially employs a subset of all the available transmit antennas. That is, k antennas, with $k \leq N_t$, are used, and the variance of the Gaussian input for each of them is $1/k$ (no transmit channel state information is assumed). The intuition behind this argument is that, in particular, for very low signal-to-noise ratios, splitting the (already very scarce) power among all the antennas may result in not being able to distinguish the signal from the noise at the receiver end. Instead, using a smaller subset of them with higher power for each transmit antenna could be a better option.

For a given signal-to-noise ratio when k of the available antennas are used in transmission, the capacity is a random variable given by

$$C(\rho) = \log\det\left(\boldsymbol{I}_{N_r} + \rho \boldsymbol{H}^H \boldsymbol{R}_x(k)\boldsymbol{H}\right), \tag{3.44}$$

where $\boldsymbol{R}_x(k) = \frac{1}{k} diag\{\underbrace{1, 1, \ldots, 1}_{k \text{ many}},\ \underbrace{0, 0, \ldots, 0}_{(N_t - k) \text{ many}}\}$.

The outage probability for a given rate of transmission R is then given by

$$P_{out} = \min_{k=1,2,\ldots,N_t} P\left(\log\det\left(\boldsymbol{I}_{N_r} + \rho \boldsymbol{H}^H \boldsymbol{R}_x(k)\boldsymbol{H}\right) \leq R\right). \tag{3.45}$$

By intuition, the number of transmit antennas employed should be smaller for higher rates (correspondingly for higher outage probabilities).

The computation of outage probabilities, hence outage capacities, can be carried out efficiently using Monte Carlo simulation techniques by simply producing many different realizations of the channel and estimating the outage probability for a given rate.

Single Transmit Antenna: Receive Diversity

For some simple cases, it is possible to analytically evaluate P_{out}. For instance for the case of a single transmit antenna, we obtain

$$P_{out} = P\left(\boldsymbol{H}\boldsymbol{H}^H < \frac{2^R - 1}{\rho}\right). \tag{3.46}$$

Noting that for independent Rayleigh fading, $\boldsymbol{H}\boldsymbol{H}^H$ is chi-square distributed with $2N_r$ degrees of freedom and mean N_r, the outage probability is given by

$$P_{out} = \frac{1}{(N_r - 1)!}\gamma\left(N_r, \frac{2^R - 1}{\rho}\right), \tag{3.47}$$

where $\gamma(a, x) = \int_0^x u^{a-1} e^{-u} du$ is the incomplete Gamma function.

Single Receive Antenna: Transmit Diversity

In a similar fashion, for the case of a single receive antenna ($N_r = 1$), when the channels are independent Rayleigh fading, we obtain

$$P_{out} = \min_{k=1,2,\ldots,N_t} \frac{1}{(k-1)!}\gamma\left(k, k\frac{2^R - 1}{\rho}\right). \tag{3.48}$$

Examples

Let us provide several examples of the outage capacity for MIMO Rayleigh fading channels. In Figure 3.11, we present the outage probability versus signal-to-noise ratio for several cases. We observe that for a fixed rate, increasing the number of transmit and/or receive antennas helps reduce the outage probability illustrating the benefits of MIMO communications. Also, when the transmission rates are increased, the outage probability also increases (if the signal-to-noise ratio is kept constant) as expected.

We now look at the problem in a different way. In Figures 3.12 and 3.13, we illustrate the outage capacity of the MIMO channels as a function of $N_t = N_r = n$ for several signal-to-noise ratios for 1% and 10% outage probability levels, respectively. We see that as in the case of ergodic channels, with the use of MIMO systems, the channel capacity increases linearly with n, which clearly demonstrates the benefits of MIMO techniques. We further note that the outage capacity is larger if the acceptable outage probability is higher. However, as for the throughput of the underlying system, one needs to compare the product of outage capacity in bits per channel use with the probability of non-outage ($1 - P_{out}$).

3.3.4 Transmit CSI for MIMO Fading Channels

In our exposition of random (fading) MIMO channels, we have assumed that the receiver has access to the channel state information (CSI), but the transmitter does not. Thus, the transmitter would use all the transmit antennas with equal power. To be precise, for some

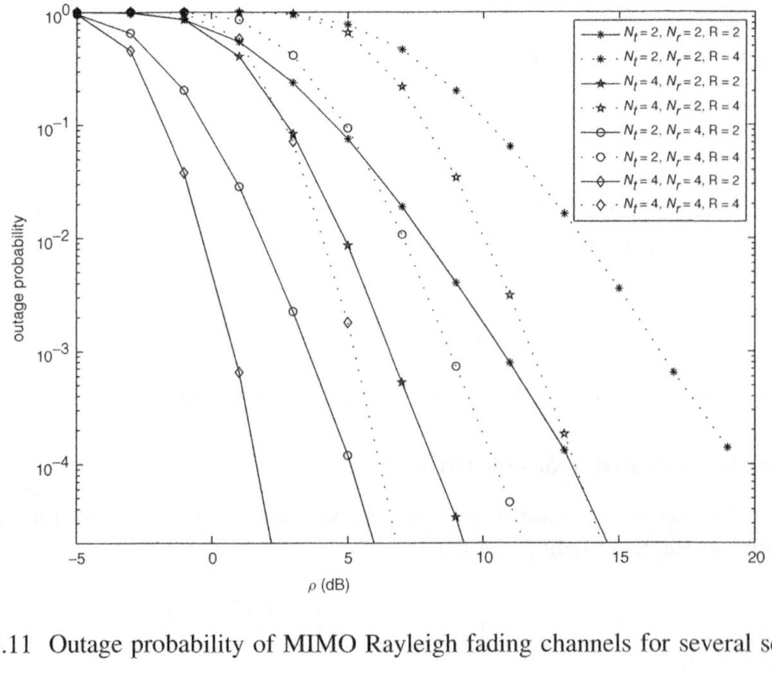

Figure 3.11 Outage probability of MIMO Rayleigh fading channels for several scenarios.

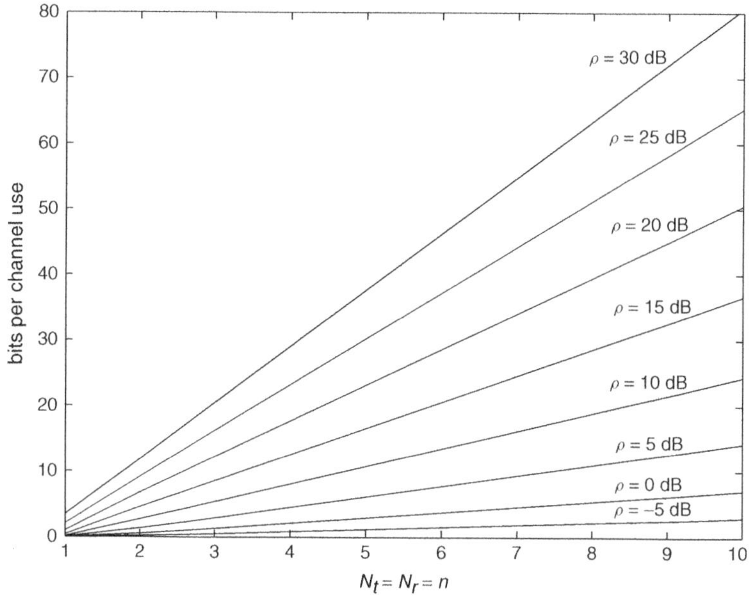

Figure 3.12 Outage capacity of MIMO Rayleigh fading channels as a function of the number of antennas for 1% outage probability.

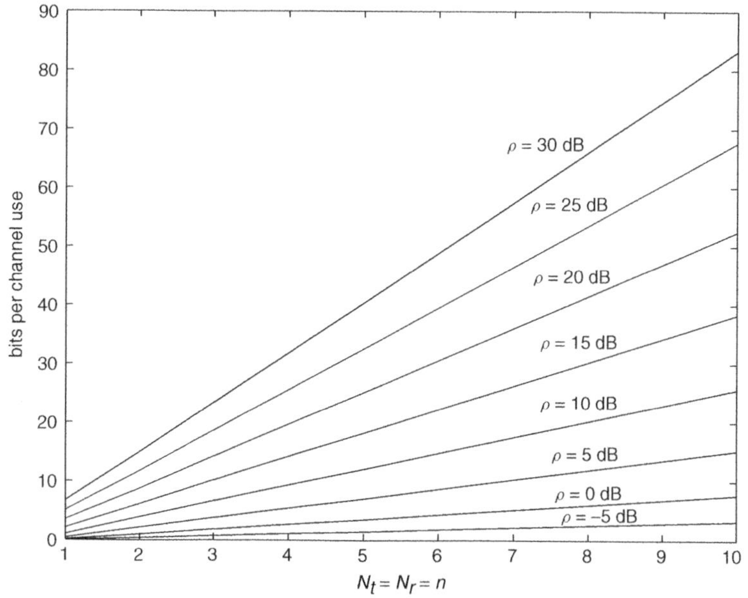

Figure 3.13 Outage capacity of MIMO Rayleigh fading channels as a function of the number of antennas for 10% outage probability.

cases, it may need to choose a randomly selected subset (for outage capacity considerations discussed) based on the average signal-to-noise ratio level (not the instantaneous CSI).

If the transmitter also has access to the channel state information via feedback from the receiver, then clearly splitting the power evenly is not optimal. In this case, "waterfilling" should be used to figure out the optimal power levels for each antenna. For quasi-static channels, the ideas would follow similar lines with the deterministic MIMO channels already discussed. If the channel is time varying, then waterfilling across time should also be used. We do not cover these scenarios here. Some pointers in this line of work include Catreux et al. (2002); Jafar and Goldsmith (2004).

3.4 Constrained Signaling for MIMO Communications

For practical wireless communications, we are usually limited to using signals from specific signal constellations such as BPSK and QAM. Since the capacity of a MIMO channel (in the presence of Gaussian noise) is achieved with Gaussian inputs, using such constellations will prevent us from achieving the channel capacity in general. For instance, if we use BPSK signaling, regardless of the signal-to-noise ratio (and the fading statistics), we can transmit at most N_t bits per channel use. However, it is important to find the limits of reliable transmission when constrained signaling is employed as well. This was rather easy to compute for the case of AWGN channels via a single integration, however for the case of fading channels, in particular for MIMO systems, it is not straightforward to do the computations analytically.

In order to compute the information rates with constrained signaling, we go back to the original capacity expression, that is,

$$C = \max_{p(\boldsymbol{X})} H(\boldsymbol{Y}) - H(\boldsymbol{N}). \tag{3.49}$$

We note that the maximization should be done using the specific input constraints. This calculation is in general difficult, however we note that

$$H(\boldsymbol{Y}) = E[-\log p(\boldsymbol{Y})], \tag{3.50}$$

where the expectation is over the statistics of the random vector $\boldsymbol{Y}$. This expectation can be easily calculated using Monte Carlo techniques, i.e., by generating a large number of realizations of the channel output (of course, using the input constraints), computing the log of the joint probability of the realization, and averaging them. This approach proves to be very useful as illustrated by examples in the following.

Let us now restrict our attention to the case of independent equally likely (independent uniformly distributed, or i.u.d.) signals picked from certain signal constellations, and present several examples. In Figure 3.14, we consider a two-transmit and two-receive antenna system with ergodic Rayleigh fading, and show the information rates achievable with BPSK, QPSK and 8-PSK signaling. Also shown in the same plot is the channel capacity (achieved with independent Gaussian inputs). We observe that for low signal-to-noise ratios, the information rates with constrained i.u.d. inputs are very close to the channel capacity. This is because, when the noise variance is very high, the channel outputs with the constrained inputs are approximately Gaussian distributed, thus the resulting mutual information is almost optimal. However, when the signal-to-noise ratio is increased, the channel capacity grows with

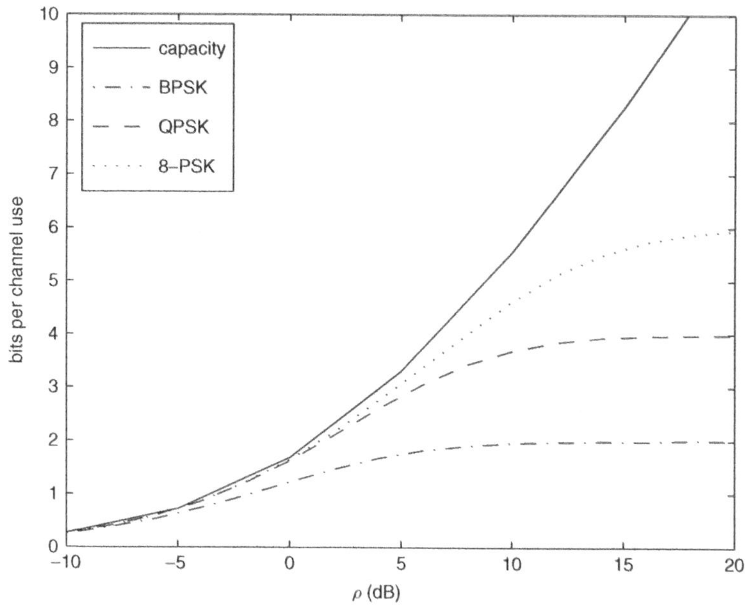

Figure 3.14 Capacity and information rates for an ergodic Rayleigh fading MIMO system with two transmit and two receive antennas.

the signal-to-noise ratio while the information rates are limited by their maximum values determined by the constellation size and the number of transmit antennas, hence the large gap between them. For the case of two-transmit antennas, the maximum transmission rate with M-PSK modulation is $2 \cdot \log_2(M)$, i.e., 2 bits per channel use for BPSK, 4 bits per channel use for QPSK and 6 bits per channel use for 8-PSK, as can be seen clearly in the figure.

The channel capacity and information rates for the $N_t = N_r = 4$ MIMO Rayleigh fading channel are shown in Figure 3.15. We make similar observations as in the 2×2 case, with the exception that the information rates approach 4 bits per channel use, 8 bits per channel use and 12 bits per channel use with BPSK, QPSK and 8-PSK inputs, respectively.

In Figures 3.16 and 3.17, we show the information rates with i.u.d. BPSK inputs for quasi-static Rayleigh fading channels for two outage levels (10% and 1%). We assume that the numbers of transmit and receive antennas are the same ($N_t = N_r = n$), and show the results for several values of signal-to-noise ratio. It is interesting to observe that, similar to the case of the channel capacity results, there is an almost linear increase of the achievable information rates (or, constrained capacity) with the number of antennas n.

3.5 Discussion: Why Use MIMO Systems?

We are now in a position to explain the main motivation behind the use of MIMO systems in a richly scattering wireless communication environment. Let us first summarize the fundamental results of this chapter. We have illustrated that the capacity of a Rayleigh fading MIMO channel increases tremendously with the use of multiple transmit and receive

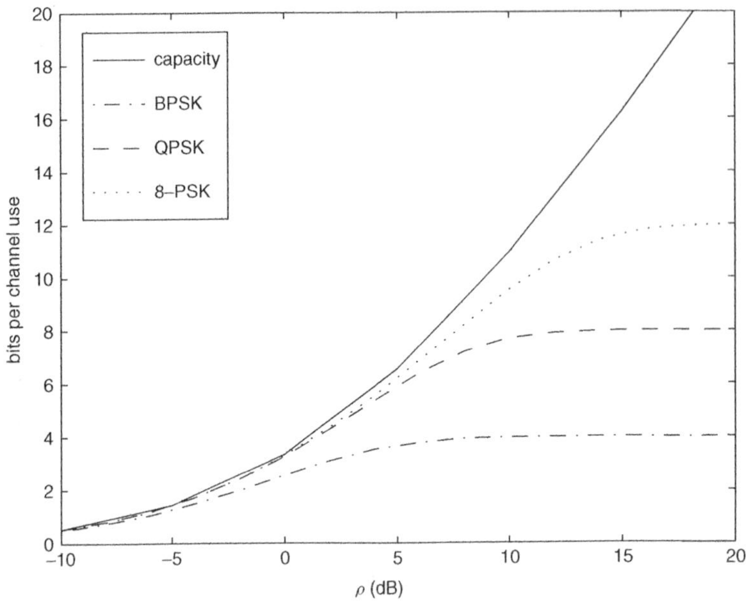

Figure 3.15 Capacity and information rates for an ergodic Rayleigh fading MIMO system with four transmit and four receive antennas.

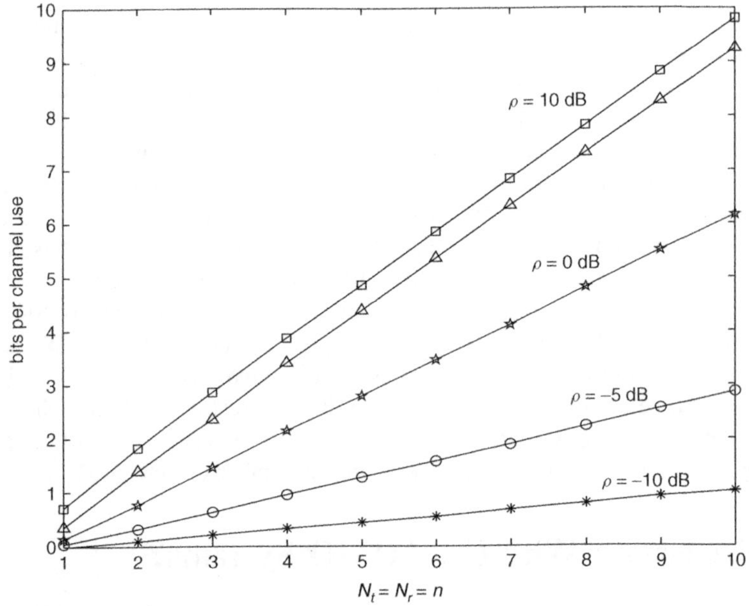

Figure 3.16 Outage information rates for $P_{out} = 10\%$ for quasi-static Rayleigh fading as a function of $N_t = N_r = n$).

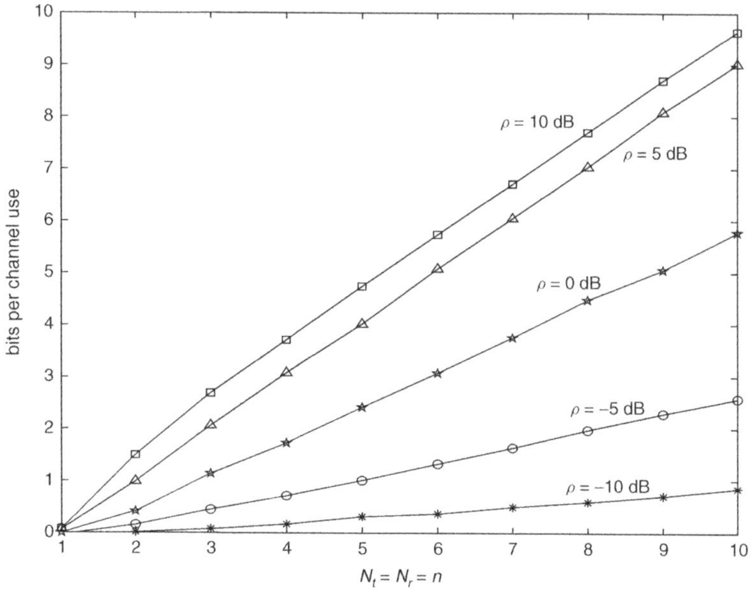

Figure 3.17 Outage information rates for $P_{out} = 1\%$ for quasi-static Rayleigh fading as a function of $N_t = N_r = n$).

antennas. While a single transmit antenna system offers only a logarithmic increase in the capacity with the number of receive antennas, if the total transmit power is kept constant, with the use of multiple transmit antennas (with $N_t = N_r$), the capacity increases linearly. Roughly speaking, the MIMO channel capacity increases linearly with $\min(N_t, N_r)$. This is a tremendous increase, and it is obtained without increasing the total transmit power or the bandwidth. Therefore, "space" or "spatial dimension" is effectively another resource that we can exploit in wireless communication system design.

The linear increase of the capacity with the number of antennas holds for a variety of cases, including ergodic and non-ergodic fading channels. It even seems to hold for the constrained information rates (obtained with practical finite constellations). In fact, as we will discuss in Chapter 9, such an increase is observed for MIMO frequency selective fading channels as well.

To exploit the significant increase in the channel capacity offered by MIMO systems, we need to design channel codes (across time and across antennas), i.e., we need practical space-time coding approaches. The rest of the book is devoted to detailing various methods for different scenarios.

3.6 Chapter Summary and Further Reading

In this chapter we have studied the ergodic (Shannon-type) and non-ergodic (outage) capacities of wireless MIMO channels. We have started with the capacity of general noisy channels, considered the case of AWGN channels and extensions to fading channels for different scenarios. We then moved on to the discussion of capacity and information rate calculations for MIMO systems. We have argued that for ergodic and non-ergodic channels, different capacity definitions apply, namely, ergodic capacity and outage capacity. Also, we have shown that there is a tremendous increase in the capacity with the use of MIMO systems over Rayleigh fading channels. We have given several examples to illustrate this point.

Capacity of noisy channels was introduced by Shannon (1948), and has been extensively studied in the literature. There are several very good books on information-theoretic exposition of communication systems, including Cover and Thomas (2006); Gallager (1968). Several fundamental works on MIMO systems from a capacity perspective include Foschini (1996); Foschini and Gans (1998); Telatar (1999). Some other important topics, such as non-coherent MIMO channel capacity, information rates for MIMO frequency selective fading channels, and channel capacity with non-idealities such as spatial channel correlations are considered in later chapters.

Problems

3.1 Consider the channel

$$y = \sqrt{\rho}hx + n,$$

where x is the channel input, y is the channel output, h is the channel gain and n is zero mean AWGN with variance $1/2$. Assume that the p.d.f. of h is $p_h(h) = p_0\delta(h) + p_1\delta(h - 1) + p_2\delta(h - 2)$. Assume that the input is real and its power is normalized to unity. The random channel gains are independent from one channel use to the next.

a) Determine the (ergodic) channel capacity assuming that only the receiver has access to the channel state information.

b) Calculate the channel capacity assuming that both the receiver and the transmitter have access to the channel state information. For this case, you will need to do waterfilling across time.

c) Compare your answers in parts (a) and (b).

3.2 Consider the same channel model as in the previous problem. But now assume that the channel is *quasi-static,* that is over a long frame of data h remains the same. Calculate the outage capacity as a function of the desired transmission rate. Assume that only the receiver knows the channel state information.

3.3 Consider a single-input single-output Rayleigh fading channel. Determine the outage probability as a function of the desired rate of transmission for the following cases (assuming that only the receiver has access to the channel state information).

a) Quasi-static fading.

b) Each codeword consists of only two independently faded blocks (note that this is still a non-ergodic channel).

c) Each codeword consists of L independently faded blocks (where L is a fixed number). Again note that this is also a non-ergodic channel. What happens if you let $L \to \infty$.

3.4 Consider a MIMO system with $N_t = 2$ and $N_r = 1$.

a) What is the channel capacity if the sub-channels are deterministic with gains $h_{1,1} = h_{2,1} = 1$ as a function of ρ?

b) What is the ergodic capacity for Rayleigh fading sub-channels (where the channel state information is available only at the receiver) as a function of ρ? Assume that $E[|h_{1,1}|^2] = E[|h_{2,1}|^2] = 1$.

c) Repeat the previous part for Rician fading, where $E[h_{1,1}] = E[h_{2,1}] = 1/2$.

d) Compare your answers in the previous parts and comment on your results.

3.5 Consider a deterministic MIMO channel (with AWGN) described by

$$H = \begin{bmatrix} j & 1-j & 1 & -0.5 \\ -0.3 & 0.4+0.1j & 1-j & j \\ 0.2j & 0 & -0.5+0.5j & 1 \\ 1 & -j & 1 & 1 \\ -j & 0.6+0.5j & 2 & -1 \end{bmatrix}.$$

a) Using the singular value decomposition, describe the equivalent representation with parallel channels.

b) Assuming no transmit channel state information, calculate the channel capacity as a function of ρ (use Matlab).

c) Redo part (b) by assuming that the channel state information is available at the transmitter.

3.6 Consider the same channel in the previous problem, but now assume that specific signal constellations are employed. Assuming no transmit CSI (i.e., no waterfilling), using Monte Carlo type simulations, estimate the achievable information rates with

a) i.u.d. BPSK inputs,

b) i.u.d. QPSK inputs.

Compare your answers with the ones in the previous problem.

3.7 Derive Equation (3.37).

3.8 Consider an $N_t \times 1$ MIMO system with quasi-static Rayleigh fading links. Assume that N_t is very large, and only the receiver knows the CSI.

a) What is the outage probability of the system for a desired rate of transmission R and a fixed ρ if the total transmit power is evenly distributed among all the transmit antennas? What is its limit as $N_t \to \infty$?

b) What is the optimal value of the number of active antennas? What is the resulting outage probability?

4

Space-Time Block Codes

In the previous chapter, we have seen that the capacity and constrained information rates over fading channels increase significantly with the employment of multiple transmit and multiple receive antennas. In this chapter, we will introduce the idea of space-time block coding which is a practical means to achieve the benefits offered by MIMO systems.

Space-time block coding is a simple yet very effective means of achieving transmit diversity when other forms of diversity may be limited or non-existent, e.g., for quasi-static fading channels. Such codes can be easily generalized to the case of multiple receive antennas as well, thus providing receive diversity in addition to transmit diversity. Furthermore, they can be decoded efficiently at the receiver by simple linear processing of the set of received signals at different receive antennas.

Our objective in this chapter is to study space-time block coding in detail. We begin with the simple case of two transmit antennas. We will detail the methods, give performance analysis results, and illustrate their performance by both analytical tools and using simulations.

The chapter is organized as follows. We first describe the Alamouti scheme which is a simple way of obtaining transmit diversity for the case of two transmit antennas. We then generalize this scheme to the case of more than two transmit antennas. For both approaches, we consider the optimal receiver structures, theoretical performance analysis in the presence of Rayleigh fading, and performance evaluation by simulations. Furthermore, we study quasi-orthogonal space-time block codes, and linear dispersion codes. Finally, we provide our conclusions and suggestions for further reading.

4.1 Transmit Diversity with Two Antennas: The Alamouti Scheme

As we have discussed in the previous chapter, it is relatively easy to obtain spatial diversity by employing multiple receive antennas. Consider, for instance, the uplink of a cellular telephony system, that is the transmission is from a mobile to the base-station. Since the base-stations can be equipped with multiple antennas with sufficient separation easily, the

Coding for MIMO Communication Systems Tolga M. Duman and Ali Ghrayeb

signal transmitted by the mobile unit can be picked up by multiple receive antennas and they can be combined using a diversity-combining technique, e.g., maximal-ratio combining, selection combining, equal-gain combining, etc., to obtain receive diversity. However, if the situation is reversed (i.e., for downlink transmission) achieving diversity gain is not that simple due to the fact that the mobile units are typically limited in size, and it is usually difficult to place multiple antennas that are separated by sufficiently large distances for reception of multiple copies of the transmitted signal through independent channels. Therefore, it is desirable to have a scheme where the benefits of (spatial) diversity are exploited through "transmit diversity". With this motivation, Alamouti (1998) introduced a way of obtaining transmit diversity when there are two transmit antennas.

4.1.1 Transmission Scheme

The Alamouti scheme is a simple transmit diversity scheme suitable for two transmit antennas. Two symbols are considered at a time, say x_1 and x_2, and they are transmitted in two consecutive time slots. In the first time slot, x_1 is transmitted from the first antenna and x_2 is transmitted from the second one. In the second time slot, $-x_2^*$ is transmitted from the first antenna, while x_1^* is transmitted from the second antenna. This process is illustrated in Figure 4.1. The signals x_1 and x_2 are picked from an arbitrary (M-ary) constellation. Since two symbols are transmitted in two time slots, the overall transmission rate is 1 symbol per channel use, or $\log_2 M$ bits per channel use.

4.1.2 Optimal Receiver for the Alamouti Scheme

Let us now derive the optimal receiver for the Alamouti scheme. We consider two separate cases, namely, a single receive antenna and multiple receive antennas.

Single Receive Antenna System

Consider the case of a single receive antenna. The received signal in the first time slot is then

$$y_1(1) = \sqrt{\rho}(h_{1,1}x_1 + h_{2,1}x_2) + n_1(1), \tag{4.1}$$

and, it is

$$y_1(2) = \sqrt{\rho}(-h_{1,1}x_2^* + h_{2,1}x_1^*) + n_1(2), \tag{4.2}$$

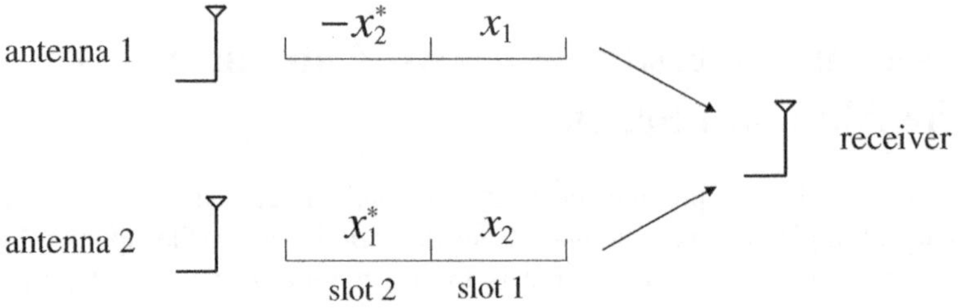

Figure 4.1 The Alamouti scheme.

in the second time slot. We assume that the channel is Rayleigh fading, i.e., $h_{1,1}$ and $h_{2,1}$ are zero mean complex Gaussian random variables with unit variance (i.e., with variance 1/2 per dimension), and they remain the same for two consecutive time intervals. We normalize the power of each constellation to be 1/2, thus the total transmit power per channel use is unity. The additive noise terms $n_1(1)$ and $n_1(2)$ are complex AWGN with variance 1/2 per dimension. With these definitions, the signal-to-noise ratio becomes ρ.

Define the vector of the received signals (where the second signal is conjugated) as

$$\boldsymbol{y} = \begin{bmatrix} y_1(1) \\ y_1^*(2) \end{bmatrix}, \tag{4.3}$$

which can be written as

$$\boldsymbol{y} = \sqrt{\rho} \begin{bmatrix} h_{1,1} & h_{2,1} \\ h_{2,1}^* & -h_{1,1}^* \end{bmatrix} \begin{bmatrix} x_1 \\ x_2 \end{bmatrix} + \begin{bmatrix} n_1(1) \\ n_1^*(2) \end{bmatrix}. \tag{4.4}$$

Assuming that the receiver has perfect knowledge of the channel state information, the optimal receiver, which minimizes the probability of error, chooses $\hat{x}_1$ and $\hat{x}_2$ as follows:

$$(\hat{x}_1, \hat{x}_2) = \arg\max_{(x_1, x_2)} P(x_1, x_2 | \boldsymbol{y}, h_{1,1}, h_{2,1}), \tag{4.5}$$

which can also be written as

$$(\hat{x}_1, \hat{x}_2) = \arg\max_{(x_1, x_2)} P(x_1, x_2 | \boldsymbol{H}^H \boldsymbol{y}, h_{1,1}, h_{2,1}), \tag{4.6}$$

where

$$\boldsymbol{H} = \begin{bmatrix} h_{1,1} & h_{2,1} \\ h_{2,1}^* & -h_{1,1}^* \end{bmatrix}.$$

We have used the fact that $\boldsymbol{H}^H \boldsymbol{y}$ (multiplication by $\boldsymbol{H}^H$) is a one-to-one transformation. Assuming that all the input symbol pairs are equally likely, and using Bayes' rule, one can equivalently write the optimal decoded symbols as

$$(\hat{x}_1, \hat{x}_2) = \arg\max_{(x_1, x_2)} P(\boldsymbol{H}^H \boldsymbol{y} | x_1, x_2, h_{1,1}, h_{2,1}). \tag{4.7}$$

Note that

$$\boldsymbol{H}^H \boldsymbol{y} = \sqrt{\rho} \begin{bmatrix} |h_{1,1}|^2 + |h_{2,1}|^2 & 0 \\ 0 & |h_{1,1}|^2 + |h_{2,1}|^2 \end{bmatrix} \begin{bmatrix} x_1 \\ x_2 \end{bmatrix} + \begin{bmatrix} n_1'(1) \\ n_1'(2) \end{bmatrix}, \tag{4.8}$$

where the new noise terms are given by

$$\begin{bmatrix} n_1'(1) \\ n_1'(2) \end{bmatrix} = \begin{bmatrix} h_{1,1}^* & h_{2,1} \\ h_{2,1}^* & -h_{1,1} \end{bmatrix} \begin{bmatrix} n_1(1) \\ n_1^*(2) \end{bmatrix}. \tag{4.9}$$

Clearly, being linear combinations of jointly Gaussian random variables, $n_1'(1)$ and $n_1'(2)$ are jointly Gaussian. Since they are also uncorrelated, they are independent. Also, they both have zero mean and a variance of $\frac{1}{2}(|h_{1,1}|^2 + |h_{2,1}|^2)$ per dimension.

Therefore, the optimal decisions $\hat{x}_1$ and $\hat{x}_2$ decouple, and simplify to the usual minimization of the Euclidean distance between the possible transmitted symbols and the respective components of the vector $\boldsymbol{H}^H \boldsymbol{y}$. That is,

$$\hat{x}_1 = \arg\min_{x_1} |h_{1,1}^* y_1(1) + h_{2,1} y_1^*(2) - \sqrt{\rho}(|h_{1,1}|^2 + |h_{2,1}|^2) x_1|, \tag{4.10}$$

$$\hat{x}_2 = \arg\min_{x_2} |h_{2,1}^* y_1(1) - h_{1,1} y_1^*(2) - \sqrt{\rho}(|h_{1,1}|^2 + |h_{2,1}|^2) x_2|. \tag{4.11}$$

We note that the above decoding rule is very useful and what makes the Alamouti scheme (and, more generally, orthogonal space-time block codes) attractive. This is because decoupling of the optimal decisions clearly reduces the search space for the optimal selection of the transmitted symbol, and thus simplifies the receiver structure considerably.

If we make further assumptions on the modulation schemes used, we can simplify the receiver structures even more. For instance, for the case of a constant energy constellation, e.g., BPSK, QPSK, 8-PSK, the optimal decision rule can be written as the usual correlation maximization. For the special case of BPSK modulation, it reduces to

$$\hat{x}_1 = \begin{cases} 1, & \text{if Re } \{h_{1,1}^* y_1(1) + h_{2,1} y_1^*(2)\} > 0 \\ 0, & \text{otherwise,} \end{cases} \tag{4.12}$$

and

$$\hat{x}_2 = \begin{cases} 1, & \text{if Re } \{h_{2,1}^* y_1(1) - h_{1,1} y_1^*(2)\} > 0 \\ 0, & \text{otherwise.} \end{cases} \tag{4.13}$$

As we will discuss later, the full-diversity of the MIMO system, which is two for the case of two transmit and one receive antenna, can be obtained with the use of the Alamouti scheme. Therefore, even if the receivers are not equipped with multiple antennas, spatial diversity can still be achieved, which makes this scheme very beneficial.

Multiple Receive Antenna System

The Alamouti scheme can easily be extended to systems with multiple receive antennas as well, resulting in receive diversity in addition to the existing transmit diversity. For this case, the available diversity order is twice the number of receive antennas, and it can also be achieved by a simple linear receiver.

Assume that the received signal during the kth time slot at the jth receive antenna is $y_j(k)$, where $k = 1, 2,\ j = 1, 2, \ldots, N_r$. The channel coefficient from the ith transmit antenna to the jth receive antenna is denoted by $h_{i,j}$. We can then write

$$y_j(1) = \sqrt{\rho}(h_{1,j} x_1 + h_{2,j} x_2) + n_j(1) \tag{4.14}$$

in the first time slot, and

$$y_j(2) = \sqrt{\rho}(-h_{1,j} x_2^* + h_{2,j} x_1^*) + n_j(2) \tag{4.15}$$

in the second time slot. Here, $n_j(k)$ is the AWGN term at time slot k and receive antenna j. Following the same line of reasoning as in the previous section, after linear combining the received signals of all the receive antennas and scaling by the factor $1/\sqrt{|h_{1,j}|^2 + |h_{2,j}|^2}$, we obtain the sufficient statistics for optimal receiver as

$$y_j(1) = \sqrt{\rho}\sqrt{|h_{1,j}|^2 + |h_{2,j}|^2}\ x_1 + n_j''(1), \tag{4.16}$$

$$y_j(2) = \sqrt{\rho}\sqrt{|h_{1,j}|^2 + |h_{2,j}|^2}\ x_2 + n_j''(2), \tag{4.17}$$

where $j = 1, 2, \ldots, N_r$. The noise terms $n_j''(k)$ $(k = 1, 2,\ j = 1, 2, \ldots, N_r)$ are all independent Gaussian random variables with variance 1/2 per dimension (with the same reasoning as in Equation (4.9)). Therefore, we basically have N_r parallel channels (each corresponding to one receive antenna) corrupted by independent Gaussian noise terms with equal variances. The optimal way to combine these to make the decisions on the symbols transmitted is to use maximum ratio combining, described in Chapter 2. That is, we scale each with their own channel gain, and add them up to obtain the decision variables of the two transmitted symbols as

$$y(k) = \sum_{j=1}^{N_r} \sqrt{|h_{1,j}|^2 + |h_{2,j}|^2}\ y_j(k) \qquad k = 1, 2. \tag{4.18}$$

These decision variables are then to be compared with the possible "clean" signals that could have been transmitted, and the one closest to $y(k)$ in the squared Euclidean sense is selected as the optimal one. Therefore, the decision rule can simply be stated as

$$\hat{x}_1 = \arg\min_{x_1} \left| \sum_{j=1}^{N_r} h_{1,j}^* y_j(1) + h_{2,j} y_j^*(2) - \sqrt{\rho}(|h_{1,j}|^2 + |h_{2,j}|^2) x_1 \right|^2, \tag{4.19}$$

$$\hat{x}_2 = \arg\min_{x_2} \left| \sum_{j=1}^{N_r} h_{2,j}^* y_j(1) - h_{1,j} y_j^*(2) - \sqrt{\rho}(|h_{1,j}|^2 + |h_{2,j}|^2) x_2 \right|^2, \tag{4.20}$$

where it is clear that the decisions for the two symbols transmitted decouple, and the minimization over the possible constellation symbols can be done separately.

Of course, the optimal decision rule can be simplified further if a constant energy constellation, such as phase shift keying, is employed. For instance, for the case of BPSK modulation, it takes the form

$$\hat{x}_1 = \begin{cases} 1, & \text{if } \mathrm{Re}\left\{\sum_{j=1}^{N_r} h_{1,j}^* y_j(1) + h_{2,j} y_j^*(2)\right\} > 0 \\ 0, & \text{otherwise,} \end{cases} \tag{4.21}$$

and

$$\hat{x}_2 = \begin{cases} 1, & \text{if } \mathrm{Re}\left\{\sum_{j=1}^{N_r} h_{2,j}^* y_j(1) - h_{1,j} y_j^*(2)\right\} > 0 \\ 0, & \text{otherwise.} \end{cases} \tag{4.22}$$

We note that this decision rule simply boils down to the earlier one derived for the case of only one receive antenna in the previous subsection when $N_r = 1$.

4.1.3 Performance Analysis of the Alamouti Scheme

Consider a $2 \times N_r$ MIMO system employing the Alamouti scheme. After linear processing (which is shown to be optimal), as argued in the previous section, the decision variable for the symbol x_k is given by (from Equation (4.18))

$$y(k) = \sqrt{\rho}\left(\sum_{i=1}^{2}\sum_{j=1}^{N_r}|h_{i,j}|^2\right)x_k + n''(k), \tag{4.23}$$

where the new noise term $n''(k)$ is given by

$$n''(k) = \sum_{j=1}^{N_r}\sqrt{|h_{1,j}|^2 + |h_{2,j}|^2}\ n''_j(k), \tag{4.24}$$

since, as argued before, the noise terms in the summation are independent Gaussian random variables with variance $1/2$ per dimension, the overall noise $n''(k)$ is also zero mean Gaussian, but with variance $\frac{1}{2}(\sum_{i=1}^{2}\sum_{j=1}^{N_r}|h_{i,j}|^2)$ per dimension.

We can easily write the probability of error in receiving the kth signal conditioned on the channel gains using results from basic digital communications. We then need to average over the statistics of the channel to obtain the average bit or symbol error rate. Before presenting the details of the probability of error computation, we would like to note that the mathematical expression above is identical to the one obtained in the Lth order diversity scheme employing maximal ratio combining at the receiver. Therefore, it is clear that the full-diversity of $2N_r$ is achieved.

Let us illustrate this procedure for the simple case of BPSK modulation. For this case, the probability of bit error conditioned on the channel gains can be easily written as (see Proakis (2001))

$$P_b(\alpha) = Q\left(\sqrt{\alpha\rho}\right), \tag{4.25}$$

where $\alpha = \sum_{i=1}^{2}\sum_{j=1}^{N_r}|h_{i,j}|^2$. For a Rayleigh fading channel, each of the terms in this summation are squared norms of complex Gaussian random variables with variance $1/2$ in each dimension. Therefore each of these terms is chi-square distributed with two degrees of freedom with parameter 1 (i.e., exponential). Assuming that all the channel gains (between different transmit and receive antenna pairs) are independent (which can be accomplished by proper separation of the transmit and receive antenna elements), the sum of N_r independent and identically distributed exponential random variables is central chi-square distributed with $4N_r$ degrees of freedom. Therefore, the probability density function of α is given by (see Papoulis and Pillai (2002))

$$f_\alpha(\alpha) = \frac{1}{(2N_r-1)!}\alpha^{2N_r-1}e^{-\alpha} \quad \text{for } \alpha \geq 0. \tag{4.26}$$

Then, the probability of error averaged over the channel gains is given by

$$P_b = E_\alpha[P_b(\alpha)] \tag{4.27}$$

$$= \int_0^\infty \frac{1}{(2N_r-1)!}\alpha^{2N_r-1}e^{-\alpha}Q\left(\sqrt{\alpha\rho}\right)d\alpha. \tag{4.28}$$

After some straightforward algebra, this integral can be simplified to

$$P_b = \left(\frac{1}{2}\left(1 - \sqrt{\frac{\rho}{2+\rho}}\right)\right)^{2N_r} \sum_{l=0}^{2N_r-1} \binom{2N_r - 1 + l}{l} \left(\frac{1}{2}\left(1 + \sqrt{\frac{\rho}{2+\rho}}\right)\right)^{l}, \tag{4.29}$$

which, for large signal-to-noise ratios, i.e., for $\rho \gg 1$, can be approximated as

$$P_b \approx \binom{4N_r - 1}{2N_r} \left(\frac{1}{2\rho}\right)^{2N_r}. \tag{4.30}$$

From this expression it is clear that the average bit error probability of the Alamouti scheme when BPSK modulation is employed decays inversely with the $(2N_r)$th power of the signal-to-noise ratio. Therefore, the diversity provided is $2N_r$ (i.e., full spatial diversity) over a Rayleigh flat fading channel.

We note that full-diversity is achieved for other modulation schemes as well, as the analysis for the BPSK modulation can be easily generalized. For example, for the general M-ary phase shift keying, the symbol error probability conditioned on the effective signal-to-noise ratio observed can be approximated by

$$P_{e,M-PSK}(\alpha) \approx 2Q\left(\sqrt{\alpha\rho}\sin\left(\frac{\pi}{M}\right)\right). \tag{4.31}$$

Following the same steps as above, by averaging over the fading statistics, it is easy to show that

$$P_{e,M-PSK} \approx 2\binom{4N_r - 1}{2N_r} \left(\frac{1}{2\sin^2(\frac{\pi}{M})}\right)^{2N_r} \left(\frac{1}{\rho}\right)^{2N_r}, \tag{4.32}$$

which also clearly shows that the full-diversity is of order $2N_r$, as expected. Similar computations can be easily carried out for other modulation schemes as well.

4.1.4 Examples

In this section, we present several error rate results for the Alamouti scheme used over Rayleigh flat fading channels based on both simulations and performance analysis carried out in the previous section.

In Figure 4.2, we present the bit error rate obtained using BPSK modulation for several cases. For comparison purposes, we also show the performance of the no-diversity case, i.e., the single-input single-output system. For the two transmit and one receive antenna case, we observe that the diversity order is two, and for the two transmit and two receive antenna case, it is four. We also observe that the simulations and theoretical results agree with each other.

In Figure 4.3, we show the symbol error rates for 8-PSK modulation obtained using simulations. In the same figure, we also provide the theoretical error probabilities. The diversity advantage of the Alamouti scheme is obvious over the system not employing transmit or receive diversity. Specifically, as expected, diversity orders of two and four are obtained for the case of a single and two receive antenna cases, respectively.

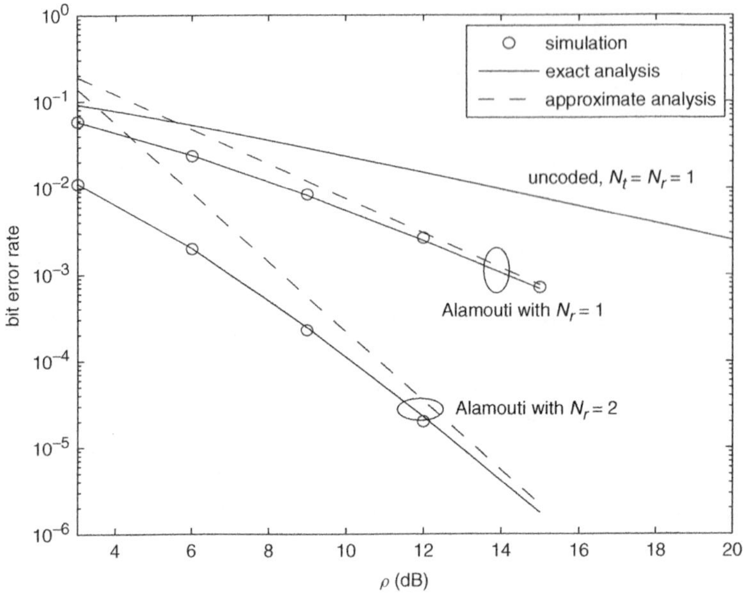

Figure 4.2 Bit error rate performance of the Alamouti scheme with BPSK modulation (simulation and bound).

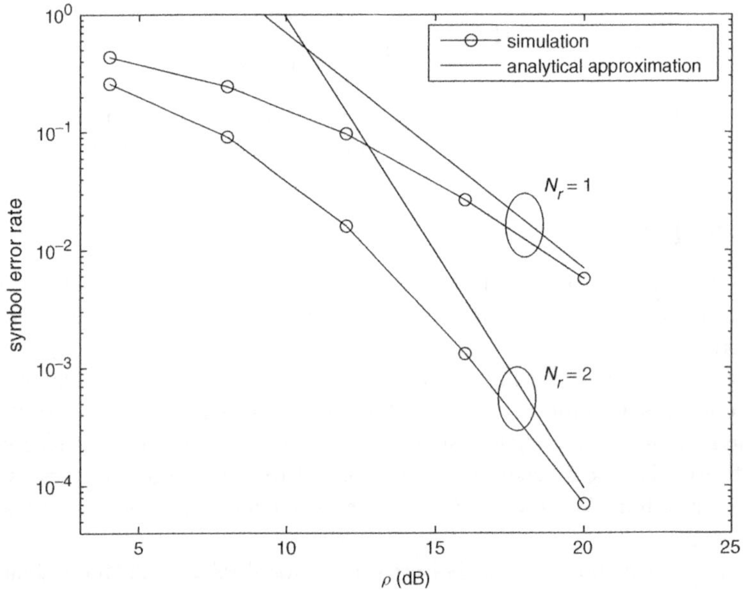

Figure 4.3 Symbol error rate performance of the Alamouti scheme with 8-PSK modulation (simulation and bound).

4.2 Orthogonal Space-Time Block Codes

The Alamouti scheme is designed for two transmit antennas. A question that naturally comes to mind is: How can this transmit diversity scheme be extended to the case of more than two transmit antennas? In this section, we describe what is known as the general space-time block codes based on the theory of orthogonal designs that precisely addresses this question (Tarokh et al. (1999)).

Consider a system with N_t transmit antennas. The objective is to design a set of $N_t \times N_t$ matrices with elements from a desired signal constellation whose columns are orthogonal to each other. The latter property is required to make sure that a linear receiver is still optimal and the decoding complexity is still kept at a minimum. Before we present the space-time block coding scheme in detail, let us give two simple examples.

Assume that $N_t = 4$ and a real signal constellation (such as M-ary pulse amplitude modulation (PAM)) is used. Assume that the constellation symbols to be transmitted are $\{x_1, x_2, x_3, x_4\}$. We form the code matrix as

$$\boldsymbol{X}_1 = \begin{bmatrix} x_1 & x_2 & x_3 & x_4 \\ -x_2 & x_1 & -x_4 & x_3 \\ -x_3 & x_4 & x_1 & -x_2 \\ -x_4 & -x_3 & x_2 & x_1 \end{bmatrix}. \tag{4.33}$$

Clearly, the columns of $\boldsymbol{X}_1$ are mutually orthogonal. For the symbol interval i ($1 \leq i \leq 4$), we transmit the ith row of this matrix using the four transmit antennas. That is, the (i, j)th element of the matrix is transmitted from antenna j during the ith time slot. This is a full-rate code example since we are able to transmit four symbols in four time slots, and as we will describe later in more detail, it achieves full-diversity.

It turns out it is not always possible to find full-rate, full-diversity space-time block codes using (square) orthogonal designs. For real constellations, such a design is possible if the number of antennas is two, four or eight. If the square orthogonal design condition is relaxed, then it is possible to find other full-rate full-diversity code examples. On the other hand, for complex signal constellations (e.g., QAM, 8 PSK), full-rate designs exist if and only if $N_t = 2$, that is, the Alamouti code is the only full-rate, full-diversity space-time block code when the signal constellation is complex (Tarokh et al. (1999)).

Let us present another example of a space-time block code. Assume that we are to transmit the four symbols $\{x_1, x_2, x_3, x_4\}$ selected from a complex signal constellation. A specific space-time block code can be described by the 8×4 matrix

$$\boldsymbol{X}_2 = \begin{bmatrix} x_1 & x_2 & x_3 & x_4 \\ -x_2 & x_1 & -x_4 & x_3 \\ -x_3 & x_4 & x_1 & -x_2 \\ -x_4 & -x_3 & x_2 & x_1 \\ x_1^* & x_2^* & x_3^* & x_4^* \\ -x_2^* & x_1^* & -x_4^* & x_3^* \\ -x_3^* & x_4^* & x_1^* & -x_2^* \\ -x_4^* & -x_3^* & x_2^* & x_1^* \end{bmatrix}. \tag{4.34}$$

At time interval i ($1 \leq i \leq 8$), the ith row of $\boldsymbol{X}_2$ is transmitted using the four transmit antennas. This code achieves full-diversity as well (this will be illustrated later); however, clearly, it is not full-rate since only four symbols are transmitted using eight time slots. This simple example clearly illustrates that if we relax the full-rate code design condition, there are many potential designs one can employ.

In order to present general space-time block codes in a more concrete framework, let us describe the important subclass of linear orthogonal designs in detail.

4.2.1 Linear Orthogonal Designs

Let us now describe the general class of linear orthogonal designs in a mathematical framework to combine different space-time block codes in a unified manner based on the description given in Xu and Kwak (2005). We denote the $N \times N_t$ code matrix that defines the space-time block code by $\boldsymbol{X}$. This matrix is used to transmit M symbols in N time slots (achieving a rate of M/N). We assume that the elements of $\boldsymbol{X}$ are linear combinations of the constellation points being transmitted or their conjugates. We impose the constraint that the columns of the code matrix $\boldsymbol{X}$ are orthogonal, that is, $\boldsymbol{X}^H\boldsymbol{X}$ is a diagonal matrix. We refer to this class of space-time block codes as linear orthogonal designs.

We can equivalently write the code matrix $\boldsymbol{X}$ in the form

$$\boldsymbol{X} = \sum_{m=1}^{M} (x_m \boldsymbol{A}_m + x_m^* \boldsymbol{B}_m), \tag{4.35}$$

where $\boldsymbol{A}_m$ and $\boldsymbol{B}_m$ are $N \times N_t$ matrices, and $\{x_m\}_{m=1}^{M}$ are the set of symbols being transmitted.

This representation is general, and thus can be used to describe different codes studied earlier. For example, for the Alamouti code, we have $N_t = T = M = 2$ with

$$\boldsymbol{A}_1 = \begin{bmatrix} 1 & 0 \\ 0 & 0 \end{bmatrix}, \quad \boldsymbol{A}_2 = \begin{bmatrix} 0 & 1 \\ 0 & 0 \end{bmatrix}, \quad \boldsymbol{B}_1 = \begin{bmatrix} 0 & 0 \\ 0 & 1 \end{bmatrix}, \quad \boldsymbol{B}_2 = \begin{bmatrix} 0 & 0 \\ -1 & 0 \end{bmatrix}. \tag{4.36}$$

As another example, for the code given in $\boldsymbol{X}_1$ (4.33), we have

$$\boldsymbol{A}_1 = \begin{bmatrix} 1 & 0 & 0 & 0 \\ 0 & 1 & 0 & 0 \\ 0 & 0 & 1 & 0 \\ 0 & 0 & 0 & 1 \end{bmatrix}, \quad \boldsymbol{A}_2 = \begin{bmatrix} 0 & 1 & 0 & 0 \\ -1 & 0 & 0 & 0 \\ 0 & 0 & 0 & -1 \\ 0 & 0 & 1 & 0 \end{bmatrix},$$

$$\boldsymbol{A}_3 = \begin{bmatrix} 0 & 0 & 1 & 0 \\ 0 & 0 & 0 & 1 \\ -1 & 0 & 0 & 0 \\ 0 & -1 & 0 & 0 \end{bmatrix}, \quad \boldsymbol{A}_4 = \begin{bmatrix} 0 & 0 & 0 & 1 \\ 0 & 0 & -1 & 0 \\ 0 & 1 & 0 & 0 \\ -1 & 0 & 0 & 0 \end{bmatrix},$$

and $\boldsymbol{B}_1 = \boldsymbol{B}_2 = \boldsymbol{B}_3 = \boldsymbol{B}_4 = \boldsymbol{0}_4$.

For the space-time block code defined by $\boldsymbol{X}_2$ in expression (4.34), it is easy to see that the matrices take the form

$$\boldsymbol{A}_1 = \begin{bmatrix} 1 & 0 & 0 & 0 \\ 0 & 1 & 0 & 0 \\ 0 & 0 & 1 & 0 \\ 0 & 0 & 0 & 1 \\ 0 & 0 & 0 & 0 \\ 0 & 0 & 0 & 0 \\ 0 & 0 & 0 & 0 \\ 0 & 0 & 0 & 0 \end{bmatrix}, \quad \boldsymbol{A}_2 = \begin{bmatrix} 0 & 1 & 0 & 0 \\ -1 & 0 & 0 & 0 \\ 0 & 0 & 0 & -1 \\ 0 & 0 & 1 & 0 \\ 0 & 0 & 0 & 0 \\ 0 & 0 & 0 & 0 \\ 0 & 0 & 0 & 0 \\ 0 & 0 & 0 & 0 \end{bmatrix},$$

$$\boldsymbol{A}_3 = \begin{bmatrix} 0 & 0 & 1 & 0 \\ 0 & 0 & 0 & 1 \\ -1 & 0 & 0 & 0 \\ 0 & -1 & 0 & 0 \\ 0 & 0 & 0 & 0 \\ 0 & 0 & 0 & 0 \\ 0 & 0 & 0 & 0 \\ 0 & 0 & 0 & 0 \end{bmatrix}, \quad \boldsymbol{A}_4 = \begin{bmatrix} 0 & 0 & 0 & 1 \\ 0 & 0 & -1 & 0 \\ 0 & 1 & 0 & 0 \\ -1 & 0 & 0 & 0 \\ 0 & 0 & 0 & 0 \\ 0 & 0 & 0 & 0 \\ 0 & 0 & 0 & 0 \\ 0 & 0 & 0 & 0 \end{bmatrix},$$

and

$$\boldsymbol{B}_1 = \begin{bmatrix} 0 & 0 & 0 & 0 \\ 0 & 0 & 0 & 0 \\ 0 & 0 & 0 & 0 \\ 0 & 0 & 0 & 0 \\ 1 & 0 & 0 & 0 \\ 0 & 1 & 0 & 0 \\ 0 & 0 & 1 & 0 \\ 0 & 0 & 0 & 1 \end{bmatrix}, \quad \boldsymbol{B}_2 = \begin{bmatrix} 0 & 0 & 0 & 0 \\ 0 & 0 & 0 & 0 \\ 0 & 0 & 0 & 0 \\ 0 & 0 & 0 & 0 \\ 0 & 1 & 0 & 0 \\ -1 & 0 & 0 & 0 \\ 0 & 0 & 0 & -1 \\ 0 & 0 & 1 & 0 \end{bmatrix},$$

$$\boldsymbol{B}_3 = \begin{bmatrix} 0 & 0 & 0 & 0 \\ 0 & 0 & 0 & 0 \\ 0 & 0 & 0 & 0 \\ 0 & 0 & 0 & 0 \\ 0 & 0 & 1 & 0 \\ 0 & 0 & 0 & 1 \\ -1 & 0 & 0 & 0 \\ 0 & -1 & 0 & 0 \end{bmatrix}, \quad \boldsymbol{B}_4 = \begin{bmatrix} 0 & 0 & 0 & 0 \\ 0 & 0 & 0 & 0 \\ 0 & 0 & 0 & 0 \\ 0 & 0 & 0 & 0 \\ 0 & 0 & 0 & 1 \\ 0 & 0 & -1 & 0 \\ 0 & 1 & 0 & 0 \\ -1 & 0 & 0 & 0 \end{bmatrix}.$$

It is straightforward to show that the space-time block code matrix $\boldsymbol{X}$ defines an orthogonal design if and only if the equalities

$$\begin{aligned} \boldsymbol{A}_i^H \boldsymbol{A}_j + \boldsymbol{B}_j^H \boldsymbol{B}_i &= \delta_{ij} \boldsymbol{D}_i, \\ \boldsymbol{A}_i^H \boldsymbol{B}_j + \boldsymbol{A}_j^H \boldsymbol{B}_i &= \boldsymbol{0}, \end{aligned} \tag{4.37}$$

are satisfied (see Xu and Kwak (2005)). Here δ_{ij} is the Kronecker delta (i.e., $\delta_{ij} = 1$ if $i = j$, and 0 otherwise), and $\boldsymbol{D}_i$ is a diagonal matrix where the diagonal elements are strictly positive.

It is easy to check that the Alamouti code, for example, satisfies the equalities above with $\boldsymbol{D}_i$ being equal to the identity matrix for all i. Similarly, we can easily demonstrate that they are also satisfied for the space-time block codes defined by $\boldsymbol{X}_1$ and $\boldsymbol{X}_2$. For $\boldsymbol{X}_1$, the diagonal entries of the corresponding $\boldsymbol{D}_i$ matrix are one, whereas they are two for the code $\boldsymbol{X}_2$.

4.2.2 Decoding of Linear Orthogonal Designs

Let us now describe the decoding of space-time block codes based on the framework presented in the previous section. Let $\boldsymbol{y}_j$ denote the $N \times 1$ received signal vector at antenna j whose kth element, $y_j(k)$, shows the received signal at the kth time interval. We have

$$\boldsymbol{y}_j = \sqrt{\rho}\boldsymbol{X}\boldsymbol{h}_j + \boldsymbol{n}_j, \tag{4.38}$$

where the $N_t \times 1$ vector $\boldsymbol{h}_j$ shows the channel coefficients from all the transmit antennas to the jth receive antenna, and the vector $\boldsymbol{n}_j$ is the complex Gaussian noise term. The channel gains and noise terms are normalized so that their variance per dimension is $1/2$ and the signal constellation is scaled so that its average energy is $1/N_t$ at each transmit antenna. Therefore, ρ denotes the signal-to-noise ratio at each receive antenna.

Define the $2N \times 1$ vectors

$$\tilde{\boldsymbol{y}}_j = \begin{bmatrix} y_j(1) \\ y_j(2) \\ \vdots \\ y_j(N) \\ y_j^*(1) \\ y_j^*(2) \\ \vdots \\ y_j^*(N) \end{bmatrix}, \quad \tilde{\boldsymbol{n}}_j = \begin{bmatrix} n_j(1) \\ n_j(2) \\ \vdots \\ n_j(N) \\ n_j^*(1) \\ n_j^*(2) \\ \vdots \\ n_j^*(N) \end{bmatrix}. \tag{4.39}$$

Based on the structure of the space-time block codes based on the orthogonal designs discussed in the previous section, we can write the equality

$$\tilde{\boldsymbol{y}}_j = \sqrt{\rho}\sum_{m=1}^{M}\left\{\begin{bmatrix} \boldsymbol{A}_m\boldsymbol{h}_j \\ \boldsymbol{B}_m^*\boldsymbol{h}_j^* \end{bmatrix} x_m + \begin{bmatrix} \boldsymbol{B}_m\boldsymbol{h}_j \\ \boldsymbol{A}_m^*\boldsymbol{h}_j^* \end{bmatrix} x_m^*\right\} + \tilde{\boldsymbol{n}}_j. \tag{4.40}$$

Since the noise is AWGN, and the channel coefficients are assumed to be known at the receiver, the optimal decision can be obtained by minimizing the squared Euclidean distance between a candidate codeword $(\boldsymbol{X}(\hat{\boldsymbol{x}}))$ and the received signal, which is given by

$$d(\hat{\boldsymbol{x}}) = \sum_{j=1}^{N_r} \|\boldsymbol{y}_j - \sqrt{\rho}\boldsymbol{X}(\hat{\boldsymbol{x}})\boldsymbol{h}_j\|^2. \tag{4.41}$$

Alternatively, we can minimize

$$2d(\hat{\boldsymbol{x}}) = \sum_{j=1}^{N_r}\left\|\tilde{\boldsymbol{y}}_j - \sqrt{\rho}\sum_{m=1}^{M}\left\{\begin{bmatrix} \boldsymbol{A}_m\boldsymbol{h}_j \\ \boldsymbol{B}_m^*\boldsymbol{h}_j^* \end{bmatrix} x_m + \begin{bmatrix} \boldsymbol{B}_m\boldsymbol{h}_j \\ \boldsymbol{A}_m^*\boldsymbol{h}_j^* \end{bmatrix} x_m^*\right\}\right\|^2. \tag{4.42}$$

It can be shown that this metric can be simplified to

$$2d(\hat{\boldsymbol{x}}) = \sum_{j=1}^{N_r} \left((1-2M)\|\tilde{\boldsymbol{y}}_j\|^2\right) + 2\sum_{m=1}^{M}\sum_{j=1}^{N_r}\left\|\tilde{\boldsymbol{y}}_j - \sqrt{\rho}\begin{bmatrix} \boldsymbol{A}_m\boldsymbol{h}_j \\ \boldsymbol{B}_m^*\boldsymbol{h}_j^* \end{bmatrix} x_m\right\|^2, \qquad (4.43)$$

which can be further simplified to

$$2d(\hat{\boldsymbol{x}}) = 2\sum_{m=1}^{M}\left(\frac{|u_m - \sqrt{\rho}v_m^2 x_m|^2}{v_m^2}\right) - 2\sum_{m=1}^{M}\left(\frac{|u_m|^2}{v_m^2}\right) + \sum_{j=1}^{N_r}\|\tilde{\boldsymbol{y}}_j\|^2, \qquad (4.44)$$

where

$$u_m = \sum_{j=1}^{N_r} \boldsymbol{h}_j^H \boldsymbol{A}_m^H \boldsymbol{y}_j + \boldsymbol{y}_j^H \boldsymbol{B}_m \boldsymbol{h}_j, \qquad (4.45)$$

and

$$v_m^2 = \sum_{j=1}^{N_r}\sum_{i=1}^{N_t} d_{i,m}|h_{i,j}|^2, \qquad (4.46)$$

where $d_{i,m}$ are the diagonal entries of the matrix $\boldsymbol{D}_m$ defined earlier.

The maximum likelihood (ML) decision rule minimizes the metric $d(\hat{\boldsymbol{x}})$ defined in (4.44). However, this is equivalent to the minimization of

$$\tilde{d}(\hat{\boldsymbol{x}}) = \sum_{m=1}^{M}\left(\frac{|u_m - \sqrt{\rho}v_m^2 x_m|^2}{v_m^2}\right), \qquad (4.47)$$

as the other terms in the original metric are independent of the transmitted symbols. We observe that the optimal decisions on the symbols transmitted can be then decoupled, and the decoding is greatly simplified. Therefore, the optimal decision rule for the mth symbol x_m is given by

$$\hat{x}_m = \arg\min_{x_m} |\sqrt{\rho}v_m^2 x_m - u_m|^2/v_m^2 \qquad (4.48)$$

where $m = 1, 2, \ldots, M$.

Using this decision rule with the code defined by the matrices $\boldsymbol{A}_1$, $\boldsymbol{A}_2$, $\boldsymbol{B}_1$ and $\boldsymbol{B}_2$ in Equation (4.36), we can easily generate the specific decision rules. For instance, consider the code represented by $\boldsymbol{X}_1$. By simplifying the above decision rule for this code, we obtain

$$u_1 = \sum_{j=1}^{N_r} h_{1,j}^* y_j(1) + h_{2,j}^* y_j(2) + h_{3,j}^* y_j(3) + h_{4,j}^* y_j(4),$$

$$u_2 = \sum_{j=1}^{N_r} h_{2,j}^* y_j(1) - h_{1,j}^* y_j(2) - h_{4,j}^* y_j(3) + h_{3,j}^* y_j(4),$$

$$u_3 = \sum_{j=1}^{N_r} h_{3,j}^* y_j(1) + h_{4,j}^* y_j(2) - h_{1,j}^* y_j(3) - h_{2,j}^* y_j(4),$$

$$u_4 = \sum_{j=1}^{N_r} h_{4,j}^* y_j(1) - h_{3,j}^* y_j(2) + h_{2,j}^* y_j(3) - h_{1,j}^* y_j(4),$$

and $v_1^2 = v_2^2 = v_3^2 = v_4^2 = \sum_{j=1}^{N_r} \sum_{i=1}^{N_t} |h_{i,j}|^2$.

Let us also consider the space-time block code defined by $\boldsymbol{X}_2$. In this case, we obtain

$$u_1 = \sum_{j=1}^{N_r} h_{1,j}^* y_j(1) + h_{2,j}^* y_j(2) + h_{3,j}^* y_j(3) + h_{4,j}^* y_j(4) + h_{1,j} y_j^*(5) + h_{2,j} y_j^*(6)$$
$$+ h_{3j} y_j^*(7) + h_{4,j} y_j^*(8),$$

$$u_2 = \sum_{j=1}^{N_r} h_{2,j}^* y_j(1) - h_{1,j}^* y_j(2) - h_{4,j}^* y_j(3) + h_{3,j}^* y_j(4) + h_{2,j} y_j^*(5) - h_{1,j} y_j^*(6)$$
$$- h_{4,j} y_j^*(7) + h_{3,j} y_j^*(8),$$

$$u_3 = \sum_{j=1}^{N_r} h_{3,j}^* y_j(1) + h_{4,j}^* y_j(2) - h_{1,j}^* y_j(3) - h_{2,j}^* y_j(4) + h_{3,j} y_j^*(5) + h_{4,j} y_j^*(6)$$
$$- h_{1,j} y_j^*(7) - h_{2,j} y_j^*(8),$$

$$u_4 = \sum_{j=1}^{N_r} h_{4,j}^* y_j(1) - h_{3,j}^* y_j(2) + h_{2,j}^* y_j(3) - h_{1,j}^* y_j(4) + h_{4,j} y_j^*(5) - h_{3,j} y_j^*(6)$$
$$+ h_{2,j} y_j^*(7) - h_{1,j} y_j^*(8),$$

and

$$v_1^2 = v_2^2 = v_3^2 = v_4^2 = \sum_{j=1}^{N_r} \sum_{i=1}^{N_t} 2|h_{i,j}|^2.$$

With this set of equations, the optimal decoding rule which is simple linear combining is obtained explicitly.

4.2.3 Performance Analysis of Space-Time Block Codes

In this section, we calculate the error rates offered by space-time block codes over Rayleigh fading channels. We again assume that the sub-channels between different antenna pairs fade independently. Assume that the sequence of symbols $\{x_1, x_2, \ldots, x_M\}$ are being transmitted. Then, the decision variable u_m computed for x_m at the receiver by linear processing is given by

$$u_m = \sum_{j=1}^{N_r} \boldsymbol{h}_j^H \boldsymbol{A}_m^H \left(\sqrt{\rho} \boldsymbol{X} \boldsymbol{h}_j + \boldsymbol{n}_j\right) + \left(\sqrt{\rho} \boldsymbol{h}_j^H \boldsymbol{X}^H + \boldsymbol{n}_j^H\right) \boldsymbol{B}_m \boldsymbol{h}_j,$$

which is obtained by substituting the received signal given in (4.38) by (4.45). By using the properties of linear orthogonal designs given by (4.37), we can simplify this expression to

$$u_m = \sqrt{\rho} x_m \sum_{j=1}^{N_r} \sum_{i=1}^{N_t} d_{i,m} |h_{i,j}|^2 + n'_m, \tag{4.49}$$

where the noise term n'_m is zero mean Gaussian with variance $\frac{1}{2}\sum_{j=1}^{N_r}\sum_{i=1}^{N_t} d_{i,m}|h_{i,j}|^2$. This is basically the equivalent single-input single-output channel model for the decoding of the space-time block codes.

Consider BPSK modulation. It is easy to observe that the probability of bit error conditioned on the channel gains can be written as

$$P_b(\alpha) = Q\left(\sqrt{\frac{2\alpha\rho}{N_t}}\right), \tag{4.50}$$

where $\alpha = \sum_{j=1}^{N_r}\sum_{i=1}^{N_t} d_{i,m}|h_{i,j}|^2$.

To simplify the rest of the derivations, at this point we assume that all the diagonal elements $d_{i,m}$ corresponding to the mth signal are identical, i.e., $d_{i,m} = d_m$. Then, $\alpha = d_m \sum_{j=1}^{N_r}\sum_{i=1}^{N_t} |h_{i,j}|^2$, and we can easily see that the parameter α is chi-square distributed with $2N_tN_r$ degrees of freedom. Following the same line of analysis as in the case of the Alamouti scheme, we can easily show that

$$P_b = \left(\frac{1}{2}(1-\beta)\right)^{N_tN_r} \sum_{l=0}^{N_tN_r-1} \binom{N_tN_r - 1 + l}{l} \left(\frac{1}{2}(1+\beta)\right)^l, \tag{4.51}$$

where

$$\beta = \sqrt{\frac{d_m\rho/N_t}{1 + d_m\rho/N_t}}.$$

This expression can be approximated for large signal-to-noise ratios as

$$P_b \approx \binom{2N_tN_r - 1}{N_tN_r} \left(\frac{N_t}{4d_m\rho}\right)^{N_tN_r}. \tag{4.52}$$

This expression clearly shows the diversity advantage provided by the space-time block coding schemes. The bit error rate for the BPSK modulation scheme decays inversely with the (N_rN_t)th power of the signal-to-noise ratio, i.e., the diversity order achieved is N_rN_t which is the full spatial diversity that can be obtained. Clearly this generalizes the result obtained for the case of the Alamouti scheme which was specific to the case of two transmit antennas.

For the case of other modulation schemes, similar expressions can also be derived. For instance, for M-PSK signaling, for an AWGN channel, the symbol error rate conditioned on the channel gains is approximately given by

$$P_{e,M-PSK}(\alpha) \approx 2Q\left(\sqrt{\frac{2\alpha d_m\rho}{N_t}} \sin\left(\frac{\pi}{M}\right)\right). \tag{4.53}$$

Averaging this expression over the statistics of α, we can easily show that

$$P_{e,M-PSK} \approx 2\binom{2N_tN_r - 1}{N_tN_r}\left(\frac{N_t}{4d_m \sin^2(\frac{\pi}{M})}\right)^{N_tN_r}\left(\frac{1}{\rho}\right)^{N_tN_r}, \tag{4.54}$$

which clearly demonstrates the diversity advantage that can be obtained.

4.2.4 Examples

We now examine the performance of the two space-time block code examples of the previous section through simulations. We first consider the case of four transmit antennas with the code defined by $\boldsymbol{X}_1$. We present the bit error rate with BPSK modulation over a Rayleigh flat fading channel in Figure 4.4. This is a full-rate code achieving one bit per transmission. We show both the simulation results and the theoretical results derived in the previous section. We observe that the diversity achieved is four for the case of one receive antenna, and it is eight for the case of two receive antennas as expected.

We consider the half rate space-time block code example given by $\boldsymbol{X}_2$ in Figure 4.5. In this case there are four transmit antennas, and we assume that QPSK is employed resulting in a transmission rate of one bit per channel use. We observe that the simulation results and the theoretical expectations clearly match, and they demonstrate a diversity order of four (since only one receive antenna is employed).

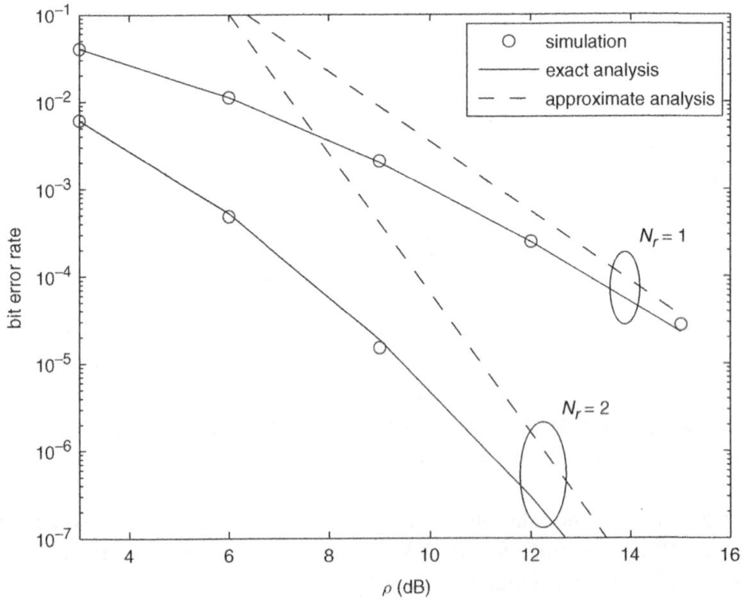

Figure 4.4 Bit error rate of $\boldsymbol{X}_1$ with BPSK modulation.

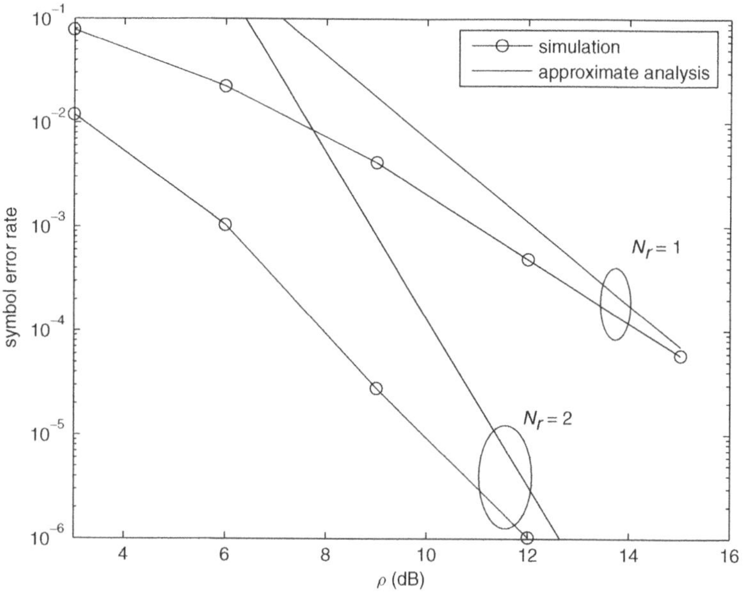

Figure 4.5 Symbol error rate of $\boldsymbol{X}_2$ with QPSK modulation.

4.3 Quasi-Orthogonal Space-Time Block Codes

As we have seen earlier in this chapter, there are no full-rate, full-diversity codes (for complex constellations) other than the Alamouti scheme, and such designs are very limited for real constellations. The quasi-orthogonal space-time block coding approach of Jafarkhani (2001) gives a way of obtaining full-rate (or increased-rate) space-time block coding designs using smaller designs as building blocks.

Let us illustrate the ideas via a simple example by presenting a full-rate design for the $N_t = 4$ scheme using the Alamouti code. We consider the transmission of four symbols, say x_1, x_2, x_3 and x_4. We use the Alamouti code to encode these symbols pairwise resulting in two 2×2 matrices of the form

$$\boldsymbol{X}_{12} = \begin{bmatrix} x_1 & x_2 \\ -x_2^* & x_1^* \end{bmatrix} \quad \text{and} \quad \boldsymbol{X}_{34} = \begin{bmatrix} x_3 & x_4 \\ -x_4^* & x_3^* \end{bmatrix}. \tag{4.55}$$

We then use these two matrices in another orthogonal design to obtain the 4×4 quasi-orthogonal space-time block code matrix as

$$\boldsymbol{X} = \begin{bmatrix} \boldsymbol{X}_{12} & \boldsymbol{X}_{34} \\ -\boldsymbol{X}_{34}^* & \boldsymbol{X}_{12}^* \end{bmatrix}, \tag{4.56}$$

where $(\cdot)^*$ denotes the complex conjugate of a matrix. Thus, we obtain

$$\boldsymbol{X} = \begin{bmatrix} x_1 & x_2 & x_3 & x_4 \\ -x_2^* & x_1^* & -x_4^* & x_3^* \\ -x_3^* & -x_4^* & x_1^* & x_2^* \\ x_4 & -x_3 & -x_2 & x_1 \end{bmatrix}. \tag{4.57}$$

This is clearly a way of obtaining a full-rate code with complex constellations for the four transmit antenna case. However, as we know it cannot achieve full-diversity as no such full-rate full-diversity design exists. In fact, it is easy to show that the diversity order achieved with this scheme is $2N_r$. We note that other code matrices can also be defined, i.e., we are not limited to only using the "Alamouti" type of structure in the definition of the overall code matrix.

Clearly not all columns of the code matrix are orthogonal. To describe the decoding process, we observe that the first and fourth columns of the code matrix are orthogonal to the second and third columns, respectively. Therefore, with appropriate linear processing, the decisions for x_1 and x_4 can be decoupled from decisions for x_2 and x_3. After that step, the pairs of codewords have to be decoded jointly, thus increasing the complexity of the decoding algorithm compared with the case of linear orthogonal designs.

We conclude this section by emphasizing that other existing smaller orthogonal designs can also be extended to a larger number of transmit antennas by a similar approach in a straightforward way, giving rise to similar conclusions.

4.4 Linear Dispersion Codes

We now describe another class of codes, called linear dispersion codes, which generalize the idea of orthogonal and quasi-orthogonal space-time block codes, as well as some other important schemes that will be described later in the book.

As discussed earlier, orthogonal space-time block codes achieve full diversity over a MIMO channel. However, the rate of transmission offered by them is not very high. To alleviate this problem, Hassibi and Hochwald (2002) introduced linear dispersion codes which basically remove the orthogonality constraint imposed before. Let us now describe this class of codes in some detail.

Assume that $x_1, x_2, \ldots, x_M$ represent M symbols to be transmitted over N time slots. The transmitted codeword is described by

$$\boldsymbol{X} = \sum_{m=1}^{M} (x_m \boldsymbol{A}_m + x_m^* \boldsymbol{B}_m), \tag{4.58}$$

where $\boldsymbol{A}$ and $\boldsymbol{B}$ are arbitrary $N \times N_t$ matrices, called dispersion matrices. This description looks identical to the one in (4.35), however we are now talking about a much broader class of codes, since the condition given in (4.37) to obtain linear orthogonal designs is not imposed here.

Clearly, linear dispersion codes include different space-time block codes described earlier in this chapter (i.e., the Alamouti scheme, linear orthogonal designs, and quasi-orthogonal space-time block codes). To give a very simple example, the code given in

Equation (4.57) can be described using the dispersion matrices

$$\boldsymbol{A}_1 = \begin{bmatrix} 1 & 0 & 0 & 0 \\ 0 & 0 & 0 & 0 \\ 0 & 0 & 0 & 0 \\ 0 & 0 & 0 & 1 \end{bmatrix}, \quad \boldsymbol{A}_2 = \begin{bmatrix} 0 & 1 & 0 & 0 \\ 0 & 0 & 0 & 0 \\ 0 & 0 & 0 & 0 \\ 0 & 0 & -1 & 0 \end{bmatrix},$$

$$\boldsymbol{A}_3 = \begin{bmatrix} 0 & 0 & 1 & 0 \\ 0 & 0 & 0 & 0 \\ 0 & 0 & 0 & 0 \\ 0 & -1 & 0 & 0 \end{bmatrix}, \quad \boldsymbol{A}_4 = \begin{bmatrix} 0 & 0 & 0 & 1 \\ 0 & 0 & 0 & 0 \\ 0 & 0 & 0 & 0 \\ 1 & 0 & 0 & 0 \end{bmatrix},$$

$$\boldsymbol{B}_1 = \begin{bmatrix} 0 & 0 & 0 & 0 \\ 0 & 1 & 0 & 0 \\ 0 & 0 & 1 & 0 \\ 0 & 0 & 0 & 0 \end{bmatrix}, \quad \boldsymbol{B}_2 = \begin{bmatrix} 0 & 0 & 0 & 0 \\ -1 & 0 & 0 & 0 \\ 0 & 0 & 0 & 1 \\ 0 & 0 & 0 & 0 \end{bmatrix},$$

$$\boldsymbol{B}_3 = \begin{bmatrix} 0 & 0 & 0 & 0 \\ 0 & 0 & 0 & 1 \\ -1 & 0 & 0 & 0 \\ 0 & 0 & 0 & 0 \end{bmatrix}, \quad \boldsymbol{B}_4 = \begin{bmatrix} 0 & 0 & 0 & 0 \\ 0 & 0 & -1 & 0 \\ 0 & -1 & 0 & 0 \\ 0 & 0 & 0 & 0 \end{bmatrix}.$$

As another example, consider the scenario where the number of symbols transmitted is $M = N\ N_t$, and the space-time codeword is given by

$$\boldsymbol{X} = \begin{bmatrix} x_1 & x_2 & \cdots & x_N \\ x_{N+1} & x_{N+2} & \cdots & x_{2N} \\ \vdots & \vdots & \ddots & \vdots \\ x_{N(N_t-1)+1} & x_{N(N_t-1)+2} & \cdots & x_{NN_t} \end{bmatrix}, \tag{4.59}$$

which basically describes a very high-rate transmission scheme where at different time slots and different antennas, different symbols are transmitted. Clearly, this is not an orthogonal design, or any other design we have seen thus far. Since the special properties that were imposed in orthogonal designs are lacking, decoding or analysis may not be simple. In fact, this scheme will be referred as the vertical Bell Labs layered space-time (VBLAST) architecture. Nevertheless, we can perceive this space-time code as a linear dispersion code. In this case, the dispersion matrices $\boldsymbol{A}_{N_t(l-1)+k}$ have a single "1" in the (l, k)th position and zeros everywhere else, and $\boldsymbol{B}_{N_t(l-1)+k} = \boldsymbol{0}_{N\times N_t}$, where $l = 1, 2, \ldots, N$ and $k = 1, 2, \ldots, N_t$.

Maximum likelihood decoding for linear dispersion codes (with parameters of interest) is out of the question due to the very high complexity requirements. Linear processing (or one of its variations) is also not feasible since there is no orthogonality constraint imposed on the dispersion matrices. On the other hand, there are sub-optimal approaches that work efficiently, such as successive nulling and cancelling, and sphere decoding (Hassibi and Hochwald (2002)). These approaches and other decoding algorithms that can be adopted for use with linear dispersion codes will be described in Chapter 6 in a different context.

Design of linear dispersion codes is a major problem. It turns out that if one selects the dispersion matrices randomly (subject to some power constraints), the resulting error rate performance may not be good. With this motivation, Hassibi and Hochwald (2002) describe a method of designing linear dispersion codes based on a criterion that maximizes the mutual information between the input and the output of the MIMO channel. The algorithm requires numerical optimization subject to certain constraints that determine the amount of complexity in the search process as well as in decoding. Based on the design algorithm, Hassibi and Hochwald (2002) report error rates that are significantly superior to those of the space-time block codes (and those of the VBLAST architecture).

4.5 Chapter Summary and Further Reading

In this chapter, we have studied the basics of space-time block codes. We have considered the case of two transmit antennas, and reviewed the Alamouti scheme, and later extended the results to the case of an arbitrary number of transmit antennas. We have presented the results in a unified framework using the idea of orthogonal designs. We have developed the optimal receiver structures which are accomplished through linear combining at the receiver, and we have conducted performance analysis over Rayleigh fading channels. Theoretical and simulation results clearly show that full spatial diversity can be obtained through the use of space-time block codes. In addition, we have summarized the idea of quasi-orthogonal space-time block codes which is a compromise between full-diversity versus full-rate designs. Finally, we have reviewed linear dispersion codes briefly.

An excellent reference on space-time block coding is the book by Jafarkhani (2005). Exact error rates of space-time block codes are given in Shin and Lee (2002). Other advances in orthogonal space-time block coding are reported in Ganesan and Stoica (2001); Li and Hu (2003); Liang (2003a,b); Lu et al. (2005); Su and Xia (2003); Tran et al. (2004). Algebraic designs of STBCs are reported in Damen and Beaulieu (2003); Damen et al. (2002); Kiran and Rajan (2005); Sethuraman et al. (2003); Xia et al. (2003). Various non-orthogonal and quasi-orthogonal space-time block codes are developed in Boariu and Ionescu (2003); Dalton and Georghiades (2005); Maddah-Ali and Khandani (2006); Sharma and Papadias (2003, 2004); Su and Xia (2004). Delay diversity and the Alamouti scheme are combined to design space-time block codes for four transmit antennas in Luo and Leib (2005), and stacked Alamouti schemes are employed for an even number of transmit antennas to develop codes that achieve the channel capacity for $N_r = 1$ by Sezgin and Jorswieck (2005). Linear dispersion code design is considered in Heath and Paulraj (2002).

Problems

4.1 Derive Equation (4.29).

4.2 Consider the space-time block code for $N_t = 3$ and $N = 4$ described by

$$\begin{bmatrix} x_1 & x_2 & x_3 \\ -x_2 & x_1 & -x_4 \\ -x_3 & x_4 & x_1 \\ -x_4 & -x_3 & x_2 \end{bmatrix},$$

where the symbols x_1, x_2, x_3, x_4 are picked from a real constellation.

a) What is the rate of this code in symbols per channel use? Is it full-rate?

b) Show that this code is an orthogonal space-time block code.

c) Assuming that the channel gains remain constant over a duration of N symbols, derive the ML optimal decision rule for BPSK modulation. Demonstrate that for the ML receiver, the decisions for each of the symbols transmitted decouple.

d) What is the diversity order achieved by this code?

4.3 Repeat Problem 4.2 for the code described by

$$\begin{bmatrix} x_1 & x_2 & x_3 & x_4 & x_5 \\ -x_2 & x_1 & x_4 & -x_3 & x_6 \\ -x_3 & -x_4 & x_1 & x_2 & x_7 \\ -x_4 & x_3 & -x_2 & x_1 & x_8 \\ -x_5 & -x_6 & -x_7 & -x_8 & x_1 \\ -x_6 & x_5 & -x_8 & x_7 & -x_2 \\ -x_7 & x_8 & x_5 & -x_6 & -x_3 \\ -x_8 & -x_7 & x_6 & x_5 & -x_4 \end{bmatrix},$$

which is also designed for a real constellation.

4.4 Design a full-rate full-diversity orthogonal space-time block code for $N = N_t = 8$ (for a real constellation).

4.5 Starting with your answer in the previous problem, design full-rate full-diversity orthogonal space-time block codes for $N = 8$ and $N_t = 5, 6, 7$. Hint: Removing columns in your original design still preserves the orthogonality.

4.6 Show the details of the decoding process for the 4×4 quasi-orthogonal space-time block code described in Section 4.3.

4.7 Consider the linear dispersion code designed for $N_t = 4$ transmit antennas

$$X = \begin{bmatrix} x_1 & x_2 & x_3 & x_4 \\ -x_2^* & x_1^* & -x_4^* & x_3^* \\ -x_7^* & -x_8^* & x_5^* & x_6^* \\ x_8 & -x_7 & -x_6 & x_5 \end{bmatrix}.$$

a) What is the rate of the code in symbols per channel use?

b) What is the corresponding rate in bits per channel use if 16-QAM constellation is employed?

c) What are the corresponding dispersion matrices?

4.8 Using Matlab, simulate the symbol error rate performance of the Alamouti scheme for $N_r = 1, 2$ (assuming Rayleigh fading with independent spatial links) using

a) BPSK modulation.

b) 4-PSK modulation.

c) 8-PSK modulation.

4.9 Using Matlab, simulate the symbol error rate performance of the codes given in Equations (4.33) and (4.34) for $N_r = 1, 2$ (assuming Rayleigh fading with independent spatial channels) using

a) BPSK modulation.

b) 8-PAM modulation.

c) 8-PSK modulation (X_2 only).

5

Space-Time Trellis Codes

In the previous chapter, we have introduced space-time block coding as a means of obtaining full spatial diversity for MIMO systems. The idea was to map several information symbols to signals transmitted simultaneously from different transmit antennas (for multiple symbol periods). We have also seen that, if properly designed, in addition to providing full spatial diversity, space-time block codes can be decoded efficiently using linear processing at the receiver.

In this chapter, we introduce space-time trellis coding which is another basic method of coding for MIMO systems. This method was originally proposed in the landmark paper by Tarokh et al. (1998). The idea is similar to the idea of convolutional coding (more precisely, trellis coded modulation) where there is a basic trellis structure that determines the (coded) symbols to be transmitted from the different antenna elements. This is in sharp contrast to the approach of the previous chapter, as space-time block codes do not have memory from one block to the next. The main benefit of space-time trellis coding is its "coding advantage" provided over the approach of space-time block coding, which comes at the cost of increased decoding complexity.

The chapter is organized as follows. We begin with a simple example of a space-time trellis code. We then consider general space-time trellis codes and the corresponding suitable decoding algorithm. We also develop the pairwise error probability expressions for Rayleigh and Rician fading channels, present basic code design principles, and give examples of good space-time trellis codes. Furthermore, we present an alternative representation of space-time trellis codes, and consider several improved design principles. Towards the end of the chapter, we analyze the performance of space-time trellis codes using a modified version of the union bound, and compare the performances of space-time block and space-time trellis codes. Finally, we conclude the chapter with a brief summary and suggestions for further reading.

5.1 A Simple Space-Time Trellis Code

To illustrate the main ideas in space-time trellis coding, let us start with a simple example of a space-time trellis code. Assume that there are $N_t = 2$ transmit antennas. To describe

Coding for MIMO Communication Systems   Tolga M. Duman and Ali Ghrayeb
© 2007 John Wiley & Sons, Ltd

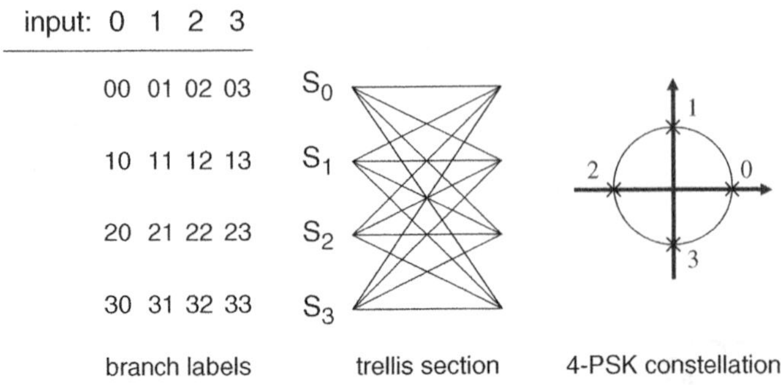

Figure 5.1 A four-state space-time trellis code example.

the code we simply need to specify what is to be transmitted from these two antennas for a frame of data. Assume that a four-state trellis whose section is shown in Figure 5.1 is used for space-time coding.

Encoding is performed as follows. The initial state of the finite state machine is assumed to be S_0. Each pair of information bits to be encoded determines the state transition, and the two coded symbols to be transmitted from the two transmit antennas simultaneously as shown in Figure 5.1. For instance, the encoded sequence corresponding to the (4-ary) information sequence

$$2, 1, 2, 3, 0, 0, 1, 3, 2, \ldots$$

is simply

$$0, 2, 1, 2, 3, 0, 0, 1, 3, \ldots$$

from the first antenna, and

$$2, 1, 2, 3, 0, 0, 1, 3, 2, \ldots$$

from the second antenna. Once all the information bits are encoded, the trellis is terminated to complete each frame of data. We observe that for this example, two information bits are transmitted per use of the channel, thus resulting in a transmission rate of 2 bits per channel use. In general to obtain a transmission rate of R bits per channel use, we need to use a trellis with 2^R branches emanating from each state.

This process is a simple extension of convolutional coding (more precisely, trellis coded modulation). That is, instead of using a trellis which determines only a single symbol to be transmitted for each trellis section (e.g., in the standard trellis coding), each branch is associated with multiple symbols that are to be transmitted from the multiple antenna elements.

5.2 General Space-Time Trellis Codes

In this section, we study general space-time trellis codes which were introduced through a simple example in the previous section. We consider a general mathematical description, establish notation, and derive the decoding algorithm for MIMO fading channels. Since spatial diversity through space-time coding is desirable for limited time diversity systems, we concentrate on quasi-static fading channels (that is, the channel fades remain constant over an entire frame of data or space-time codeword). Furthermore, we assume that the channel state information is known at the receiver (which can be accomplished through the use of pilot tones).

5.2.1 Notation and Preliminaries

Consider a MIMO system with N_t transmit and N_r receive antennas. An n_s state space-time trellis code with transmission rate R is defined by an n_s state trellis diagram where there are 2^R branches emanating from each state. For every branch there are N_t branch labels chosen from a certain signal constellation (e.g., 4-PSK, 16-QAM). We will discuss the optimal selection of the branch labels in later sections when we consider the code design principles. We assume that at the beginning of each frame, we start with the S_0 state, and at the end, the trellis is terminated to that state as well.

Assuming that $x_i(k)$ is the transmitted signal from antenna i at time k, the received signal at the jth antenna corresponding to this time interval is given by

$$y_j(k) = \sqrt{\rho}\sum_{i=1}^{N_t} h_{i,j}x_i(k) + n_j(k)$$

where $i = 1, 2, \ldots, N_t$, $j = 1, 2, \ldots, N_r$, $t = 1, 2, \ldots, N$, and N is the frame length; $h_{i,j}$ denotes the complex Gaussian channel coefficient between the ith transmit and jth receive antennas. Thus the channel is Rayleigh or Rician fading where the channel gains are constant over an entire frame of length N (i.e., quasi-static fading), and are independent across different sub-channels; $n_j(k)$ is the additive Gaussian noise sample at the jth receive antenna at time k. The noise samples are assumed to be spatially and temporally white. We also assume that both the channel gains and additive noise samples are normalized such that their variance per dimension is $1/2$, and the constellation energy is normalized to $1/N_t$ so that ρ denotes the signal-to-noise ratio at each receive antenna.

The input-output relationship can be written in the matrix form as

$$\boldsymbol{Y} = \sqrt{\rho}\boldsymbol{X}\boldsymbol{H} + \boldsymbol{N}, \tag{5.1}$$

where $\boldsymbol{X}$ denotes the $N \times N_t$ transmitted codeword (whose (k, i)th element is $x_i(k)$), $\boldsymbol{H}$ is the $N_t \times N_r$ matrix of channel coefficients (whose (i, j)th element is $h_{i,j}$), $\boldsymbol{Y}$ is the $N \times N_r$ received matrix consisting of $y_j(k)$ as its entries, and $\boldsymbol{N}$ is the $N \times N_r$ noise matrix.

Quasi-static fading, which is assumed here, is often encountered in practice. For instance, if we assume that the Doppler spread is a few hertz, the coherence time of the wireless channel will be in the order of 100 milliseconds. Since typically the bit/symbol durations are very short, this coherence time over which the channel effectively remains constant corresponds to many symbols, e.g., hundreds, or even thousands. If a code of a frame length shorter than the coherence time of the channel is used (which is the case for many applications, such as real-time speech communications), we will effectively observe quasi-static fading. MIMO communications and space-time trellis codes are well motivated for such channels, as obtaining spatial diversity is very important since there is no time diversity.

Throughout this chapter, we assume that the exact channel state information (i.e., $\boldsymbol{H}$) is available at the receiver, but not at the transmitter. Since in most practical wireless systems, the channel fading is very slow, the estimation of $\boldsymbol{H}$ can be easily accomplished through the use of pilot tones with little overhead. We will discuss ways of estimating these channel coefficients, and the effects of estimation errors on the system performance in Chapter 10.

5.2.2 Decoding of Space-Time Trellis Codes

Assuming that the receiver has access to the channel coefficients, the optimal decision rule minimizing the probability of error is given by

$$\hat{\boldsymbol{X}} = \arg\min_{\boldsymbol{X}} P(\boldsymbol{X}|\boldsymbol{Y}, \boldsymbol{H}). \tag{5.2}$$

This is the maximum a posteriori (MAP) decoding rule. If all the symbols are equally likely, then it is equivalent to the ML decoding rule, and it is given by

$$\hat{\boldsymbol{X}} = \arg\min_{\boldsymbol{X}} P(\boldsymbol{Y}|\boldsymbol{X}, \boldsymbol{H}), \tag{5.3}$$

which is easier to manipulate. Given the transmitted signal matrix and the set of channel coefficients, the elements of the received matrix are jointly Gaussian since the additive noise is Gaussian, and temporally and spatially white. Therefore, this likelihood function is proportional to the negative of the squared Euclidean distance between the received matrix and the transmitted matrix (multiplied by the channel coefficient matrix), resulting in the optimal decoding rule

$$\hat{\boldsymbol{X}} = \arg\min_{\boldsymbol{X}} \|\boldsymbol{Y} - \sqrt{\rho}\boldsymbol{X}\boldsymbol{H}\|^2, \tag{5.4}$$

where $\|\cdot\|^2$ denotes the sum of the norm squares of the elements of its matrix argument (i.e., the square of the Frobenius norm of a matrix). Clearly, the resulting decision rule can also be written as

$$\hat{\boldsymbol{X}} = \arg\min_{\boldsymbol{X}} \sum_{k=1}^{N} \sum_{j=1}^{N_r} \left| y_j(k) - \sqrt{\rho} \sum_{i=1}^{N_t} h_{i,j} x_i(k) \right|^2. \tag{5.5}$$

By way of construction, the space-time trellis codewords are paths through the code trellis. Therefore, the minimization above is nothing but the process of finding a path through the space-time code trellis with the minimum (Euclidean) distance from the received signal. Also, observing that the metric is additive at each time step, it is clear that one can use the well-known Viterbi algorithm to perform the decoding very efficiently.

Let us describe the Viterbi decoding algorithm in this context in a bit more detail. At time $k = 0$, we assume that the encoder is in state S_0. We extend the paths emanating from this state, and record the value of the path metric computed using

$$\sum_{j=1}^{N_r} \left| y_j(1) - \sqrt{\rho} \sum_{i=1}^{N_t} h_{i,j} x_i(1) \right|^2.$$

At time k, we have one path through the trellis for each state of the encoder together with the corresponding value of the accumulated path metric. To extend each of these paths by one more step, for each state, at time $k + 1$, all the paths that merge with that particular state are considered as candidates. The possible path metrics are computed by adding

$$\sum_{j=1}^{N_r} \left| y_j(k+1) - \sqrt{\rho} \sum_{i=1}^{N_t} h_{i,j} x_i(k+1) \right|^2$$

to the current path metrics. All of these extensions except the one with the minimum accumulated path metric are discarded, and the time index is incremented. Typically the trellis is terminated in the final step(s) back to state S_0. Therefore, for these steps, only the extensions of the paths that lead to trellis termination are considered, and at the end of the frame, the path through the trellis that is closest to the received vector is declared as the maximum likelihood codeword. For more details on the Viterbi algorithm, the reader is referred to textbooks on digital communications, such as the one by Proakis (2001).

5.3 Basic Space-Time Code Design Principles

In this section, we develop the code design principles for space-time trellis codes for quasi-static Rayleigh fading channels. To derive the principles, we start with the pairwise error probability expression, and examine it closely to derive the code design criteria.

5.3.1 Pairwise Error Probability

Assume that two space-time trellis codewords are given by $\boldsymbol{X}_1$ and $\boldsymbol{X}_2$. The pairwise error probability $P(\boldsymbol{X}_1 \to \boldsymbol{X}_2)$ is the probability that the received signal vector is closer to the erroneous codeword $\boldsymbol{X}_2$ given that the codeword $\boldsymbol{X}_1$ is transmitted. Conditioned on the instantaneous channel realization, the pairwise error probability is given by

$$P(\boldsymbol{X}_1 \to \boldsymbol{X}_2|\boldsymbol{H}) = P(\|\boldsymbol{Y} - \sqrt{\rho}\boldsymbol{X}_1\boldsymbol{H}\|^2 > \|\boldsymbol{Y} - \sqrt{\rho}\boldsymbol{X}_2\boldsymbol{H}\|^2), \quad (5.6)$$

where $\boldsymbol{Y} = \sqrt{\rho}\boldsymbol{X}_1\boldsymbol{H} + \boldsymbol{N}$. This can equivalently be written as

$$P(\boldsymbol{X}_1 \to \boldsymbol{X}_2|\boldsymbol{H}) = P(\|\boldsymbol{N}\|^2 > \|\boldsymbol{N} - \sqrt{\rho}\boldsymbol{D}\|^2), \quad (5.7)$$

where $\boldsymbol{D} = (\boldsymbol{X}_2 - \boldsymbol{X}_1)\boldsymbol{H}$. Noting that the norm used in the above expression is the Frobenius norm, and writing the argument of the probability expression explicitly, we have

$$P(\boldsymbol{X}_1 \to \boldsymbol{X}_2|\boldsymbol{H}) = P\left(\sum_{k=1}^{N}\sum_{j=1}^{N_r} |n_j(k)|^2 > \sum_{k=1}^{N}\sum_{j=1}^{N_r} |n_j(k) - \sqrt{\rho} d_j(k)|^2\right), \quad (5.8)$$

where $d_j(k)$ is the (k, j)th element of the matrix $\boldsymbol{D}$. Cancelling the common terms, we get

$$P(\boldsymbol{X}_1 \to \boldsymbol{X}_2|\boldsymbol{H}) = P\left(\sum_{k=1}^{N}\sum_{j=1}^{N_r} 2\,\mathrm{Re}\{(d_j(k))^* n_j(k)\} > \sum_{k=1}^{N}\sum_{j=1}^{N_r} \sqrt{\rho}|d_j(k)|^2\right). \quad (5.9)$$

Since the noise is spatially and temporally white and Gaussian with zero mean and variance $1/2$ per dimension, the left-hand side of the argument in (5.9) is a Gaussian random variable with zero mean and variance $2\|\boldsymbol{D}\|^2$. Therefore, we obtain

$$P(\boldsymbol{X}_1 \to \boldsymbol{X}_2|\boldsymbol{H}) = Q(\sqrt{\rho/2}\|\boldsymbol{D}\|), \tag{5.10}$$

which can be simply be upper-bounded as

$$P(\boldsymbol{X}_1 \to \boldsymbol{X}_2|\boldsymbol{H}) \leq \exp\left(-\frac{\rho\|\boldsymbol{D}\|^2}{4}\right). \tag{5.11}$$

Denoting the columns of the channel coefficient matrix by $\boldsymbol{h}_j$, we can rewrite the matrix $\boldsymbol{D}$ as

$$\boldsymbol{D} = [(\boldsymbol{X}_1 - \boldsymbol{X}_2)\boldsymbol{h}_1 \quad (\boldsymbol{X}_1 - \boldsymbol{X}_2)\boldsymbol{h}_2 \quad \cdots \quad (\boldsymbol{X}_1 - \boldsymbol{X}_2)\boldsymbol{h}_{N_r}]. \tag{5.12}$$

Therefore, its Frobenius norm can be written as

$$\|\boldsymbol{D}\|^2 = \sum_{j=1}^{N_r} \|(\boldsymbol{X}_1 - \boldsymbol{X}_2)\boldsymbol{h}_j\|^2, \tag{5.13}$$

$$= \sum_{j=1}^{N_r} \boldsymbol{h}_j^H (\boldsymbol{X}_1 - \boldsymbol{X}_2)^H (\boldsymbol{X}_1 - \boldsymbol{X}_2)\boldsymbol{h}_j. \tag{5.14}$$

Since the matrix $\boldsymbol{A} = (\boldsymbol{X}_1 - \boldsymbol{X}_2)^H(\boldsymbol{X}_1 - \boldsymbol{X}_2)$ is Hermitian, it can be diagonalized using a unitary matrix $\boldsymbol{U}$, i.e., $\boldsymbol{U}^H\boldsymbol{A}\boldsymbol{U}$ is a diagonal matrix whose diagonal elements are the eigenvalues of the matrix $\boldsymbol{A}$. Denoting the eigenvalues of $\boldsymbol{A}$ by λ_i (where $i = 1, 2, \ldots, N_t$), and defining

$$\boldsymbol{\Lambda} = \text{diag}\{\lambda_1, \lambda_2, \ldots, \lambda_{N_t}\}, \tag{5.15}$$

we obtain

$$\|\boldsymbol{D}\|^2 = \sum_{j=1}^{N_r} \boldsymbol{h}_j^H \boldsymbol{U}^H \boldsymbol{\Lambda} \boldsymbol{U} \boldsymbol{h}_j. \tag{5.16}$$

Since $\boldsymbol{U}$ is a unitary matrix, and the vector $\boldsymbol{h}_j$ consists of independent complex Gaussian random variables as its entries, the vector $\boldsymbol{U}\boldsymbol{h}_j$ has the same distribution. Denoting the ith element of $\boldsymbol{U}\boldsymbol{h}_j$ by $\beta_{i,j}$, we can write

$$\|\boldsymbol{D}\|^2 = \sum_{j=1}^{N_r}\sum_{i=1}^{N_t} \lambda_i |\beta_{i,j}|^2. \tag{5.17}$$

Using this expression, the conditional pairwise error probability that we require can be written as

$$P(\boldsymbol{X}_1 \to \boldsymbol{X}_2|\boldsymbol{H}) \leq \exp\left(-\frac{\rho}{4}\sum_{j=1}^{N_r}\sum_{i=1}^{N_t} \lambda_i |\beta_{i,j}|^2\right), \tag{5.18}$$

$$= \prod_{j=1}^{N_r}\prod_{i=1}^{N_t} \exp\left(-\frac{\rho\lambda_i|\beta_{i,j}|^2}{4}\right). \tag{5.19}$$

Rayleigh Fading

Assuming that the wireless channel is modeled by Rayleigh fading, the random variables $\beta_{i,j}$ are zero-mean complex Gaussian with variance $1/2$ per dimension. Therefore, their norm square is exponential with parameter 1. Thus, we can easily average over the channel variations to obtain an upper bound on the (unconditional) pairwise error probability as

$$P(\boldsymbol{X}_1 \to \boldsymbol{X}_2) \leq \prod_{j=1}^{N_r} \prod_{i=1}^{N_t} \int_0^{\infty} \exp\left(-\frac{\rho\lambda_i z}{4}\right) e^{-z} dz, \tag{5.20}$$

$$= \prod_{j=1}^{N_r} \prod_{i=1}^{N_t} \frac{1}{1 + \frac{\rho\lambda_i}{4}}, \tag{5.21}$$

$$= \left(\frac{1}{\prod_{i=1}^{N_t}\left(1 + \frac{\rho\lambda_i}{4}\right)}\right)^{N_r}. \tag{5.22}$$

Examining this upper bound carefully, we can develop basic code design principles for space-time codes over quasi-static MIMO Rayleigh fading channels.

Rician Fading

For the sake of completeness, let us consider the case of Rician fading as well. In this case, the elements $\beta_{i,j}$ have non-zero means, i.e., their norms are Rician distributed. Assuming that $K_{i,j} = |E(\beta_{i,j})|^2$, by averaging the conditional pairwise error probability expression over the channel statistics, it can be shown that (Tarokh et al. (1998))

$$P(\boldsymbol{X}_1 \to \boldsymbol{X}_2) \leq \prod_{j=1}^{N_r} \left(\prod_{i=1}^{N_t} \frac{1}{1 + \frac{\rho\lambda_i}{4}} \exp\left\{ -\frac{K_{i,j}\frac{\rho}{4}\lambda_i}{1 + \frac{\rho\lambda_i}{4}} \right\} \right). \tag{5.23}$$

Clearly, the earlier expression on the pairwise error probability derived for Rayleigh fading is a special case of this one for $K_{i,j} = 0$.

5.3.2 Space-Time Code Design Principles

In this section, we consider Rayleigh fading since it represents practically the worst case scenario among different types of fading as noted above. Assuming that the first r of the N_t eigenvalues of the matrix $\boldsymbol{A}$ are non-zero (they have to be positive as the matrix is Hermitian), and the remaining are zeros, we can further upper-bound the pairwise error probability expression as

$$P(\boldsymbol{X}_1 \to \boldsymbol{X}_2) \leq \left(\prod_{i=1}^{r} \lambda_i\right)^{N_r} (\rho/4)^{-rN_r}, \tag{5.24}$$

which is a tight bound on the pairwise error probability for large signal-to-noise ratios.

In light of the above results, we now make the following observations. The pairwise error probability decays with the inverse of the (rN_r)th power of the signal-to-noise ratio ρ. Therefore, the diversity order obtained is simply rN_r. Since $r \leq N_t$, clearly the diversity order provided by the space-time code is upper-bounded by N_tN_r, and this upper bound is achieved if there are no codeword pairs in the space-time code for which some of the eigenvalues of the corresponding $\boldsymbol{A}$ matrix are zeros. We note that since there are only N_tN_r different independent fading coefficients, the maximum available diversity over this channel is N_tN_r regardless of the code used. We also observe that the eigenvalues of the matrix $\boldsymbol{A}$ determine the "coding gain" that can be achieved with the space-time code, and that the larger the product of the non-zero eigenvalues, the lower the pairwise error probability at high signal-to-noise ratios.

Based on these observations, the basic code design principles for space-time codes over quasi-static Rayleigh fading channels are as follows:

- **Rank Criterion:** The maximum diversity is achieved if the matrix

$$\boldsymbol{A} = (\boldsymbol{X}_1 - \boldsymbol{X}_2)^H(\boldsymbol{X}_1 - \boldsymbol{X}_2) \tag{5.25}$$

  is full rank for all the pairs of distinct codewords $\boldsymbol{X}_1$ and $\boldsymbol{X}_2$. Otherwise, if the minimum rank of $\boldsymbol{A}$ among all the codeword pairs, r, is smaller than N_t, a diversity of order rN_r is achieved.

- **Determinant Criterion:** Assume that full rank is obtained using the criterion in the first part – clearly, this is the more interesting case. To obtain the maximum coding advantage possible, the minimum of the product of the eigenvalues of $\boldsymbol{A}$ over all pairs of distinct codewords (i.e., the minimum determinant of possible $\boldsymbol{A}$ matrixes) should be maximized.

Clearly, the first criterion is the more important code design principle as it is more beneficial to get the full diversity advantage. Therefore, among the codes that satisfy the rank criterion, one can search for the codes that satisfy the determinant criterion. In other words, there is no need to try to optimize the determinant criterion without making sure that the rank criterion is satisfied. The effect of the minimum determinant of the matrix $\boldsymbol{A}$ is simply a shift of the probability of error curve without changing the slope, thus, it can be used to obtain the optimal coding gain.

We also note that, although we concentrate in this chapter on space-time trellis codes, the above design rules are general, that is, they can be used to design other coding schemes that do not have a trellis structure as well, as long as the channel is quasi-static Rayleigh fading with independent fading coefficients across different antenna pairs. For instance, the rank criterion above can be readily applied to the case of space-time block codes to prove that they will provide full rank. Let us consider the Alamouti scheme as a simple example. In this case, the pair of distinct codewords are of the form

$$\boldsymbol{X}_1 = \begin{bmatrix} x_{1,1} & x_{1,2} \\ -x_{1,2}^* & x_{1,1}^* \end{bmatrix}, \tag{5.26}$$

and

$$\boldsymbol{X}_2 = \begin{bmatrix} x_{2,1} & x_{2,2} \\ -x_{2,2}^* & x_{2,1}^* \end{bmatrix}, \tag{5.27}$$

therefore, the corresponding $\boldsymbol{A}$ matrix is given by

$$\boldsymbol{A} = \begin{bmatrix} |x_{1,1} - x_{2,1}|^2 + |x_{1,2} - x_{2,2}|^2 & 0 \\ 0 & |x_{1,1} - x_{2,1}|^2 + |x_{1,2} - x_{2,2}|^2 \end{bmatrix}, \tag{5.28}$$

Clearly, since the codewords are distinct, the rank of the above matrix is definitely two, and full diversity is achieved.

We also note that the minimum of the determinants among all codeword pairs for the Alamouti scheme is simply the fourth power of the minimum Euclidean distance of the underlying signal constellation, which can be very small. Therefore, this code does not take advantage of the second code design criterion above, and it does not yield a coding gain. We will elaborate on this point, and compare these codes later in the chapter.

To give another example, consider the 4×4 orthogonal space-time block code given in Equation (4.33). In this case, the two distinct codewords are of the form

$$\boldsymbol{X}_1 = \begin{bmatrix} x_{1,1} & x_{1,2} & x_{1,3} & x_{1,4} \\ -x_{1,2} & x_{1,1} & -x_{1,4} & x_{1,3} \\ -x_{1,3} & x_{1,4} & x_{1,1} & -x_{1,2} \\ -x_{1,4} & -x_{1,3} & x_{1,2} & x_{1,1} \end{bmatrix}, \tag{5.29}$$

and

$$\boldsymbol{X}_2 = \begin{bmatrix} x_{2,1} & x_{2,2} & x_{2,3} & x_{2,4} \\ -x_{2,2} & x_{2,1} & -x_{2,4} & x_{2,3} \\ -x_{2,3} & x_{2,4} & x_{2,1} & -x_{2,2} \\ -x_{2,4} & -x_{2,3} & x_{2,2} & x_{2,1} \end{bmatrix}. \tag{5.30}$$

Therefore, the corresponding $\boldsymbol{A}$ matrix is given by

$$\boldsymbol{A} = \left((x_{1,1} - x_{2,1})^2 + (x_{1,2} - x_{2,2})^2 + (x_{1,3} - x_{2,3})^2 + (x_{1,4} - x_{2,4})^2\right) \boldsymbol{I}_4, \tag{5.31}$$

which clearly shows that as long as the symbols being transmitted (at least one of them) are different, the rank of the matrix is four, meaning that the full diversity is achieved. However, as in the Alamouti example, there is no coding gain, i.e., the minimum determinant can be very small.

Clearly, it is easy to use the same argument for other orthogonal space-time block codes to demonstrate that the full diversity is obtained without resorting to the more complicated analysis performed in the previous chapter.

5.3.3 Examples of Good Space-Time Codes

Based on the rank and determinant criteria, a number of space-time trellis codes are developed in the literature. The basic idea is to start with a certain trellis and a given signal constellation, and perform a computer search for the branch labels for which the rank criterion and the determinant criterion presented in the previous section are satisfied. Clearly, searching through all possible codeword pairs is formidable. However, by limiting the examination to the set of codeword pairs that differ only in one "short" error event would be sufficient for all intents and purposes.

Using a computer search, Tarokh et al. (1998) designed various space-time trellis codes. Here, we consider a few of those. We consider the case of two transmit antennas at this point.

The first example is a four-state space-time trellis code using a QPSK signal constellation which was already described in Figure 5.1. This code is designed for two transmit antennas and it achieves a transmission rate of 2 bits per channel use. Examples of eight-state and 16-state space-time trellis codes with 2 bits per channel use transmission rate using the same number of transmit antennas and the same signal constellation are given in Figure 5.2.

All the three examples considered above achieve a rate of 2 bits per channel use. An example of a code employing 8-PSK modulation that achieves a higher bandwidth efficiency is given in Figure 5.3. For code examples with larger number of states, i.e., higher complexity levels but with better performance, the reader is referred to Tarokh et al. (1998).

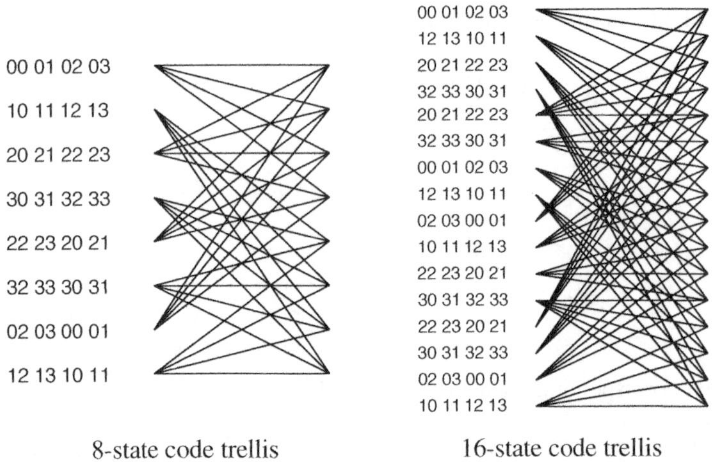

Figure 5.2 Eight-state and 16-state space-time trellis codes using 4-PSK modulation (2 bits per channel use).

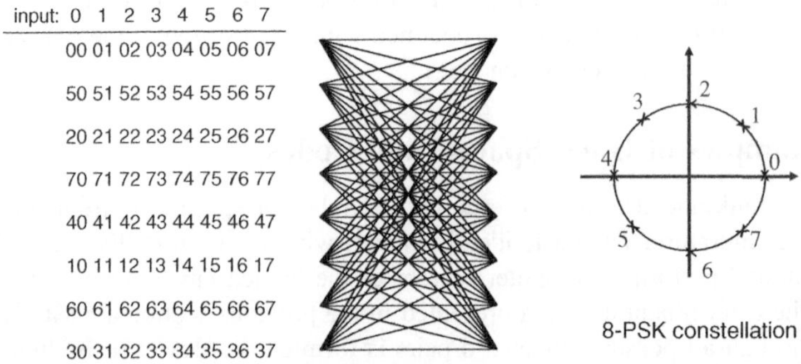

Figure 5.3 Eight-state space-time trellis code using 8-PSK modulation (3 bits per channel use).

It is certainly possible to design codes for rates higher than 2 bits per channel use, however, the complexity of the decoder (implementing the Viterbi algorithm on the code trellis) is exponential in the transmission rate R. Therefore, for systems requiring higher spectral efficiencies, it is more appropriate to use other space-time coding techniques (e.g., space-time block codes discussed in the previous chapter, or spatial multiplexing schemes that will be described in Chapter 6). The same is true for a larger number of transmit antennas. Both the code design (which requires computer search) and decoding complexities increase considerably if more than two transmit antennas are employed (in an attempt to increase the spectral efficiencies of the MIMO system). Therefore for a larger number of transmit antennas, other space-time coding alternatives may be more suitable.

Let us give several examples illustrating the performance of the codes whose structures are described in this section. We consider a two transmit antenna system over a quasi-static Rayleigh fading channel. We assume that a frame length of $N = 130$ is selected, and perfect channel state information is available at the receiver. Figure 5.4 illustrates the frame error rates of the four-, eight-, and 16-state space-time trellis codes that achieve a transmission rate of 2 bits per channel use when only a single receive antenna is used. We observe that all three schemes achieve a diversity order of two, i.e., full spatial diversity, as expected. We also see that the coding gain improves by using a more complicated trellis (with a larger number of states), but this comes at the expense of some increase in the decoding complexity.

It is interesting to compare these results with the channel capacity. Since the channel is quasi-static (i.e., non-ergodic), in this case the suitable channel capacity is the "outage" capacity as described in Chapter 3. For two transmit and one receive antenna Rayleigh flat

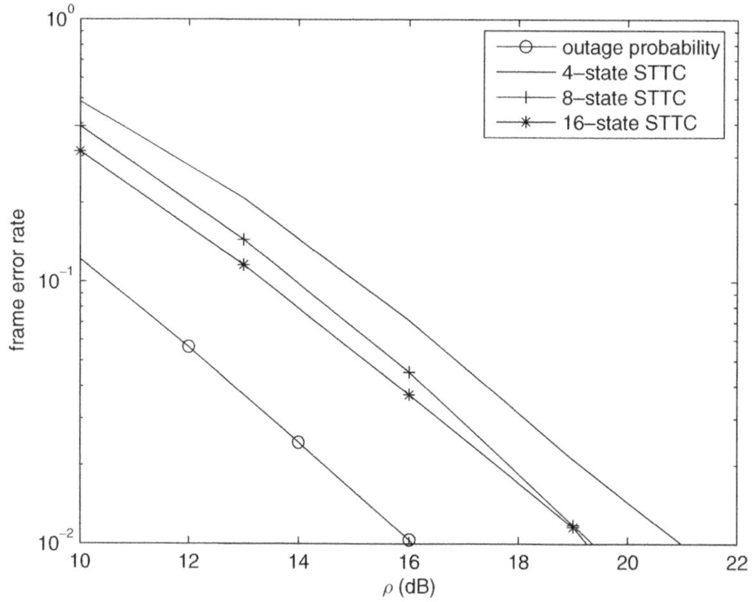

Figure 5.4 Outage probability and frame error rates of several space-time trellis codes with QPSK (quasi-static fading, two transmit and one receive antennas).

fading channels, the 10% outage capacity results state that we need about 10 dB signal-to-noise ratio to achieve a transmission rate of 2 bits per channel use. If we compare this signal-to-noise ratio with the signal-to-noise ratio needed for 10% frame error rate (about 13.5 dB for the 16-state code), we see that the performance of the practical space-time trellis code is only about 3.5 dB away. This gap can be closed further by using more complicated trellis structures. For instance, the results in Tarokh et al. (1998) show that the frame error rate performance further improves by more than 0.5 dB by employing a 64-state code, bringing the code performance within 3 dB of the outage capacity limit. Similar observations can be made for other frame error rates selected as well. For instance, for 1% outage level, the performance difference is about 3–4 dB.

Frame error rates for the same three codes and the corresponding outage capacity results for the case of the two receive antennas are shown in Figure 5.5. Clearly, in this case too, the full spatial diversity of order four is attained. Also, similar to the previous example, the code performance is close to the outage capacity limits.

Let us also present the bit error rate results for the four-state space-time trellis code with one and two receive antennas in Figure 5.6. We see that the observations on the diversity order achieved are the same as before, i.e., for the two transmit and one receive antenna system, the diversity order is two, and for the $N_t = 2$ and $N_r = 2$ system, it is four.

5.3.4 Space-Time Trellis Codes for Fast Fading Channels

So far, all of our discussion has been focused on the case of quasi-static fading channels, where the channel remains constant for the entire frame of data. Let us now consider the

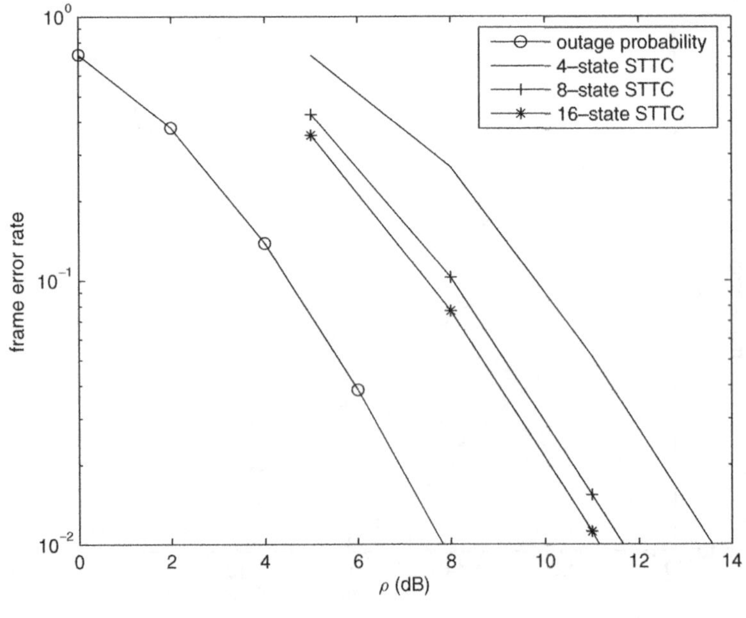

Figure 5.5 Outage probability and frame error rates of several space-time trellis codes with QPSK (quasi-static fading, two transmit and two receive antennas).

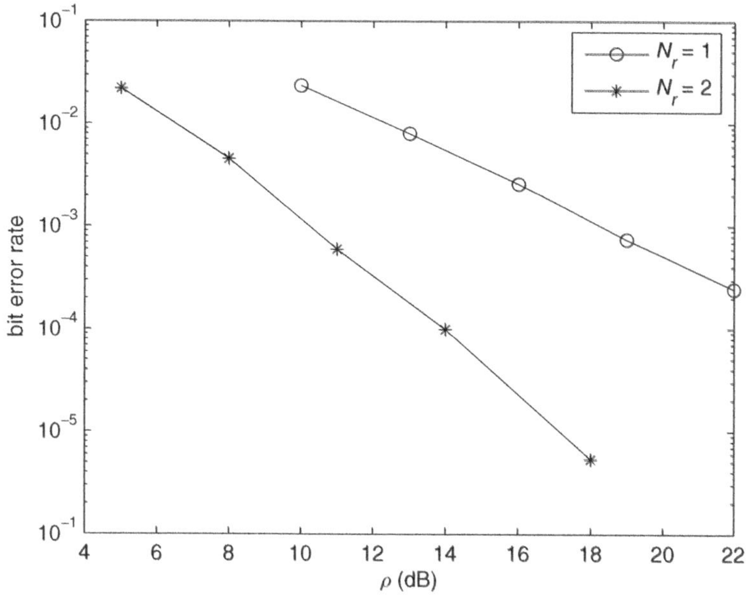

Figure 5.6 Bit error rates for the four-state code in quasi-static fading (two transmit and two receive antennas).

case where fading is more rapid. With proper interleaving, it is possible to convert this into a fast-fading one, i.e., a fully interleaved MIMO channel. Although this may be a less practical scenario, we would like to comment on the performance of space-time trellis codes over these channels for completeness of exposition.

We note that the design principles, based on the rank and the determinant criteria were developed assuming that the channel does not change over an entire frame. Therefore, they are not applicable to this case. In fact, the overall diversity order available in the system is greatly increased as well (due to the available time diversity). A space-time code should exploit this diversity if it is to perform well.

Let us give a simple example of a space-time trellis code performance over a fully interleaved Rayleigh flat fading channel. Consider the case with two transmit antennas. The bit error rate of the four, eight and 16-state codes described in Section 5.3.3 are presented in Figure 5.7 assuming that perfect channel state information is available at the receiver. We observe that the diversity order for the three codes is more than two, and it is improved as the number of states is increased. This is because, in addition to the spatial diversity, we also have time diversity available. Since the error event lengths on the code trellis become larger with a larger number of states, the number of independent path gains observed increases, thereby increasing the diversity order. For instance, a four-state code designed using the code trellis we have used in this example can provide a diversity order of up to four (as the shortest error event length is two providing a potential time diversity advantage of two). Despite the increase in the diversity order using these codes, considering that the (ergodic) channel capacity states that 2 bits per channel use

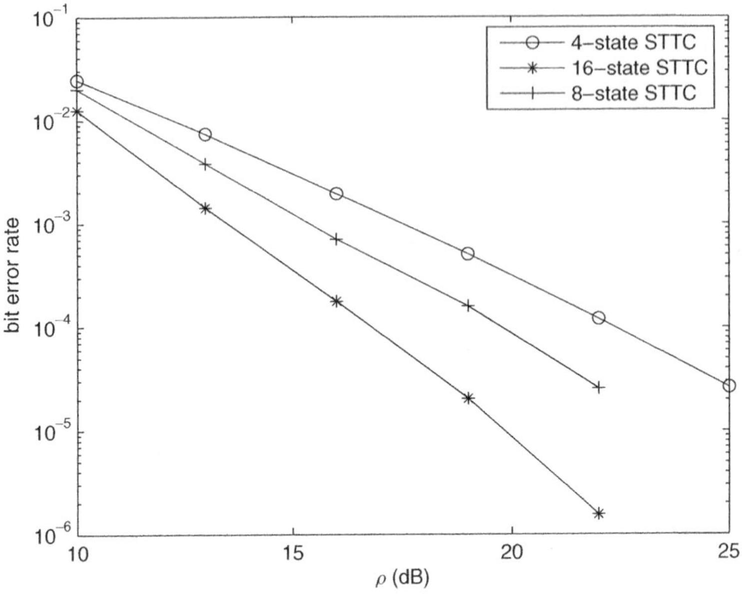

Figure 5.7 Bit error rate results for four-, eight- and 16-state space-time trellis codes over fully interleaved Rayleigh fading channels (two transmit and one receive antennas).

should be achievable for a signal-to-noise ratio of $\rho = 5.5$ dB (for Gaussian inputs) or $\rho = 6.3$ dB (for i.i.d. QPSK inputs), we see that the code performance, even for the more complicated 16-state code, is poor, that is, there is a difference of over 10 dB with the Shannon limit – if we take probability of error of 10^{-4} as a reliable communication level.

Obviously, better classes of codes can be used to improve the space-time code performance that can efficiently exploit the available time diversity in the system. One such class of codes is developed in Tarokh et al. (1998) named "smart greedy" space-time codes. A better alternative is to use other capacity-approaching coding schemes, such as turbo codes or low-density parity check codes, adopted for MIMO systems (see Chapter 7).

The situation is similar if an intermediate scenario is considered. For instance, consider the case of a block fading channel, which refers to the case where the coefficients are constant over a block of data, but change from one block to another, and a coded frame consists of several such blocks. A direct application of the space-time trellis codes does not provide a good performance. Even when a channel interleaver is employed to make sure the space-time trellis code takes some advantage of the channel variations in time, space-time trellis coding usually is not the best choice. For this case too, better codes should be developed using this new channel model which may include the use of turbo-based coding schemes.

5.4 Representation of Space-Time Trellis Codes for PSK Constellations

The space-time trellis codes given by Tarokh et al. (1998) are described using a trellis diagram. Such a representation certainly provides a clear understanding of the code structure. However, it may not be the best choice when the objective is to search for optimal codes. Let us present a different representation of such codes (suitable for PSK signal constellations) using the approach of Baro et al. (2000), which can be considered to be analogous to the generator matrix representation of convolutional codes.

5.4.1 Generator Matrix Representation

For a code with transmission rate R and a frame length of N, there are a total of RN information bits to be transmitted in each frame of data. Let us denote these information bits by the sequence $\{u_n\}$. The bits $u_{Rk+R-1}, \ldots, u_{Rk-s}$ are used to compute the kth set of transmitted symbols. In vector form, we can write this set of bits as

$$\boldsymbol{u}_k = [u_{Rk-s}\ u_{Rk-s+1}\ \cdots u_{Rk}\ \ \cdots\ \ u_{Rk+R-1}]. \tag{5.32}$$

The coded symbols transmitted from the N_t antennas at the kth stage of the trellis can be written using matrix multiplication as

$$\boldsymbol{x}_k = [x_1(k)\ x_2(k)\ \cdots\ x_{N_t}(k)] \tag{5.33}$$

$$= \boldsymbol{u}_k \boldsymbol{G}, \tag{5.34}$$

where the operation is modulo 2^R, s is a parameter that depends on the trellis memory (for instance, $s = 0$ will be the case with no memory; a simple mapping, i.e., uncoded transmission), and $\boldsymbol{G}$ is a matrix that describes the space-time trellis code.

A simple example is in order. Consider the four-state space-time trellis code described in the previous section. It can be observed that this code can be generated using the parameters $R = 2$, $s = 2$ and

$$\boldsymbol{G} = \begin{bmatrix} 2 & 0 \\ 1 & 0 \\ 0 & 2 \\ 0 & 1 \end{bmatrix}.$$

To see how the encoding is performed, consider the same sequence used in Section 5.1, i.e., 2, 1, 2, 3, 0, 0, 1, 3, 2. Expressing this in terms of bits, we have $\boldsymbol{u} = [100110110000011\ 110]$. Therefore, the symbols to be transmitted are given by (note that the first symbol considers two 0's for initialization, i.e., the bits with negative indexes are taken as 0's)

$$\boldsymbol{x}_0 = [0\ 0\ 1\ 0]\boldsymbol{G} = [0\ 2],$$

$$\boldsymbol{x}_1 = [1\ 0\ 0\ 1]\boldsymbol{G} = [2\ 1],$$

$$\boldsymbol{x}_2 = [0\ 1\ 1\ 0]\boldsymbol{G} = [1\ 2],$$

$$x_3 = [1\ 0\ 1\ 1]G = [2\ 3],$$
$$x_4 = [1\ 1\ 0\ 0]G = [3\ 0],$$
$$x_5 = [0\ 0\ 0\ 0]G = [0\ 0],$$
$$x_6 = [0\ 0\ 0\ 1]G = [0\ 1],$$
$$x_7 = [0\ 1\ 1\ 1]G = [1\ 3],$$
$$x_8 = [1\ 1\ 1\ 0]G = [3\ 2],$$

which are the same as the ones produced by the trellis representation used above.

5.4.2 Improved Space-Time Code Design

Baro et al. (2000) note that the generator matrix representation of the previous section is useful when searching for optimal space-time codes. With this representation, using exhaustive computer search, several codes that are superior to the ones given by Tarokh et al. (1998) are produced. The resulting code matrices for four- ($s = 1$), eight- ($s = 2$) and 16-state ($s = 2$) space-time trellis codes with 4-PSK modulation are

$$G_{4\text{-}state} = \begin{bmatrix} 2 & 2 \\ 0 & 2 \\ 1 & 0 \\ 3 & 1 \end{bmatrix}, \quad G_{8\text{-}state} = \begin{bmatrix} 2 & 2 \\ 0 & 1 \\ 2 & 0 \\ 1 & 0 \\ 2 & 2 \end{bmatrix}, \quad G_{16\text{-}state} = \begin{bmatrix} 1 & 2 \\ 2 & 0 \\ 2 & 1 \\ 0 & 2 \\ 0 & 2 \\ 2 & 0 \end{bmatrix}. \tag{5.35}$$

These codes provide gains over the codes given in the previous section. The gains, however, are only significant when a larger number of antennas are employed.

In general, it is not an easy task to perform space-time trellis code design for more than two transmit antennas. However, such a representation (or, equivalently, an implementation using shift registers) makes it easier to systematically search for good codes, and thus it helps us design codes for three or four transmit antenna systems as well (see Chen et al. (2002)). We note, however, that the complexity of decoding space-time trellis codes increases with the number of transmit antennas. Thus, for large N_t, other coding schemes, e.g., space-time block codes, or spatial multiplexing schemes (described in Chapter 6) would be more appropriate.

Blum (2002) derives some analytical tools that enable easier searches for optimal codes by deriving some properties of the pairwise coding gain, i.e., to obtain codes that satisfy the determinant criterion (assuming that full rank is already achieved). These tools allow a relatively easy search for optimal codes that achieve the best coding gains possible. Using these ideas, Yan and Blum (2002) design a comprehensive set of space-time trellis codes that outperform the codes given in the previous section. Code design examples for more than two transmit antennas are also provided. We note that, as with most other improved design efforts, improvements reported are more pronounced if there are more than one receive antennas.

A number of approximations were made to reach the rank criterion and the determinant criterion of the previous section, and also a computer search was necessary to find the "optimal" branch labels to design the best possible codes. It is in fact possible to improve these code design principles further, thereby leading to codes with better performance. Using the upper bound on the pairwise error probability in Equation (5.22) instead of the simplified one in Equation (5.24), we can improve the space-time trellis codes designed (Tao and Cheng (2001)). Clearly, this expression is dependent on the signal-to-noise ratio, and will only give different results if we are considering moderate or low values of ρ. However, for practical applications, this may be important. For instance, by assuming that $\rho/4$ is approximately unity, the design rule becomes the maximization of the minimum value of $\det(\boldsymbol{I} + \boldsymbol{A})$, over all codeword pairs (where $\boldsymbol{A}$ is as defined earlier). This rule is used by Tao and Cheng (2001) to design several codes that show an improved performance over the codes given in the previous section. However, as in the codes designed by Baro et al. (2000), the resulting improvements are significant only for the case of multiple receive antennas.

5.5 Performance Analysis for Space-Time Trellis Codes

The performance results (on bit error rates and frame error rates) of space-time trellis codes we have presented so far were solely based on computer simulations. Obviously, it is also of interest to provide theoretical performance results as done for the case of space-time block codes in Chapter 4. However, this is a significantly more challenging task in this case.

The standard method for computing theoretical upper bounds on the probability of error of a coded system is to use union bounding techniques. However, when directly applied, the union bound over quasi-static fading channels turns out to be very loose, rendering itself useless (Stefanov (2001)). This is true for both single-antenna and multi-antenna systems.

The union bound is obtained by summing the pairwise error probabilities over all possible erroneously received codewords. Clearly, this sum may be counting some of the possible errors multiple times (by summing/integrating over many different regions of the multi-dimensional space). This may not be a big problem when the bound is applied to a non-fading Gaussian channel as each of the pairwise error probability terms is an exponentially decaying function of the signal-to-noise ratio and the squared Euclidean distance between the codewords considered, and the overall error probability will be dominated by only a few error events. As a result, the bound will be tight (at least at high signal-to-noise ratios).

The situation is different for fading channels, in particular, when the available diversity is limited. For instance, for quasi-static flat fading channels, the only form of diversity available is due to the use of multiple transmit and receive antennas (given by $N_r N_t$). Therefore, each of the pairwise error probability terms (at best) will decay inversely with this power of the signal-to-noise ratio – as opposed to an exponential decay as in the case of non-fading channels. For instance, for an $N_t = 2$ and $N_r = 1$ system, we will be adding many terms that are proportional to $\frac{1}{\rho^2}$, and there will not be a few terms that dominate the overall bound, making the bound loose. The situation is not as severe if time diversity is also available (i.e., the fading is not slow).

Motivated by the fact that a direct application of the union bound on the error probability for space-time trellis codes over quasi-static fading channels will be useless, in this section, we provide a modified performance bound based on the work of Stefanov and Duman (2003a).

As another approach to the analysis of space-time trellis codes, Aktas and Fitz (2003) provide a distance spectrum interpretation of space-time trellis codes. This interpretation is a useful extension of the earlier analysis techniques that use pairwise error probabilities only. The authors employ the idea of expurgation of the union bound based on the work of Verdu (1987). However, they observe that this technique by itself is not sufficient to obtain tight error rate bounds over quasi-static fading channels. Uysal and Georghiades (2000) derive useful performance bounds (based on the union bound technique) over fully interleaved Rayleigh fading channels. However, this bound does not extend to the case of quasi-static fading channels.

Our objective here is to derive performance bounds for space-time trellis codes over both fully interleaved and quasi-static Rayleigh fading channels. The bound for the fully interleaved fading case is similar to the one by Uysal and Georghiades (2000). We start with the usual union bound, and we observe that it is sufficient for the fast fading case. However, significant modifications are needed for the quasi-static fading channels as will described in detail.

5.5.1 Union Bound for Space-Time Trellis Codes

Let us start our investigation with a naive application of the union bounding technique for a space-time trellis coded system. The union bounding technique is commonly used to upper bound the error rates of coded or uncoded communication systems when the derivation of the exact error rates is difficult, or impossible. To illustrate the general idea, consider a digital communication system where the possible transmitted signals are $\boldsymbol{X}_1, \boldsymbol{X}_2, \ldots, \boldsymbol{X}_{N'}$. The exact error probability for this communication system is given by

$$P_e = \sum_{i=1}^{N'} P(\boldsymbol{X}_i) P\left(\bigcup_{j=1, j\neq i}^{N'} E_{i,j} \right), \tag{5.36}$$

where $P(\boldsymbol{X}_i)$ is the probability that the ith codeword is transmitted, and $E_{i,j}$ denotes the event that the received signal is closer to the jth codeword than to the ith codeword, given that $\boldsymbol{X}_i$ is transmitted. For example, for the case of an additive white Gaussian noise channel, the proper metric is the squared Euclidean distance. Typically, the evaluation of the probability of the union of these events is very difficult for coded systems. Therefore, we resort to upper bounding this quantity with the sum of the probabilities of the individual events, i.e., we have

$$P_e \leq \sum_{i=1}^{N'} \sum_{j=1, j\neq i}^{N'} P(\boldsymbol{X}_i) P(E_{i,j}), \tag{5.37}$$

which is referred as the union bound. This bound is usually tractable as it only involves the computation of pairwise error probabilities $P(E_{i,j})$, i.e., the probability that the received vector is closer to the jth codeword of the signal set when the ith codeword is transmitted, as opposed to the union of these events. It is possible to simplify this further if the signal set is geometrically uniform (e.g., for linear codes with BPSK modulation), by removing the outer summation as the conditional error probabilities for all the elements of the signal set will be identical.

In the context of space-time trellis codes, the union bound on the frame error rate is given by

$$P_f \leq \sum_{i=1}^{2^{RN}} P(\boldsymbol{X}_i) \sum_{j=1, j\neq i}^{2^{RN}} P(\boldsymbol{X}_i \rightarrow \boldsymbol{X}_j), \tag{5.38}$$

where R is the transmission rate of the system, and N is the frame length, hence 2^{RN} is the total number of different possible codewords, $\boldsymbol{X}_i$ is the ith $N \times N_t$ space-time codeword matrix, and $P(\boldsymbol{X}_i \rightarrow \boldsymbol{X}_j)$ is the pairwise error probability. Since all the codewords are equally likely, we can write

$$P_f \leq 2^{-RN} \sum_{i=1}^{2^{RN}} \sum_{j=1, j\neq i}^{2^{RN}} P(\boldsymbol{X}_i \rightarrow \boldsymbol{X}_j). \tag{5.39}$$

The union bound on the bit error probability can also be written in a similar fashion as

$$P_b \leq 2^{-RN} \sum_{i=1}^{2^{RN}} \sum_{j=1, j\neq i}^{2^{RN}} \frac{k_{i,j}}{RN} P(\boldsymbol{X}_i \rightarrow \boldsymbol{X}_j), \tag{5.40}$$

where $k_{i,j}$ is the number of information (uncoded) bit differences corresponding to the codewords $\boldsymbol{X}_i$ and $\boldsymbol{X}_j$. The weights in front of the pairwise error probability terms simply show the percentage of errors that would be made if $\boldsymbol{X}_j$ is incorrectly decoded when $\boldsymbol{X}_i$ is transmitted.

Let us evaluate the union bound on the frame error and bit error rates of space-time trellis codes for the case of fully interleaved and quasi-static Rayleigh fading channels separately.

Fully Interleaved Rayleigh Fading Channels

In the pairwise error probability derivation earlier in the chapter, we have assumed that the channel is quasi-static fading. However, it is straightforward to extend the result to the case of fully interleaved channels, in particular, for the case of Rayleigh fading, it can be shown that

$$P(\boldsymbol{X}_i \rightarrow \boldsymbol{X}_j) \leq \prod_{k=1}^{N} \left(1 + \frac{d_k^2 \rho}{4}\right)^{N_r}, \tag{5.41}$$

where $d_k^2 = \sum_{l=1}^{N_t} |x_{i,l}(k) - x_{j,l}(k)|^2$ with $x_{i,l}(k)$ representing the symbol transmitted from the lth antenna at time k for the codeword $\boldsymbol{X}_i$.

Substituting this upper bound in the general union bound expression gives the union bound on the frame error rate as

$$P_e \leq 2^{-RN} \sum_{i=1}^{2^{RN}} \sum_{j=1, j\neq i}^{2^{RN}} \prod_{k=1}^{N} \left(1 + \frac{d_k^2 \rho}{4}\right)^{N_r}, \tag{5.42}$$

and the bit error rate as

$$P_b \leq 2^{-RN} \sum_{i=1}^{2^{RN}} \sum_{j=1, j\neq i}^{2^{RN}} \frac{k_{i,j}}{RN} \prod_{k=1}^{N} \left(1 + \frac{d_k^2 \rho}{4}\right)^{N_r}. \tag{5.43}$$

We note that, depending on the code structure, each of the pairwise error probability terms is proportional to $1/\rho^{N_t N_r L}$ where L is the amount of the time diversity available, i.e., L is the number of positions that the codewords differ in time (i.e., additional gain due to the time diversity available). Since these terms could be relatively small, the overall union bound computed in this manner may be useful. It may not be very tight as the decay of each term of the summation is still proportional to only a power of the signal-to-noise ratio, as opposed to being exponential as in an AWGN channel.

Let us illustrate the union bound for the fast fading case by an example. We consider a two transmit and one receive antenna system employing the four-state space-time trellis code described before. Assuming that the frame length is 130, we obtain the union bound on the frame error rate of the system in Figure 5.8. Also shown in the same figure are the simulation results. We observe that the union bound and the simulation results on the frame error rates are close to each other (there is only a difference of about 3 dB), and the diversity order is predicted by the bound. This example clearly shows that routine application of the union bound for the fully interleaved Rayleigh fading channel may be sufficient for analysis purposes.

Quasi-static Fading Channels

The situation is quite different for the quasi-static Rayleigh fading channel case. Using the bounds on the pairwise error probability derived earlier, we obtain the union bound on the

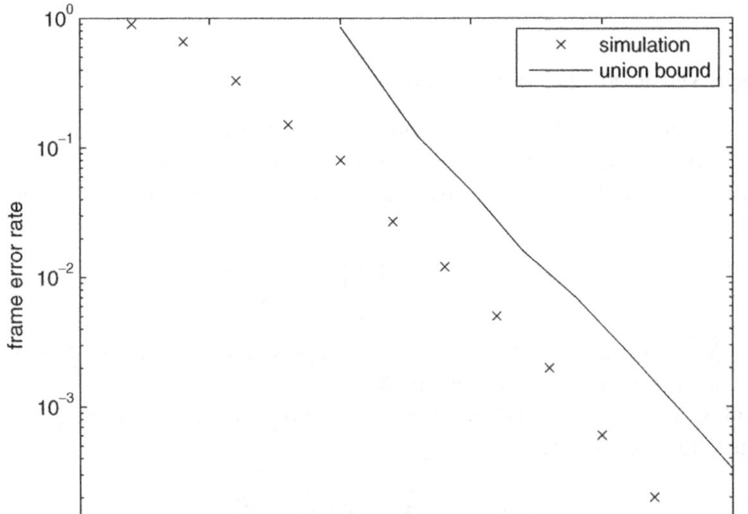

Figure 5.8 Frame error rate bound and simulation results for fast fading (two transmit and one receive antennas).

frame error rate as

$$P_e \leq 2^{-RN} \sum_{i=1}^{2^{RN}} \sum_{j=1, j \neq i}^{2^{RN}} \prod_{i=1}^{N_t} \left(1 + \lambda_i \frac{\rho}{4}\right)^{-N_r}, \tag{5.44}$$

and the union bound on the bit error probability as

$$P_b \leq 2^{-RN} \sum_{i=1}^{2^{RN}} \sum_{j=1, j \neq i}^{2^{RN}} \frac{k_{i,j}}{RN} \prod_{i=1}^{N_t} \left(1 + \lambda_i \frac{\rho}{4}\right)^{-N_r}. \tag{5.45}$$

We have seen in the previous subsection that the simple union bound is useful for the case of fast fading. However, this is not the case for the quasi-static fading scenario, particularly when the number of transmit and receive antennas are small. For instance, although it is not shown here, straightforward evaluation will result in an error rate bound exceeding unity for a wide range of signal-to-noise ratios (say within 10–20 dB) of the range of interest. This is clearly unacceptable, and it means that a straightforward union bound for this case is useless. The reason behind this behavior is that the summation on the right-hand side of (5.44) and (5.45) is not dominated by a few codeword pairs (unlike the case of fast fading). Instead, a very large number of terms contribute significantly to the bound, resulting in a very loose expression.

5.5.2 Useful Performance Bounds for Space-Time Trellis Codes

One way to try to remedy the looseness of the union bound for space-time trellis codes is to employ the idea of expurgation (Aktas and Fitz (2003)). However, this alone is not sufficient to obtain a tight upper bound on the error probabilities. Our analysis here uses a different approach developed by Stefanov and Duman (2003a), and is based on two main ideas. The first idea is to employ expurgation of some of the terms in the summation while still making sure that the expression remains a valid upper bound (as in Aktas and Fitz (2003); Verdu (1987)). The second (more important) idea computes the conditional union bound given the channel gains, upper bounding this value with unity if it exceeds unity, and averaging over the channel gains at the end. Such an approach was first used by Malkamaki and Leib (1999) in the context of convolutional codes. Let us present the details of these ideas separately.

Expurgation to Tighten the Union Bound

It can be shown that in computing the union bound, the only terms that need to be kept are the pairs of codewords that diverge and merge only once for the entire frame. We refer to such error events as "simple," and the ones that diverge from and merge with the correct path more than once as "compound" error events, as illustrated in Figure 5.9. With this terminology, the terms involving "compound" error events should be omitted from the summation, making the bounds tighter.

The reasoning behind expurgation can be explained as follows. The region of integration in calculating the pairwise error probability for a compound error event is a subset of the unions of the regions corresponding to a number of simple error events. Therefore, in the error probability expression, it is already accounted for, and need not be included again. Details of this argument are provided in Stefanov and Duman (2003a).

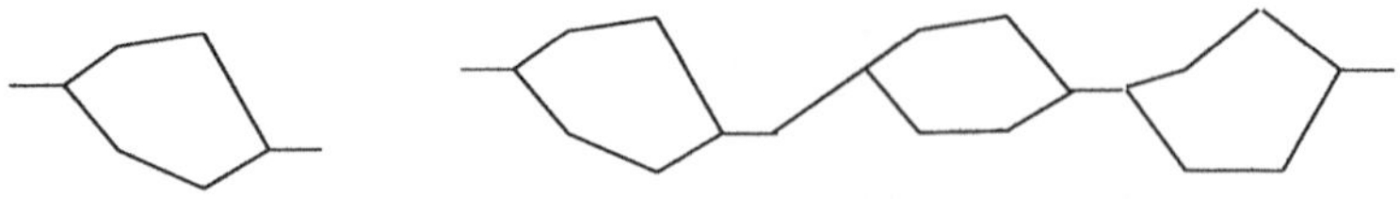

Figure 5.9 Simple versus compound error events.

Improving the Union Bound for Quasi-Static Fading

Although it is certainly useful, expurgating some of the terms in the union bound expression alone does not suffice to give a useful bound over quasi-static fading channels. To gain further insight into the problem, let us write the union bound on the frame error rate for quasi-static fading channels in a slightly different form as

$$P_e \leq E_{\boldsymbol{H}} \left[2^{-RN} \sum_{i=1}^{2^{RN}} \sum_{j=1, j \neq i}^{2^{RN}} P(\boldsymbol{X}_i \to \boldsymbol{X}_j | \boldsymbol{H}) \right], \tag{5.46}$$

where the pairwise error probability is computed conditioned on the channel coefficients $\boldsymbol{H}$ and the expectation is over their joint probability density function.

Earlier expressions are obtained by changing the order of the integration and the summations, and using the average pairwise error probability terms already computed. The main problem in the union bound is that, for some channel realizations, the fading channel coefficients are so small that the individual (conditional) pairwise error probabilities become large (close to 1), and thus when many such terms are added up, the right-hand side of the union bound expression gives results much larger than unity, which is clearly useless as a bound on the error probability. Averaging over the fading channel statistics before the summation is computed hides this fact, but the final result does not change.

With this observation, to develop a useful bound, we can upper bound the conditional union bound with a useful term (an obvious choice is unity) if it turns out to be large, and then we can average over the fading channel statistics. The resulting frame error rate bound is given by

$$P_e \leq E_{\boldsymbol{H}} \left[\min \left\{ 1, 2^{-RT} \sum_{i=1}^{2^{RT}} \sum_{j=1, j \neq i}^{2^{RT}} P(\boldsymbol{X}_i \to \boldsymbol{X}_j | \boldsymbol{H}) \right\} \right]. \tag{5.47}$$

Conditioned on the fading channel coefficients $\boldsymbol{H}$, the conditional pairwise error probability term can be written using the squared Euclidean distance between the two space-time codewords (for a given set of channel coefficients) as

$$P(\boldsymbol{X}_i \to \boldsymbol{X}_j | \boldsymbol{H}) = Q\left(\sqrt{\frac{\rho \, d^2(\boldsymbol{X}_i, \boldsymbol{X}_j | \boldsymbol{H})}{2}} \right), \tag{5.48}$$

where

$$d^2(\boldsymbol{X}_i, \boldsymbol{X}_j|\boldsymbol{H}) = \sum_{k=1}^{N}\sum_{k'=1}^{N_r}\left|\sum_{l=1}^{N_t} h_{l,k'}(x_{i,l}(k) - x_{j,l}(k))\right|^2. \tag{5.49}$$

Using this expression, we can write the improved bound on the frame error rate as

$$P_e \leq E_{\boldsymbol{H}}\left[\min\left\{1, 2^{-RT}\sum_{i=1}^{2^{RT}}\sum_{j=1, j\neq i}^{2^{RT}} Q\left(\sqrt{\frac{\rho\, d^2(\boldsymbol{X}_i, \boldsymbol{X}_j|\boldsymbol{H})}{2}}\right)\right\}\right], \tag{5.50}$$

and the one on the bit error rate as

$$P_b \leq E_{\boldsymbol{H}}\left[\min\left\{1, 2^{-RT}\sum_{i=1}^{2^{RT}}\sum_{j=1, j\neq i}^{2^{RT}} \frac{k_{i,j}}{RT} Q\left(\sqrt{\frac{\rho\, d^2(\boldsymbol{X}_i, \boldsymbol{X}_j|\boldsymbol{H})}{2}}\right)\right\}\right]. \tag{5.51}$$

It is obvious that these bounds will not diverge (due to limiting before averaging over the fading statistics). However, two issues remain: the first one is the evaluation of the overall expectation (i.e., a multi-dimensional integral); the second one is the evaluation of the double summation by going over all the possible codeword pairs (i.e., by using some sort of counting process). It is also not clear if the bounds will be useful eventually to predict the performance of the space-time trellis coded system.

The minimization inside the integral above makes it intractable to evaluate the bounds analytically. However, it is possible to resort to numerical integration techniques, or Monte Carlo techniques for evaluation. In particular, the Monte Carlo type evaluation is attractive since it is a simple matter to generate a large number of fading channel coefficients using the proper joint probability distribution, and to estimate the value of the integral by averaging the resulting values. We note that, as an added benefit, we can also evaluate the upper bound on the error rates for correlated channels as well, i.e., if the different sub-channels between the transmitter and the receiver elements are correlated, the computation/numerical evaluation of the integral does not become more difficult. The effects of channel correlation are discussed in Chapter 10.

To address the second issue, i.e., evaluation of the double summation for a fixed channel realization, we need to find a method to count all the possibilities and add their contributions to the overall sum. To understand the problem further, let us look into the squared Euclidean distance (for a fixed channel matrix) between the two codewords more closely. As an example, assume that we have two transmit antennas; in this case, we can easily write

$$\begin{aligned} d^2(\boldsymbol{X}_i, \boldsymbol{X}_j|\boldsymbol{H}) &= \sum_{k=1}^{N}\sum_{k'=1}^{N_r}\left|\sum_{l=1}^{2} h_{l,k'}(x_{i,l}(k) - x_{j,l}(k))\right|^2, \\ &= \left(\sum_{k'=1}^{N_r}|h_{1,k'}|^2\right)\sum_{k=1}^{N}|x_{i,1}(k) - x_{j,1}(k)|^2 \\ &\quad + \left(\sum_{k'=1}^{N_r}|h_{2,k'}|^2\right)\sum_{k=1}^{N}|x_{i,2}(k) - x_{j,2}(k)|^2 \end{aligned} \tag{5.52}$$

$$+2\,\mathrm{Re}\left\{\left(\sum_{k'=1}^{N_r} h_{1,k'}h_{2,k'}^*\right)\sum_{k=1}^{N}(x_{i,1}(k)-x_{j,1}(k))(x_{i,2}(k)-x_{j,2}(k))^*\right\}, \quad (5.53)$$

$$= \left(\sum_{k'=1}^{N_r}|h_{1,k'}|^2\right)A_1(i,j) + \left(\sum_{k'=1}^{N_r}|h_{2,k'}|^2\right)A_2(i,j)$$

$$+2\,\mathrm{Re}\left\{\left(\sum_{k'=1}^{N_r} h_{1,k'}\,h_{2,k'}^*\right)B(i,j)\right\}, \quad (5.54)$$

where

$$A_1(i,j) = \sum_{k=1}^{N}|x_{i,1}(k)-x_{j,1}(k)|^2, \quad (5.55)$$

$$A_2(i,j) = \sum_{k=1}^{N}|x_{i,2}(k)-x_{j,2}(k)|^2, \quad (5.56)$$

and

$$B(i,j) = \sum_{k=1}^{N}(x_{i,1}(k)-x_{j,1}(k))(x_{i,2}(k)-x_{j,2}(k))^*. \quad (5.57)$$

Therefore, the squared Euclidean distance between the two space-time codewords conditioned on $\boldsymbol{H}$ can be expressed using the channel fading coefficients (which are independent of the pair of codewords under consideration), and the terms $A_1(i, j)$, $A_2(i, j)$ and $B(i, j)$ (which depend on the specific codewords). We note that if these three terms are the same for another pair of codewords, the (conditional) squared Euclidean distances are identical. Therefore, these two different pairs of codewords contribute the same amount to the double sum in the argument of the union bound expressions (given in (5.50) and (5.51)). Consequently, the double sum over the codeword pairs can be performed in a different and more efficient way. That is, if we can count the multiplicities of codeword pairs that result in a specific triplet (A_1, A_2, B), we can evaluate the conditional union bound. This idea should be compared with the "weight enumerating function" usually needed in the evaluation of the union bound for linear block codes.

We note that the explanation above is valid for the calculation of the frame error rate bound. If we are interested in the bit error rates, we need to keep track of the total number of information bit differences between the two codewords ($k_{i,j}$) as well. We also note that the situation will be more complicated if we have more than two transmit antennas, since we will need to keep track of many more variables (not just three). To complete the derivation of the bound, let us describe the counting procedure, i.e., the calculation of the relevant weight-enumerating function, following the approach of Stefanov and Duman (2003a) (again for $N_t = 2$).

Distance Spectrum Calculation

As mentioned above, in order to evaluate the union bound, we need to know the possible values of the triples (A_1, A_2, B), and their multiplicities, i.e., the distance spectrum of

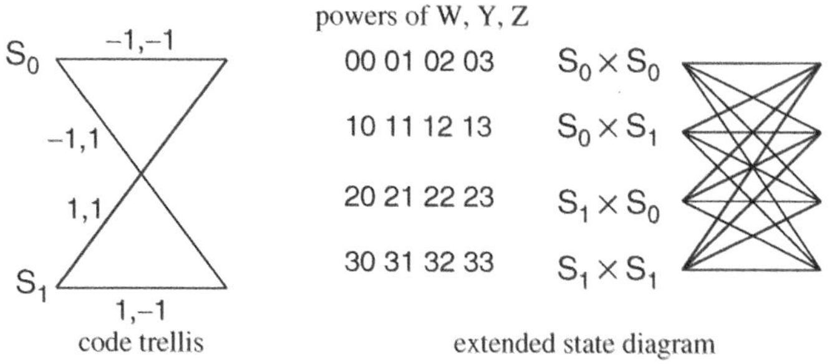

Figure 5.10 Two-state space-time trellis code with BPSK modulation and its extended state diagram for bound computation.

the space-time trellis code. Since, B is in general complex, we in fact need to keep track of two variables for B, one for the real part and one for the imaginary part. Since the space-time trellis codes are described by trellis diagrams, the distance spectrum can be computed using the idea of (extended) state diagrams. This procedure is best illustrated by an example.

Let us consider the two-state space-time trellis code given in Figure 5.10. Since the code is employing BPSK modulation, the relevant cross-term is real, hence we keep track of only three values. The resulting extended state diagram that counts the multiplicities of the triples (A_1, A_2, B) is also given in the same figure. The extended state diagram has four states, each denoting a possible pair of states for the two codewords that are being compared. Its branch labels show the contribution of the particular state transition at a trellis section to each of the terms A_1, A_2 and B.

The one-step state transition matrix for the extended state diagram can be written as (see Biglieri et al. (1991))

$$S = \begin{bmatrix} W^0Y^0Z^0 & W^0Y^2Z^0 & W^0Y^2Z^0 & W^0Y^0Z^0 \\ W^2Y^2Z^2 & W^2Y^0Z^0 & W^2Y^0Z^0 & W^2Y^2Z^{-2} \\ W^2Y^2Z^2 & W^2Y^0Z^0 & W^2Y^0Z^0 & W^2Y^2Z^{-2} \\ W^0Y^0Z^0 & W^0Y^2Z^0 & W^0Y^2Z^0 & W^0Y^0Z^0 \end{bmatrix}. \tag{5.58}$$

In each entry of the matrix S, the exponents of terms W, Y and Z denote the contribution to the particular pair of state transitions to the summation (over the entire frame) that computes the values of A_1, A_2 and B. For instance, for the first entry of the state transition matrix, all these powers are simply zeros, since this corresponds to the case where both codewords are making a transition from a state to the same one (they produce the same labels), and there is no contribution to the squared Euclidean distance.

To compute the distance spectrum, we can use the standard technique of simply raising this matrix to the Nth power (where N is the frame length), and reading off the resulting powers of the dummy variables W, Y and Z to find A_1, A_2 and B. If the trellis of the code is not terminated, we can simply sum the first row of S^N to accomplish this.

If we employ the expurgation technique mentioned earlier in order to tighten the bounds, we can simply extend this procedure by adding a new state (that is basically reached when the two paths that diverged before are merging for the first time). Therefore, the same ideas are applicable (for details, see Stefanov and Duman (2003a)).

We further note that if the code trellis is terminated (i.e., the codewords are driven back to the initial all-zero state at the end of the frame), we can take this into account in the counting procedure as well. All we need to do is to figure out the one-step state transition matrix for the termination process, multiply the Nth power of $\boldsymbol{S}$ with this matrix, and read off the first term of the resulting matrix (since both codewords will be back at the same state they have started).

5.5.3 Examples

To illustrate the use of the resulting performance bound for space-time trellis codes over quasi-static fading channels, we consider two different examples. The first one is the four-state space-time trellis code considered earlier in the chapter, and the second one is the improved four-state code by Yan and Blum (2002) (whose trellis diagram is given in Figure 5.11). The frame length is assumed to be $N = 130$.

Assuming a quasi-static Rayleigh fading channel, the frame error rate bounds as well as the simulation results for both codes are given in Figure 5.12 for the case of two transmit and two receive antennas. These bounds are computed using the idea of limiting the conditional union bound before averaging, and can be considered to be tight – they are only about 3 dB away from the simulation results, and they correctly identify the diversity order achievable by the codes (in this case it is four). Direct evaluation of the union bound is not given, but it will be several orders of magnitude larger than the true error rate values, and thus it is useless. The frame error rates for the same two codes for the case of two transmit and three receive antennas are provided in Figure 5.13. Similar observations are valid for this case as well.

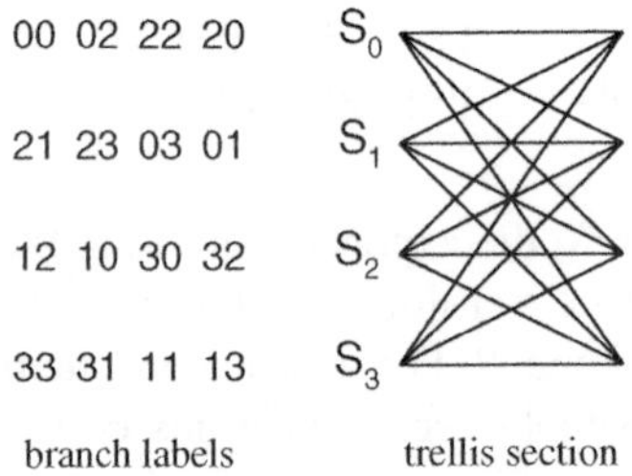

Figure 5.11 Four-state space-time trellis code with 4-PSK modulation (by Yan and Blum).

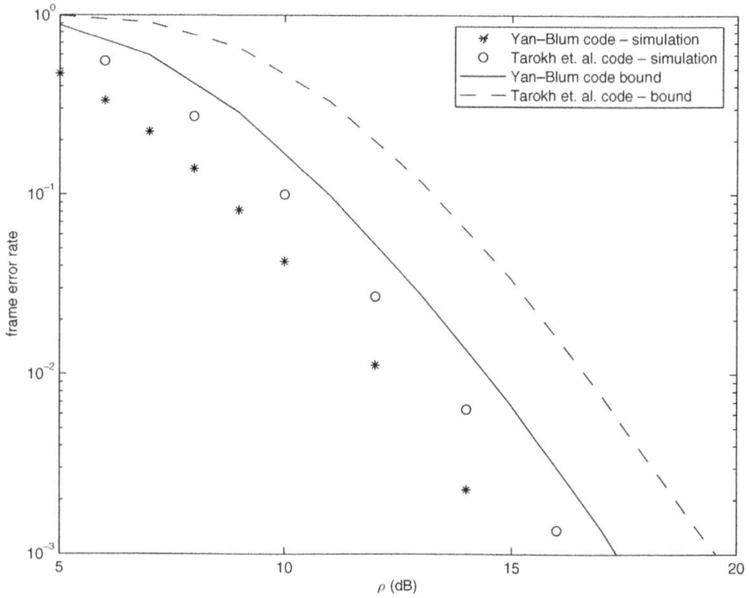

Figure 5.12 Frame error rate bound and simulation results for quasi-static fading with a frame length of 130 (two transmit and two receive antennas).

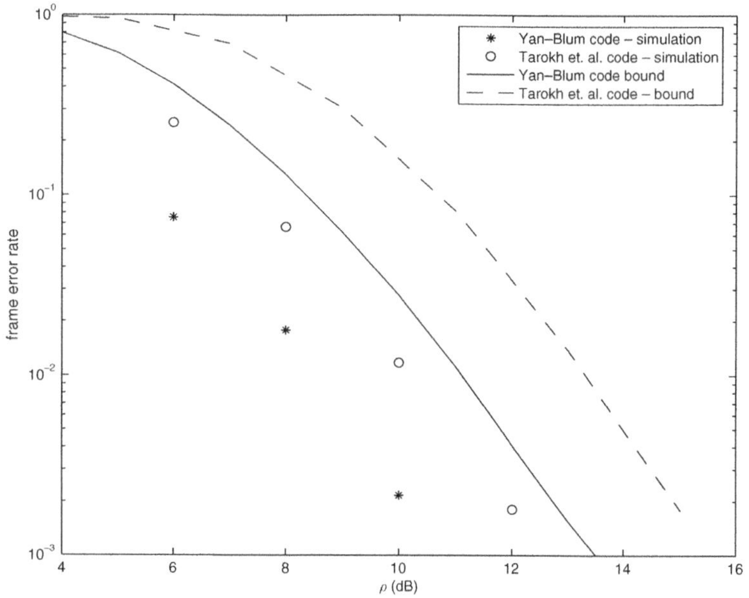

Figure 5.13 Frame error rate bound and simulation results for quasi-static fading with a frame length of 130 (two transmit and three receive antennas).

5.6 Comparison of Space-Time Block and Trellis Codes

It is of interest to compare and contrast the two approaches for space-time coding in this chapter and in Chapter 4, namely, space-time block coding versus space-time trellis coding. Both space-time block codes and space-time trellis codes are designed to achieve full diversity advantage over the MIMO wireless channels. However, there are some basic differences in their use. For instance, space-time block codes are very easy to encode and decode (which is achieved using the simple linear processing receivers), whereas space-time trellis codes require more complicated trellis-based decoders. Also, it is relatively easy to find and employ space-time block codes for more than two transmit antennas (although there may be a rate loss penalty), this is not the case for space-time trellis codes – they are most widely used for the case of two transmit antennas. These are clear advantages for space-time block coding. On the other hand, the resulting error rates of the space-time trellis codes are generally better than those of the space-time block codes.

As an example, in Figure 5.14, we show the error rates for a two transmit and one receive antenna system (over a quasi-static Rayleigh fading channel) for several space-time trellis codes, and the Alamouti scheme (assuming that the frame length is $N = 130$). All the schemes are designed to achieve 2 bits per channel use transmission rate. We observe that the space-time trellis codes outperform the Alamouti scheme, particularly when the number of states is increased. This is because space-time trellis codes provide a coding advantage in addition to providing full diversity when properly designed. However, there is no coding

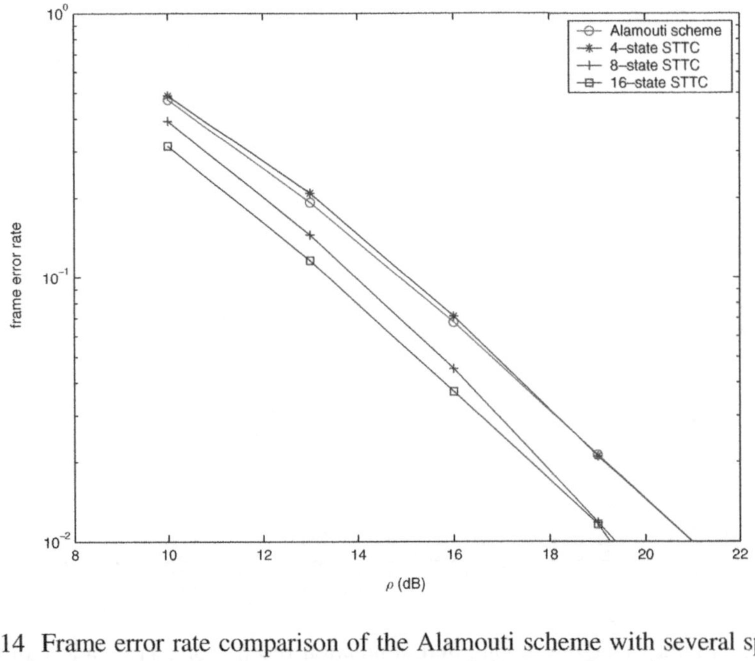

Figure 5.14 Frame error rate comparison of the Alamouti scheme with several space-time trellis codes (two transmit and one receive antennas, 2 bits per channel use).

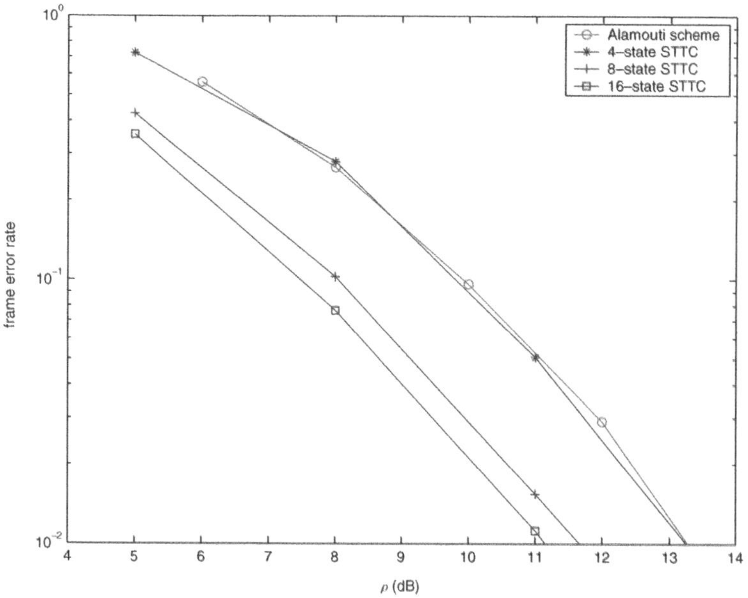

Figure 5.15 Frame error rate comparison of the Alamouti scheme with several space-time trellis codes (two transmit and two receive antennas, 2 bits per channel use).

advantage provided by space-time block codes. In fact, as we have discussed earlier, the performance of the space-time trellis codes with a larger number of states is not far from the outage capacity of the channel, thus they are near optimal over quasi-static fading channels.

The frame error rate comparisons of the Alamouti scheme and the same space-time trellis codes for the case of two receive antennas are provided in Figure 5.15. The observations are similar to the case of one receive antenna, except that the coding gains provided by the space-time trellis codes are even larger.

5.7 Chapter Summary and Further Reading

In this chapter, we have described space-time trellis codes. We derived the basic design principles for such codes over quasi-static Rayleigh fading channels, and gave several examples of good codes. We have also derived the corresponding performance bounds over quasi-static fading channels. We have also given an extensive set of performance results for different channel models, and compared the space-time trellis and space-time block codes.

There are a number of other important results on space-time trellis codes that have not been addressed in this chapter. For example, algebraic code design principles are derived by Hammons and El Gamal (2000); Liu et al. (2002a). Work on performance analysis of space-time coded systems includes Stefanov and Duman (2001a,b, 2002, 2003a,b). A new class of space-time trellis codes is proposed by Jafarkhani and Seshadri (2003).

Problems

5.1 Using the rank criterion described in this chapter, find the diversity orders offered by the following linear dispersion codes.

a) $$\begin{bmatrix} x_1 & x_2 & x_3 & x_4 \\ -x_2^* & x_1^* & -x_4^* & x_3^* \\ -x_3^* & -x_4^* & x_1^* & x_2^* \\ x_4 & -x_3 & -x_2 & x_1 \end{bmatrix},$$

b) $$\begin{bmatrix} x_1 & x_2 & x_3 & x_4 \\ -x_2^* & x_1^* & -x_4^* & x_3^* \\ -x_7^* & -x_8^* & x_5^* & x_6^* \\ x_8 & -x_7 & -x_6 & x_5 \end{bmatrix}.$$

5.2 Consider the four-state STTC given in Figure 5.1. Apply the rank and determinant criteria to the space-time codewords corresponding to the following pairs of transmitted (information) symbols.

a) 0, 0, 0, 0, 0 and 2, 1, 2, 1, 0,

b) 0, 0, 0 and 1, 2, 0,

c) 1, 2, 1, 3 and 3, 2, 1, 3.

5.3 Find the generator matrix representation of the eight-state 8-PSK code given in Figure 5.3.

5.4 Write a Matlab code to estimate the frame error rates for the eight-state 8-PSK code given in Figure 5.3 for $N_t = 2$ and $N_r = 1, 2$. Consider quasi-static Rayleigh fading and use a frame length of 100 symbols. Compare your results with the corresponding outage capacities.

5.5 Derive Equation (5.41).

5.6 Prove that there is no need to consider a compound error event in the calculation of the union bound (on the codeword or bit error probability) if all the simple error events (comprising this compound error event) are taken into account.

6

Layered Space-Time Codes

In MIMO systems with N_t transmit and N_r receive antennas, as was shown in previous chapters, STTCs and STBCs can achieve a diversity order of $N_t N_r$ over quasi-static fading channels, rendering such codes very effective in combatting the adverse effects of fading. However, these coding schemes achieve a spatial rate of unity or less (in terms of the number of symbols per channel use) without offering flexibility in trading diversity for rate. Providing such a trade-off is essential to accommodate a wide range of wireless applications, especially those applications that have high data rate requirements.

The notion of layered space-time (LST) coding, first introduced by Foschini (1996), has emerged since then as a powerful architecture suitable for applications with high data rates. Several LST coding architectures exist, including Horizontal Bell Laboratories Layered Space-Time (HBLAST), Vertical BLAST (VBLAST), and Diagonal BLAST (DBLAST) architectures. A common feature of these architectures is that N_t independent substreams are transmitted simultaneously from the available N_t transmit antennas. Consequently, these LST architectures achieve a spatial rate of $R_c b N_t$ where R_c denotes the rate of the channel code employed and 2^b is the signal constellation size. These BLAST schemes are sometimes referred to as spatial multiplexing schemes. Other LST schemes also exist, including the multilayered space-time coding scheme proposed by Tarokh et al. (1999) and the threaded space-time coding scheme proposed by El Gamal and Hammons (2001). These schemes combine channel coding and LST coding, resulting in a trade-off between diversity and rate.

In this chapter, we discuss in detail various architectures of LST coding. We begin our discussion with the VBLAST since it is the simplest, followed by the HBLAST and the single code BLAST (SCBLAST), and then the DBLAST. For each scheme, we describe the encoding mechanism, and highlight its merits relative to the other schemes. We also describe in great detail the detection algorithms used for each scheme, with emphasis on the ones based on the zero-forcing and minimum mean squared error detection criteria. Then, we move on to the multilayered space-time coding scheme, where we also describe its encoding mechanism, as well as suitable detection methods. We also discuss the threaded space-time scheme. This involves describing how the layers are formed, how the algebraic space-time codes are designed, and what iterative detection techniques can be used. The

Coding for MIMO Communication Systems Tolga M. Duman and Ali Ghrayeb

chapter is concluded with a brief summary and a few suggestions for further reading on the subject.

6.1 Basic Bell Laboratories Layered Space-Time (BLAST) Architectures

In this section, we consider the VBLAST, HBLAST and SCBLAST architectures as they all have similar encoder structures. The DBLAST architecture is treated later in a separate section since its encoder structure is somewhat different from the above three.

6.1.1 VBLAST/HBLAST/SCBLAST

The VBLAST encoder is depicted in Figure 6.1. As shown in the figure, the message bit stream is demultiplexed into N_t parallel substreams. Each substream is modulated using a 2^b-ary constellation, interleaved and then assigned to a transmit antenna. As such, the number of layers is N_t and the spatial rate is bN_t. Since each layer is associated with a fixed transmit antenna, this architecture can accommodate applications with possibly different data rates and/or different users. The spatial diversity achieved by this scheme varies between one and N_r, depending on the detection scheme employed at the receiver. For instance, when interference cancellation and suppression is used, the first layer detected will have a spatial diversity of $N_r - N_t + 1$ because the other layers are suppressed where they are treated as interference (Loyka and Gagnon (2004)). The last layer detected, on the other hand, will have a spatial diversity of N_r since the $N_t - 1$ previously detected layers are subtracted from the last layer, i.e., there is no suppression but rather cancellation.

When channel coding is involved, often in conjunction with interleaving, the VBLAST scheme becomes the HBLAST, as shown in Figure 6.2. As can be seen from the figure, each layer is encoded by a separate channel code, yielding flexibility in accommodating different users and/or different data rates. Additional diversity may be possible to achieve by this scheme because of the presence of channel coding and interleaving. This is true provided that the interleaving depth is larger than the coherence time of the channel. When

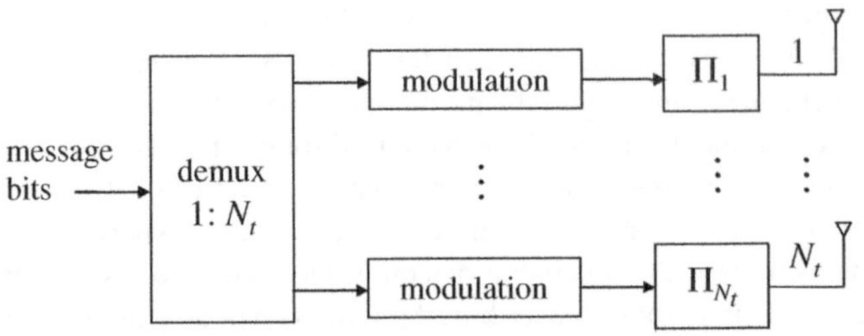

Figure 6.1 VBLAST encoder structure (Π_i denotes the interleaver corresponding to the ith layer).

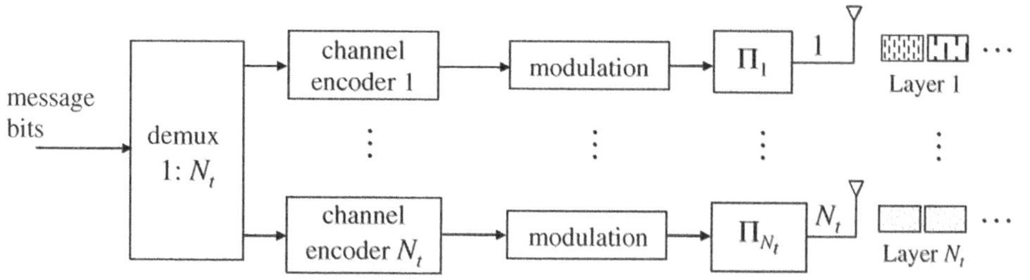

Figure 6.2 HBLAST encoder structure.

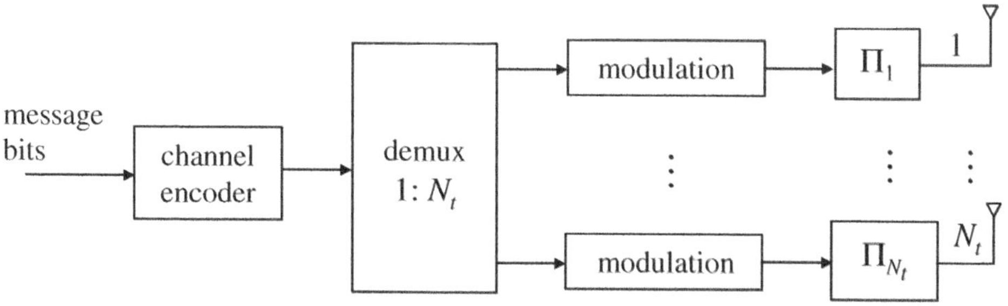

Figure 6.3 SCBLAST encoder structure.

the channel is quasi-static fading, however, interleaving has no impact on the performance and thus can be dropped.

Another way of applying channel coding is to use a single channel code for all layers as shown in Figure 6.3. This scheme is referred to as the SCBLAST architecture. It is clear that this scheme is simpler than the HBLAST scheme in the sense that only one decoder is used for the spatially N_t-dimensional symbols. Also the SCBLAST scheme could be superior to HBLAST in terms of spatial diversity in a slow or block fading environment because the presence of the demultiplexing (at the transmitter) and multiplexing (at the receiver) can be viewed as spatial interleaving where they together help in breaking some of the correlation in the received signal. Both the HBLAST and SCBLAST achieve a transmission rate of $R_c b N_t$ where R_c is the channel code rate.

6.1.2 Detection Algorithms for Basic BLAST Architectures

The optimal detector for the BLAST signals is the maximum likelihood detector that is matched to the combined trellis of the LST code and the channel code (for the HBLAST and SCBLAST schemes). However, the complexity of this detector grows exponentially in the number of transmit antennas and the number of states of the channel code. When interleaving is employed, the complexity grows even further. Even for a small number of transmit antennas, the associated complexity is still too high, rendering this detection approach not an option.

Other less complex but suboptimal approaches are commonly used for detecting BLAST signals, including the detection algorithms based on the zero forcing (ZF) and the minimum mean squared error (MMSE) criteria. In the ZF criterion, when a layer is detected, the interference coming from undetected layers is suppressed, whereas in the MMSE criterion, a compromise between interference suppression and noise reduction is achieved. For both detection schemes, interference suppression can be combined with interference cancellation to achieve further improvement in performance. In the rest of the chapter, we shall consider this combination of interference suppression and cancellation. Other detection schemes suitable for BLAST will be discussed briefly at the end of the chapter.

While both detection approaches are asymptotically equivalent, the ZF approach is less practical than the MMSE approach because the complete interference suppression achieved by ZF comes at the expense of enhancing the noise power, which leads to performance degradation. Another difference between the two schemes is that the constraint $N_r \geq N_t$ that is required for the ZF detector can be relaxed for the MMSE detector.

Zero Forcing Detection

Let $x_i(k)$ denote the symbol transmitted from antenna i at time k (see Figure 6.2). As such, the substreams transmitted from all transmit antennas can be expressed in a compact form as

$$\boldsymbol{X} = \begin{bmatrix} x_1(1) & x_2(1) & \cdots & x_{N_t}(1) \\ x_1(2) & x_2(2) & \cdots & x_{N_t}(2) \\ \vdots & \vdots & \ddots & \vdots \\ x_1(N) & x_2(N) & \cdots & x_{N_t}(N) \end{bmatrix}, \tag{6.1}$$

where N denotes the length of the transmitted sequence per transmit antenna.

The received signal, denoted by $\boldsymbol{Y}$, can be expressed in matrix form as

$$\boldsymbol{Y} = \sqrt{\rho}\boldsymbol{X}\boldsymbol{H} + \boldsymbol{N}, \tag{6.2}$$

where $\boldsymbol{H}$ is a matrix of size $N_t \times N_r$ whose (i, j)th entry represents the fading coefficient between the ith transmit and jth receive antenna, and $\boldsymbol{N}$ is a noise matrix of size $N \times N_r$. The entries of $\boldsymbol{H}$ and $\boldsymbol{N}$ are modeled as complex i.i.d. Gaussian random variables with zero mean and variance 1/2 per dimension. The signal constellation is scaled so that its average energy per transmit antenna is $1/N_t$, i.e., $\text{trace}\{\boldsymbol{X}^H\boldsymbol{X}\} = N_t$. Thus, ρ represents the average signal-to-noise ratio per receive antenna.

Assuming quasi-static fading and that $N_r \geq N_t$, $\boldsymbol{H}$ can be expressed using the *QR factorization* principle (Horn and Johnson (1985)) as

$$\boldsymbol{H} = \boldsymbol{R}\boldsymbol{Q}, \tag{6.3}$$

where $\boldsymbol{R}$ is an $N_t \times N_t$ lower triangular matrix and $\boldsymbol{Q}$ is an $N_t \times N_r$ unitary matrix. The rows of $\boldsymbol{Q}$ are mutually orthonormal. Consequently, $\boldsymbol{Q}$ has the property that $\boldsymbol{Q}\boldsymbol{Q}^H = \boldsymbol{I}_{N_t}$ where $\boldsymbol{I}_{N_t}$ is the $N_t \times N_t$ identity matrix. There are several fast algorithms available in the literature that can be used to compute these vectors. If $\boldsymbol{H}$ is nonsingular, which is the case for Rayleigh fading channels (with probability 1), the diagonal elements of $\boldsymbol{R}$ are all positive. Representing $\boldsymbol{H}$ as such is essential in suppressing inter-layer interference. To elaborate, let us right-multiply $\boldsymbol{Y}$ in (6.2) by $\boldsymbol{Q}^H$. Let $\tilde{\boldsymbol{Y}}$ denote the matrix resulting from this multiplication, that is,

$$\tilde{\boldsymbol{Y}} = \sqrt{\rho}\boldsymbol{X}\boldsymbol{H}\boldsymbol{Q}^H + \boldsymbol{N}\boldsymbol{Q}^H \tag{6.4}$$

$$= \sqrt{\rho}\boldsymbol{X}\boldsymbol{R} + \tilde{\boldsymbol{N}}, \tag{6.5}$$

where $\tilde{\boldsymbol{N}} \triangleq \boldsymbol{N}\boldsymbol{Q}^H$. Equation (6.5) is obtained by plugging the expression for $\boldsymbol{H}$ given by (6.3) into (6.4) and knowing that $\boldsymbol{Q}\boldsymbol{Q}^H = \boldsymbol{I}_{N_t}$.

The (k, i)th element of $\tilde{\boldsymbol{Y}}$, denoted by $\tilde{y}_i(k)$, represents the received signal corresponding to the symbol transmitted from the ith transmit antenna at time k, which can be expressed as

$$\tilde{y}_i(k) = \sqrt{\rho}\sum_{j=i}^{N_t} r_{j,i}x_j(k) + \tilde{n}_i(k), \tag{6.6}$$

$$= \underbrace{\sqrt{\rho}r_{i,i}x_i(k)}_{\text{desired symbol}} + \underbrace{\sqrt{\rho}\sum_{j=i+1}^{N_t} r_{j,i}x_j(k) + \tilde{n}_i(k)}_{\text{interference}}, \tag{6.7}$$

where $r_{j,i}$ is the (j, i)th element of $\boldsymbol{R}$, $x_j(k)$ is the (k, j)th element of $\boldsymbol{X}$ and $\tilde{n}_i(k)$ is the (k, i)th element of $\tilde{\boldsymbol{N}}$. The lower limit on j in (6.6) is i because $\boldsymbol{R}$ is lower triangular, i.e., $r_{j,i} = 0$ for $j = 1, 2, \ldots, i-1$. It is clear from (6.7) that the interference coming from layers $1, \ldots, i-1$ has been suppressed. As for the interference coming from the remaining layers (the last term in (6.7)), since these layers have already been detected, it can be cancelled easily.

From the above discussion, it is clear that the detection should start with layer $i = N_t$. In this case, $\tilde{y}_{N_t}(k)$ is expressed as

$$\tilde{y}_{N_t}(k) = \sqrt{\rho}r_{N_t,N_t}x_{N_t}(k) + \tilde{n}_{N_t}(k). \tag{6.8}$$

It is clear from (6.8) that there is no interference from other layers, and hence an estimate of the signal corresponding to this layer, denoted as $\hat{x}_{N_t}(k)$, can be obtained. Next, layer $i = N_t - 1$ is detected and this involves suppressing the interference coming from layers $j = 1, 2, \ldots, N_t - 2$ and canceling the interference coming from layer N_t. This process continues until the last layer is detected. In general, when layer i is to be detected, the interference coming from layers $j = 1, 2, \ldots, i-1$ is suppressed and the interference coming from layers $j = i+1, i+2, \ldots, N_t$ is cancelled. The interference to be cancelled from layer i can be expressed as $\sum_{j=i+1}^{N_t} r_{j,i}\hat{x}_j(k)$. As such, the soft-decision statistics for the ith layer, namely $\tilde{y}_i(k)$ can be written as

$$\tilde{y}_i(k) = \sqrt{\rho}r_{i,i}x_i(k) + \sqrt{\rho}\sum_{j=i+1}^{N_t} r_{j,i}\left[x_j(k) - \hat{x}_j(k)\right] + \tilde{n}_i(k). \tag{6.9}$$

The expression in (6.9) suggests that, when the hard decisions for all previously detected layers are all correct, the next layer to be detected is interference-free, although in practice this assumption is somewhat optimistic.

Table 6.1 The ZF-IC algorithm for the VBLAST/HBLAST/SCBLAST schemes

decompose $\boldsymbol{H}$ according to (6.3)
compute $\tilde{\boldsymbol{Y}}$ according to (6.4)
detect layer N_t according to (6.8)
obtain the corresponding hard decisions
set $i = N_t - 1$
while $i \geq 1$
    subtract the interference from layer i according to (6.9)
    detect layer i and obtain hard decisions using $\tilde{y}_i(k)$ defined by (6.9)
    $i = i - 1$
end of the while loop

The foregoing discussion applies in the exact same way to the VBLAST and SCBLAST schemes. It also applies to the HBLAST scheme with possibly a slight variation, particularly in the way hard decisions are obtained from the soft-decision statistics. For example, when the HBLAST is considered, one may opt to pass the soft-decision statistics to the corresponding channel decoder, encode the output of the channel decoder again, and use the coded sequence in the interference cancellation process. By this, these hard decisions will be more reliable than those that would have been obtained by just applying threshold detection to the soft-decision statistics.

A summary of the zero forcing with interference cancellation (ZF-IC) algorithm for the VBLAST/HBLAST/SCBLAST schemes is given in Table 6.1.

Sorted Zero Forcing Detection

By examining (6.9) more closely, we can see the level of data reliability differs from one layer to the other. Specifically, the first detected layer has the lowest reliability, whereas the last detected layer has the best reliability. As mentioned before, in the former case, the diversity order achieved is $N_r - N_t + 1$ because the contribution from all other layers is suppressed at the time of detection of this layer. On the other hand, the diversity order of the last detected layer is N_r since the interference coming from all previously detected layers is cancelled and not suppressed for the last detected layer. In general, the diversity order of the ith layer is $N_r - N_t + i$ (Loyka and Gagnon (2004)). This may not be desirable in certain applications.

One approach to alleviate this problem is to sort the received sequences from strongest to weakest in terms of power and start the detection with the strongest one. This involves sorting the rows of $\boldsymbol{H}$ according to their squared norms. In particular, the row with the highest squared norm becomes the (N_t)th row of $\boldsymbol{H}$ and the row with the lowest squared norm becomes the first row of $\boldsymbol{H}$. The rest of the ZF detection proceeds as outlined in the previous section. The only impact of the sorting process is the order in which the different layers are detected. For example, without sorting, the first detected layer corresponds to the signal transmitted from the (N_t)th transmit antenna and the last detected layer corresponds to the signal transmitted from the first transmit antenna, whereas, with sorting, the first detected layer could correspond to any of the transmitted signals. In fact, the order of the detected layers follows the order of the rows of the sorted $\boldsymbol{H}$.

Another approach is to assign different transmit powers to the transmitted sequences. As such, the layer that has the largest diversity order is assigned the lowest transmit power and the one with the lowest diversity order is assigned the highest transmit power. One may also consider employing both approaches simultaneously. After sorting and/or new power allocation is done, the algorithm outlined in Table 6.1 is applied.

Minimum Mean Squared Error Detection

As mentioned before, the MMSE criterion minimizes the expected value of the difference between the transmitted signal and a linear combination of the received ones, that is,

$$\mathcal{D}^2 = \min_{\boldsymbol{W}} E\left\{\|\boldsymbol{X} - \boldsymbol{Y}\boldsymbol{W}\|^2\right\}, \tag{6.10}$$

where $\|\boldsymbol{A}\|^2$ is the squared Frobenius norm of matrix $\boldsymbol{A}$, and $\boldsymbol{W}$ is an $N_r \times N_t$ matrix over which $\mathcal{D}^2$ is minimized. The optimal solution to (6.10), denoted by $\boldsymbol{W}_{opt}$, is the well-known Wiener solution given by (Kay (1993))

$$\boldsymbol{W}_{opt} = \sqrt{\frac{\rho}{N_t}}\boldsymbol{H}^H\left[\frac{\rho}{N_t}\boldsymbol{H}\boldsymbol{H}^H + \boldsymbol{I}_{N_t}\right]^{-1}, \tag{6.11}$$

where $\boldsymbol{I}_{N_t}$ is the $N_t \times N_t$ identity matrix.

By right multiplying $\boldsymbol{Y}$ by $\boldsymbol{W}_{opt}$, one may obtain the soft-decision statistics for the transmitted symbols, that is,

$$\tilde{\boldsymbol{Y}} = \boldsymbol{Y}\boldsymbol{W}_{opt}. \tag{6.12}$$

The (k, i)th element of $\tilde{\boldsymbol{Y}}$ corresponds to the symbol transmitted at time k from antenna i, which can be expressed in vector notation as

$$\tilde{y}_i(k) = \boldsymbol{y}_k\boldsymbol{w}_i, \tag{6.13}$$

where $\boldsymbol{y}_k$ is the kth row of $\boldsymbol{Y}$ which is of size $1 \times N_r$, and $\boldsymbol{w}_i$ is the ith column of $\boldsymbol{W}_{opt}$ which is of size $N_r \times 1$. From (6.2), $\boldsymbol{y}_k$ can be expressed as

$$\boldsymbol{y}_k = \sqrt{\rho}\boldsymbol{x}_k\boldsymbol{H} + \boldsymbol{n}_k, \tag{6.14}$$

where $\boldsymbol{n}_k$ is the kth row of the noise matrix $\boldsymbol{N}$, and $\boldsymbol{x}_k$ is the kth row of $\boldsymbol{X}$. By plugging (6.14) into (6.13) and after simple manipulations, $\tilde{y}_i(k)$ can be written as

$$\tilde{y}_i(k) = \sqrt{\rho}\sum_{j=1}^{N_r} w_{j,i}\left(\boldsymbol{x}_k\boldsymbol{h}_j\right) + \tilde{n}_i(k), \tag{6.15}$$

where $\boldsymbol{h}_j$ is the jth column of $\boldsymbol{H}$. Since $\tilde{y}_i(k)$ corresponds to $x_i(k)$, which is the kth component of the ith layer at time k, it is clear from (6.15) that symbols from all other layers interfere with $x_i(k)$. This does not imply, however, that no interference is suppressed, but rather, according to the MMSE criterion, some of the interference from other layers is compromised as a trade-off for not enhancing the noise power, unlike the ZF case.

Let us assume, without loss of generality, that we are interested in detecting layer N_t, which is the first layer to be detected, i.e., layers $1, 2, \ldots, N_t - 1$ have not been detected yet. From (6.15), $\tilde{y}_{N_t}(k)$ can be expressed as

$$\tilde{y}_{N_t}(k) = \sqrt{\rho} \sum_{j=1}^{N_r} w_{j,N_t} \left[\underbrace{x_{N_t}(k) h_{N_t,j}}_{\text{desired symbol}} + \underbrace{\sum_{m=1}^{N_t-1} x_m(k) h_{m,j}}_{\text{interference}} \right] + \tilde{n}_{N_t}(k). \tag{6.16}$$

Expression (6.16) suggests that the desired symbol suffers from interference from the other layers, but this interference is minimized according to the MMSE criterion. Estimates of $x_{N_t}(k)$ for $k = 1, 2, \ldots, N$, denoted by $\hat{x}_{N_t}(k)$, are then obtained by applying threshold detection to $\tilde{y}_{N_t}(k)$ defined by (6.16).

Then layer $N_t - 1$ is detected, but the interference coming from layer N_t has to be cancelled first. Let us write $\tilde{y}_{N_t-1,k}$ as

$$\tilde{y}_{N_t-1}(k) = \sum_{j=1}^{N_r} w_{j,N_t-1} \left[\left(\sum_{m=1}^{N_t-2} x_m(k) h_{m,j} \right) + x_{N_t-1}(k) h_{N_t-1,j} + x_{N_t}(k) h_{N_t,j} \right]$$
$$+ \tilde{n}_{N_t-1}(k). \tag{6.17}$$

The first term in the brackets in (6.17) represents the interference coming from the undetected layers, the middle term represents the desired symbol, and the last term represents the interference coming from layer N_t, which has already been detected. Thus the interference contribution coming from layer N_t, expressed as

$$\hat{\boldsymbol{x}}_{N_t} \boldsymbol{h}_{N_t} = \begin{bmatrix} \hat{x}_{N_t}(1) \\ \hat{x}_{N_t}(2) \\ \vdots \\ \hat{x}_{N_t}(N) \end{bmatrix} \begin{bmatrix} h_{N_t,1} & h_{N_t,2} & \cdots & h_{N_t,N_r} \end{bmatrix},$$

where $\boldsymbol{h}_{N_t}$ represents the (N_t)th row of $\boldsymbol{H}$, can be cancelled before layer $N_t - 1$ is detected. This interference cancellation involves updating the received matrix $\boldsymbol{Y}$.

Let $\boldsymbol{Y}^{(N_t)}$ denote the received signal matrix defined by (6.2) before any interference cancellation takes place. When layer $N_t - 1$ is to be detected, the corresponding received symbol matrix, denoted by $\boldsymbol{Y}^{(N_t-1)}$, is updated as

$$\boldsymbol{Y}^{(N_t-1)} = \boldsymbol{Y}^{(N_t)} - \sqrt{\rho} \hat{\boldsymbol{x}}_{N_t} \boldsymbol{h}_{N_t}. \tag{6.18}$$

In general, the detection of layer $i - 1$ is accomplished after subtracting the interference from layer i, i.e.,

$$\boldsymbol{Y}^{(i-1)} = \boldsymbol{Y}^{(i)} - \sqrt{\rho} \hat{\boldsymbol{x}}_i \boldsymbol{h}_i. \tag{6.19}$$

It is important to note that $\boldsymbol{W}_{opt}$ defined by (6.11) has to be re-calculated every time a layer is cancelled, and this involves updating $\boldsymbol{H}$ by deleting its ith row, i.e., $\boldsymbol{h}_i$. The detection process for the other layers repeats exactly in the same way until the last layer is detected. Note that the interference cancellation is done recursively as, for any given

Table 6.2 The MMSE-IC algorithm for the VBLAST/HBLAST/SCBLAST schemes

---

set $i = N_t$
compute $\boldsymbol{W}_{opt}^H$ using (6.11)
while $i \geq 1$
    compute $\tilde{y}_i(k)$, for $k = 1, 2, \ldots, N$ using (6.13)
    obtain hard decisions on $\tilde{y}_i(k)$, $k = 1, 2, \ldots, N$ denoted above by $\hat{\boldsymbol{x}}_i$
    compute $\boldsymbol{Y}^{(i-1)}$ using (6.19)
    update $\boldsymbol{H}$ by removing its ith row, $\boldsymbol{h}_i$
    compute $\boldsymbol{W}_{opt}^H$ using (6.11) with the updated $\boldsymbol{H}$
    $i = i - 1$
end of the while loop

---

layer, only the interference contribution from the previously detected layer is cancelled. Similar to the ZF case, different layers have different reliabilities here as well. Therefore, the approaches used in the ZF case can be applied here to remedy this problem. That is, after ordering the layers according to their signal-to-noise ratios, the detection process may start with the layer that has the largest signal-to-noise ratio and end with the one that has the lowest signal-to-noise ratio.

A summary of the MMSE with interference cancellation (MMSE-IC) algorithm for the VBLAST/HBLAST/SCBLAST schemes is given in Table 6.2.

6.1.3 Examples

In this section, we study the performance of the above-mentioned architectures, namely, the VBLAST, HBLAST and SCBLAST decoded using the ZF and MMSE detectors with and without interference cancellation. We consider a MIMO system with $N_t = N_r = 4$. Consequently, there are four layers. The length of the codeword coming out of each transmit antenna is set to $N = 2000$. The underlying channel is quasi-static fading, and thus the fading coefficients remain the same for a frame of 2000 consecutive symbols and change independently from one frame to the next (this applies to all four layers). BPSK modulation is used. Interleaving is not employed in all examples. In addition, where applicable, hard decision Viterbi decoding is used.

The bit error rate performance results for the VBLAST scheme are shown in Figure 6.4 for the following detection criteria: ZF, ZF-IC, MMSE, and MMSE-IC. We also plot in the figure, as a baseline, the interference-free bound (IFB), which can be approximated as (Proakis (2001))

$$P_b \approx \binom{2N_r - 1}{N_r} (4\rho)^{-N_r} . \tag{6.20}$$

This expression corresponds to a single-transmit, N_r-receive antenna system, and thus the diversity order achieved in this case is N_r.

From the figure, we can see that the performance of the MMSE detection criterion is better than that of the ZF detection criterion. We can also see that additional performance improvements can be obtained when interference cancellation is used. However, these results are still far away from the IFB. This is attributed to the fact that the overall

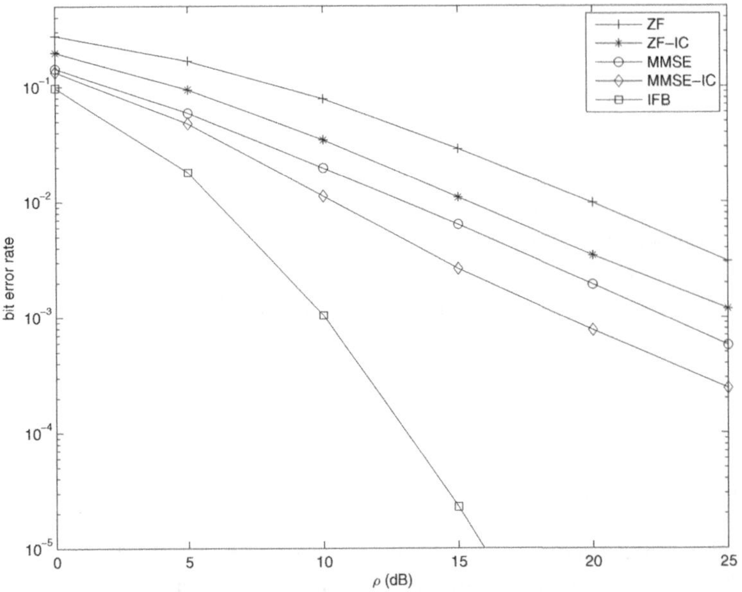

Figure 6.4 VBLAST bit error rate performance with various detection criteria for $N_t = N_r = 4$ with uncoded BPSK.

performance is dominated by the performance of the first layer detected, which is normally the worst among the other layers since it has the least amount of diversity.

To illustrate the above point further, we plot in Figure 6.5 the performance of the four layers separately using the ZF-IC detector when all the detected symbols from previous layers are correct. This implies that when a layer is detected, the interference coming from previously detected layers is cancelled perfectly. We observe from the figure how the diversity order increases for successively detected layers. Specifically, for the example under consideration, the diversity for the first detected layer is one, whereas it is two, three and four for the second, third and last detected layers, respectively. These results clearly suggest that sorting and/or non-uniform power allocation is much needed to achieve equal (or near equal) performances across all layers.

To examine the impact of applying sorting on the performance of ZF-IC, we plot in Figure 6.6 the bit error rate performance of the VBLAST scheme employing the ZF-IC detector with and without sorting. We observe from the figure the improvement achieved due to sorting is about 2 dB at a bit error rate of 10^{-3}.

We plot in Figure 6.7 the bit error rate performance results for the HBLAST scheme for various detection criteria. The channel codes used for all layers were identical, with each code being a rate $1/2$ convolutional code with generator polynomials $(171, 133)_{octal}$. Similar observations are made here. That is, the MMSE detector outperforms the ZF detector, and further performance improvements are achieved with interference cancellation for both detectors.

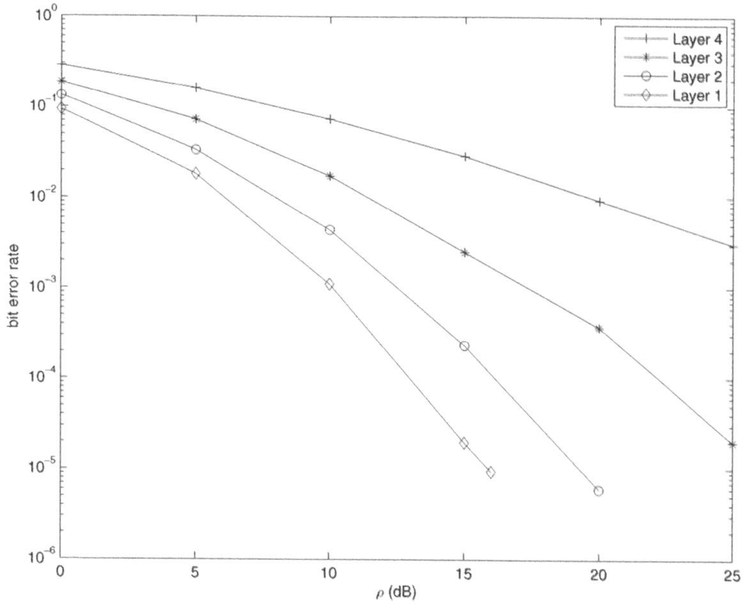

Figure 6.5 VBLAST bit error rate performance for $N_t = N_r = 4$ for the four layers using the ZF-IC criterion (for uncoded BPSK).

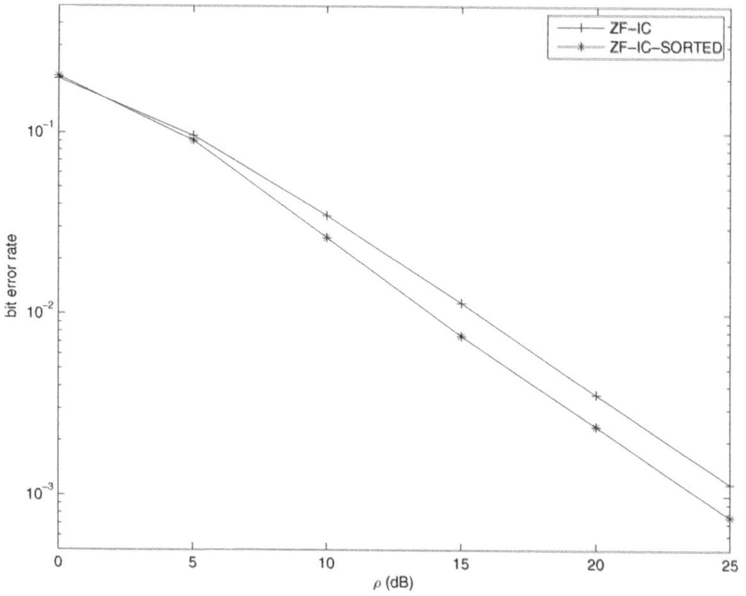

Figure 6.6 VBLAST bit error rate performance for $N_t = N_r = 4$ using the ZF-IC criterion with and without sorting (for uncoded BPSK).

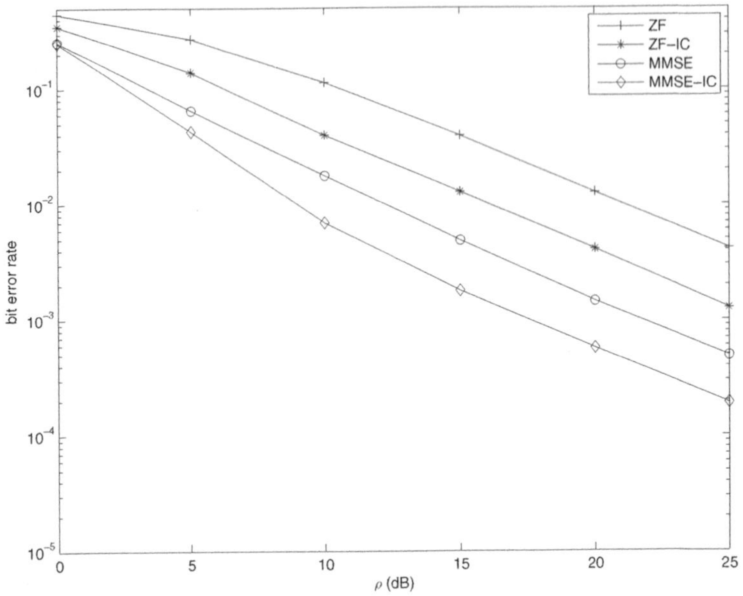

Figure 6.7 HBLAST bit error rate performance with various detection criteria for $N_t = N_r = 4$ and BPSK signaling.

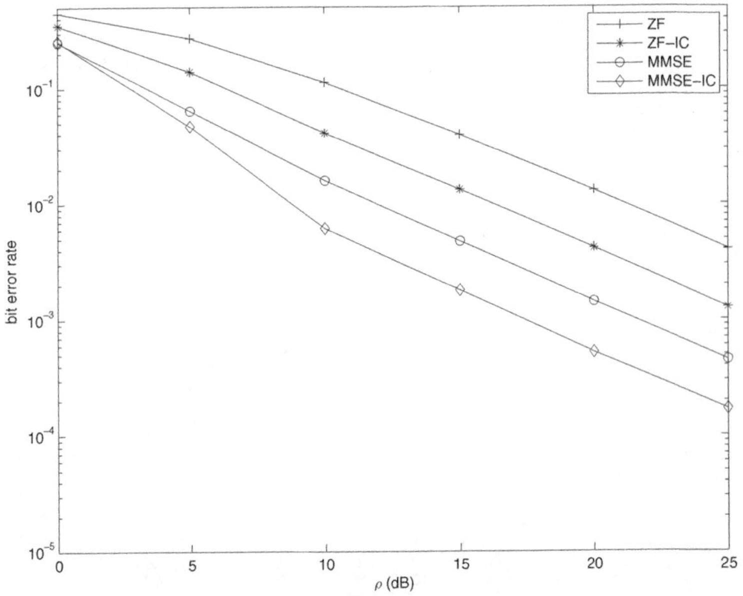

Figure 6.8 SCBLAST bit error rate performance with various detection criteria for $N_t = N_r = 4$ and BPSK signaling.

The bit error rate performance results for the SCBLAST scheme are plotted in Figure 6.8. By comparing the results in this figure with their counterparts in Figure 6.7, we can see that the SCBLAST outperforms the HBLAST and VBLAST. This should be expected because the SCBLAST employs a single channel encoder that is used for all layers, which is similar to the use of time diversity, and hence the improved performance.

6.2 Diagonal BLAST (DBLAST)

The DBLAST encoder structure is shown in Figure 6.9. As can be seen from the figure, the DBLAST and HBLAST schemes have the same structure except in the way the substreams are assigned to the transmit antennas. In particular, in the DBLAST scheme, each codeword is broken down into smaller blocks. These blocks are then transmitted in a cyclic fashion where N_t consecutive blocks belonging to the same codeword are transmitted from different antennas. This results in diagonally layering the signal in space and time, which is equivalent to block-wise spatial interleaving of the incoming substreams. The motivation behind this transmission scheme is to introduce some additional spatial diversity that the HBLAST scheme lacks.

To illustrate the DBLAST encoding process more clearly, we consider an example in which $N_t = 4$, i.e., there are four layers. Accordingly, each codeword is divided into four blocks. The number of blocks in a codeword should equal N_t. This constraint is imposed by the decoding process, as we will discuss later. The transmission process proceeds as shown in Figure 6.10. First, $\boldsymbol{a}_1$ is transmitted from the first transmit antenna. Second, $\boldsymbol{d}_1$ and $\boldsymbol{a}_2$ are transmitted simultaneously from the first and second antennas, respectively.

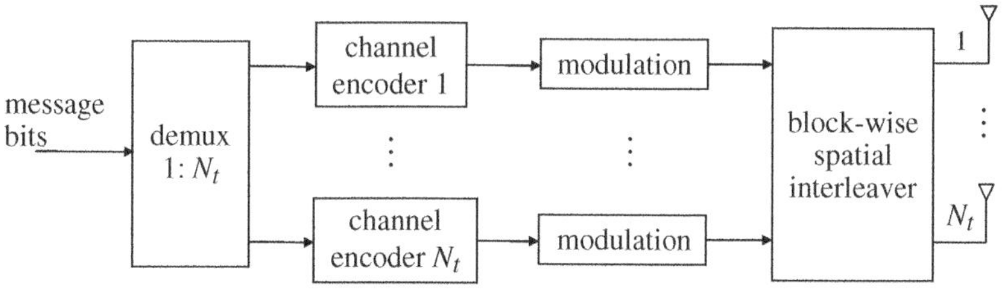

Figure 6.9 DBLAST encoder structure.

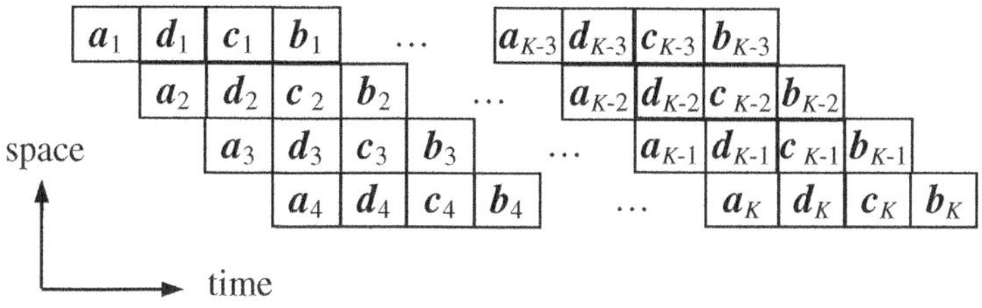

Figure 6.10 Illustration of the DBLAST transmission.

Third, c_1, d_2 and a_3 are also transmitted simultaneously from the first, second and third antennas, respectively. This process continues until b_K is transmitted from the last antenna. It is implied here that there are $K/4$ codewords per layer.

Since some of the transmit antennas remain idle at the beginning and end of each transmission period, the resulting transmission rate is slightly lower than $R_c b N_t$, but this is a minor issue especially when the transmission period is relatively long. On the other hand, the spatial diversity here can be higher than that of the HBLAST due to the presence of spatial interleaving. However, this depends on several factors, including the detection scheme employed at the receiver and the underlying fading channel model. For example, when the channel is block fading, the spatial diversity would be comparable in the HBLAST and DBLAST cases. Additionally, it is implied here that the substreams at the input of the spatial interleaver are assumed to be of equal rate, rendering this scheme less flexible in this respect as compared to the HBLAST scheme.

6.2.1 Detection Algorithms for DBLAST

In principle, the ZF and MMSE detection algorithms (with or without interference cancellation) discussed above for the other BLAST architectures extend to the DBLAST architecture. However, the way in which the signals are transmitted in the DBLAST case, which is different from the earlier BLAST schemes, introduces some minor differences in the detection process.

Let us consider the same example used in Figure 6.10 in which $N_t = 4$. Accordingly, the corresponding transmitted signal matrix, arranged with respect to the transmission time and transmitting antenna, can be expressed as

$$X = \begin{bmatrix} a_1^T & 0 & 0 & 0 \\ d_1^T & a_2^T & 0 & 0 \\ c_1^T & d_2^T & a_3^T & 0 \\ b_1^T & c_2^T & d_3^T & a_4^T \\ \vdots & b_2^T & c_3^T & d_4^T \\ a_{K-3}^T & \vdots & b_3^T & c_4^T \\ d_{K-3}^T & a_{K-2}^T & \vdots & b_4^T \\ c_{K-3}^T & d_{K-2}^T & a_{K-1}^T & \vdots \\ b_{K-3}^T & c_{K-2}^T & d_{K-1}^T & a_K^T \\ 0 & b_{K-2}^T & c_{K-1}^T & d_K^T \\ 0 & 0 & b_{K-1}^T & c_K^T \\ 0 & 0 & 0 & b_K^T \end{bmatrix}, \tag{6.21}$$

where a_i^T is the transpose of a_i and the rest of the entries are similarly defined. Note that X is of size $N \times N_t$, and its entries, based on the way it is defined, are blocks of symbols whose lengths depend on the codeword length at the output of each channel encoder, as shown in Figure 6.10. For this particular example, consider the codewords at the output of the top channel encoder. The first codeword is divided into four equal-length blocks, namely, a_1, a_2, a_3 and a_4. The second codeword is divided into a_5, a_6, a_7 and a_8. This continues until the last codeword which is divided into $\boldsymbol{a}_{K-3}$, $\boldsymbol{a}_{K-2}$, $\boldsymbol{a}_{K-1}$ and $\boldsymbol{a}_K$. The same procedure applies to the other channel encoder outputs. Note that, due to constraints imposed by the decoding method for this architecture, all of these blocks across all layers are of equal length. To simplify the presentation, we denote the length of each data block by m. We now describe the ZF and MMSE detection algorithms for DBLAST based on the $N_t = 4$ example. Extensions of these algorithms to an arbitrary number of transmit antennas are straightforward.

Zero Forcing Detection

After obtaining $\boldsymbol{Y}$, defined by (6.2), calculating the *QR factorization* of $\boldsymbol{H}$, right multiplying $\boldsymbol{Y}$ by $\boldsymbol{Q}^H$, the matrix

$$\tilde{\boldsymbol{Y}} = \sqrt{\rho}\boldsymbol{X}\boldsymbol{R} + \tilde{\boldsymbol{N}} \tag{6.22}$$

can be written in an expanded form as

$$\tilde{\boldsymbol{Y}} = \begin{bmatrix} \overbrace{\tilde{\boldsymbol{y}}_1^T\Big|_1^m}^{\text{corresponds to } \boldsymbol{a}_1} & \text{only noise} & \text{only noise} & \text{only noise} \\ \tilde{\boldsymbol{y}}_1^T\Big|_{m+1}^{2m} & \overbrace{\tilde{\boldsymbol{y}}_2^T\Big|_{m+1}^{2m}}^{\text{corresponds to } \boldsymbol{a}_2} & \text{only noise} & \text{only noise} \\ \tilde{\boldsymbol{y}}_1^T\Big|_{2m+1}^{3m} & \tilde{\boldsymbol{y}}_2^T\Big|_{2m+1}^{3m} & \overbrace{\tilde{\boldsymbol{y}}_3^T\Big|_{2m+1}^{3m}}^{\text{corresponds to } \boldsymbol{a}_3} & \text{only noise} \\ \tilde{\boldsymbol{y}}_1^T\Big|_{3m+1}^{4m} & \tilde{\boldsymbol{y}}_2^T\Big|_{3m+1}^{4m} & \tilde{\boldsymbol{y}}_3^T\Big|_{3m+1}^{4m} & \overbrace{\tilde{\boldsymbol{y}}_4^T\Big|_{3m+1}^{4m}}^{\text{corresponds to } \boldsymbol{a}_4} \\ \vdots & \vdots & \vdots & \vdots \end{bmatrix}, \tag{6.23}$$

where

$$\tilde{\boldsymbol{y}}_i^T\Big|_{k_1}^{k_2} = \begin{bmatrix} \tilde{y}_i(k_1) \\ \tilde{y}_i(k_1+1) \\ \vdots \\ \tilde{y}_i(k_2) \end{bmatrix},$$

which belongs to the ith column of $\tilde{\boldsymbol{Y}}$, and

$$\boldsymbol{R} = \begin{bmatrix} r_{1,1} & 0 & 0 & 0 \\ r_{2,1} & r_{2,2} & 0 & 0 \\ r_{3,1} & r_{3,2} & r_{3,3} & 0 \\ r_{4,1} & r_{4,2} & r_{4,3} & r_{4,4} \end{bmatrix}.$$

Note that the matrix entries in (6.23) with *only noise* are not functions of the transmitted signals, i.e., they are purely noise (see (6.22)).

The task now is to detect the transmitted codewords one by one, starting with the codeword $\boldsymbol{a}_1, \ldots, \boldsymbol{a}_4$, followed by the codeword $\boldsymbol{d}_1, \ldots, \boldsymbol{d}_4$, and so on. As shown in (6.23), the first m elements of the first row of $\tilde{\boldsymbol{Y}}$ are the soft-decision statistics corresponding to the symbols comprising $\boldsymbol{a}_1$, and they can be expressed as

$$\tilde{y}_1(k) = \sqrt{\rho} r_{1,1} x_1(k) + \tilde{n}_1(k), \qquad k = 1, \ldots, m. \tag{6.24}$$

Similarly, the soft-decision statistics corresponding to the symbols within blocks $\boldsymbol{a}_2$, $\boldsymbol{a}_3$ and $\boldsymbol{a}_4$ can be written, respectively, as

$$\tilde{y}_2(k) = \sqrt{\rho} r_{2,2} x_2(k) + \tilde{n}_2(k), \qquad k = m+1, \ldots, 2m, \tag{6.25}$$

$$\tilde{y}_3(k) = \sqrt{\rho} r_{3,3} x_3(k) + \tilde{n}_3(k), \qquad k = 2m+1, \ldots, 3m, \tag{6.26}$$

and

$$\tilde{y}_4(k) = \sqrt{\rho} r_{4,4} x_4(k) + \tilde{n}_4(k), \qquad k = 3m+1, \ldots, 4m. \tag{6.27}$$

Expressions (6.24)–(6.27) suggest that these soft-decision statistics have no interference from other layers, and thus they are ready to be used to obtain hard decisions on the codeword comprising $\boldsymbol{a}_1, \ldots, \boldsymbol{a}_4$. These hard decisions are in turn used in the interference cancellation process before the next codeword is detected. There are two ways to obtain these hard decisions. One way is to apply threshold detection directly to these decision statistics. The other way is to first decode these decision statistics by the corresponding channel decoder (either hard or soft decision decoding, depending on the type of code used), encode again to obtain an estimate of the corresponding transmitted codeword, and use these estimates in interference cancellation. The latter method is supposed to yield more reliable estimates of the transmitted codewords, which comes of course at the expense of some latency and possibly additional complexity. The importance of getting as reliable as possible estimates lies in their role in detecting the subsequent codewords when interference cancellation is considered so as to avoid or reduce error propagation.

In detecting the next codeword, i.e., $\boldsymbol{d}_1, \ldots, \boldsymbol{d}_4$, the interference coming from $\boldsymbol{a}_1, \ldots, \boldsymbol{a}_4$ should be cancelled first. To accomplish this, matrix $\boldsymbol{Y}$ should be updated such that the interference coming from $\boldsymbol{a}_1, \ldots, \boldsymbol{a}_4$ is cancelled. Once $\boldsymbol{Y} = \sqrt{\rho}\boldsymbol{X}\boldsymbol{H} + \boldsymbol{N}$ is written in an expanded form, one can easily see that the first m rows of $\boldsymbol{Y}$ should be deleted because they correspond to $\boldsymbol{a}_1$ whose symbols do not interfere with any other symbols. Rows $m+1$ through $2m$ should be updated by canceling the interference coming from $\boldsymbol{a}_2$, which is expressed as

$$\sqrt{\rho} \begin{bmatrix} \hat{x}_2(m+1) \\ \hat{x}_2(m+2) \\ \vdots \\ \hat{x}_2(2m) \end{bmatrix} \begin{bmatrix} h_{2,1} & h_{2,2} & \cdots & h_{2,N_r} \end{bmatrix}, \tag{6.28}$$

where $\hat{x}_2(m+1), \hat{x}_2(m+2), \ldots, \hat{x}_2(2m)$ are hard decisions on the symbols comprising block $\boldsymbol{a}_2$. Furthermore, rows $2m+1$ through $3m$ should be updated by subtracting the interference coming from $\boldsymbol{a}_3$, which is expressed as

$$\sqrt{\rho} \begin{bmatrix} \hat{x}_3(2m+1) \\ \hat{x}_3(2m+2) \\ \vdots \\ \hat{x}_3(3m) \end{bmatrix} \begin{bmatrix} h_{3,1} & h_{3,2} & \cdots & h_{3,N_r} \end{bmatrix}, \tag{6.29}$$

and finally rows $3m+1$ through $4m$ should be updated by subtracting the interference coming from $\boldsymbol{a}_4$, which is expressed as

$$\sqrt{\rho}\begin{bmatrix} \hat{x}_4(3m+1) \\ \hat{x}_4(3m+2) \\ \vdots \\ \hat{x}_4(4m) \end{bmatrix} \begin{bmatrix} h_{4,1} & h_{4,2} & \cdots & h_{4,N_r} \end{bmatrix}. \tag{6.30}$$

Once $\boldsymbol{Y}$ is updated, as per the foregoing discussion, the new $\tilde{\boldsymbol{Y}}$ can then be calculated by right multiplying the updated $\boldsymbol{Y}$ by $\boldsymbol{Q}^H$. The new $\tilde{\boldsymbol{Y}}$ obviously has the same structure as the one given by (6.23) except that the first m elements of the first row of $\tilde{\boldsymbol{Y}}$ now correspond to $\boldsymbol{d}_1$, and so on. The above process repeats exactly in the same way in detecting the rest of the codewords.

A summary of the ZF-IC algorithm for the DBLAST scheme is given in Table 6.3.

Minimum Mean Squared Error Detection

The MMSE detection of DBLAST signals with interference cancellation involves obtaining $\boldsymbol{Y}$ defined by (6.2), computing $\boldsymbol{W}_{opt}^H$ using (6.11), and computing $\tilde{\boldsymbol{Y}}$ according to (6.12). Following the above example, by writing $\boldsymbol{Y}$ in an expanded form, one can easily see that the first set of m rows of $\tilde{\boldsymbol{Y}}$ are functions of $\boldsymbol{a}_1$, the second set of m rows correspond to the linear combination of $\boldsymbol{a}_2$ and $\boldsymbol{d}_1$, and so on. Thus, estimates of the elements of $\boldsymbol{a}_1$, i.e., $\hat{x}_1(1), \ldots, \hat{x}_1(m)$, can be obtained by using $\tilde{y}_1(1), \ldots, \tilde{y}_1(k)$; estimates of the elements of $\boldsymbol{a}_2$ can be obtained by using $\tilde{y}_2(m+1), \ldots, \tilde{y}_2(2m)$; estimates of the elements of $\boldsymbol{a}_3$ can be obtained by using $\tilde{y}_3(2m+1), \ldots, \tilde{y}_3(3m)$; and finally estimates of the elements of $\boldsymbol{a}_4$ can be obtained by using $\tilde{y}_4(3m+1), \ldots, \tilde{y}_4(4m)$.

Once the codeword $\boldsymbol{a}_1, \ldots, \boldsymbol{a}_4$ is detected, the next codeword to be detected is $\boldsymbol{d}_1, \ldots, \boldsymbol{d}_4$, but the interference coming from $\boldsymbol{a}_1, \ldots, \boldsymbol{a}_4$ should be cancelled first. Note that the estimates of $\boldsymbol{a}_1$ are not used in the interference cancellation process because they do not interfere with any other codeword. Thus, the first m rows of $\boldsymbol{Y}$, which correspond to $\boldsymbol{a}_1$, should be deleted. Now to detect $\boldsymbol{d}_2$, the interference coming from $\boldsymbol{a}_2$, expressed by

Table 6.3 The ZF-IC detection algorithm for the DBLAST scheme

obtain $\boldsymbol{Y}$, as per (6.2)
set $i = K$
while $i \geq 1$
    compute $\tilde{\boldsymbol{Y}}$ according to (6.22)
    make hard decisions using the decision statistics given by (6.24)–(6.27)
    update $\boldsymbol{Y}$, that is,
        delete row $1, 2, \ldots, m$
        subtract the interference from rows $m+1, m+2, \ldots, 2m$ using (6.28)
        subtract the interference from rows $2m+1, 2m+2, \ldots, 3m$ using (6.29)
        subtract the interference from rows $3m+1, 3m+2, \ldots, 4m$ using (6.30)
    $i = i - 1$
end of the while loop

Table 6.4 The MMSE-IC detection algorithm for the DBLAST scheme

```
set i = K
let m be the number of blocks per codeword
obtain Y
compute W^H_opt
while i ≥ 1
    compute Ỹ
    detect the first set of m rows of Y
    update Y, that is,
        delete rows 1, 2, ..., m
        subtract the interference from rows m + 1, m + 2, ..., 2m using (6.28)
        subtract the interference from rows 2m + 1, 2m + 2, ..., 3m using (6.29)
        subtract the interference from rows 3m + 1, 3m + 2, ..., 4m using (6.30)
    i = i − 1
end of the while loop
```

(6.28), should be cancelled from the second set of m rows of $\boldsymbol{Y}$. Similarly, the interference coming from $\boldsymbol{a}_3$, expressed by (6.29), should be cancelled from the third set of m rows of $\boldsymbol{Y}$. Finally, $\boldsymbol{d}_4$ is detected after the interference coming from $\boldsymbol{a}_4$, expressed by (6.30), is cancelled from the fourth set of m rows of $\boldsymbol{Y}$.

In computing the new $\tilde{\boldsymbol{Y}} = \boldsymbol{Y}\boldsymbol{W}_{opt}$, with $\boldsymbol{Y}$ updated by removing its first set of m rows and the interference coming from $\boldsymbol{a}_2$, $\boldsymbol{a}_3$, and $\boldsymbol{a}_4$ is cancelled, $\boldsymbol{W}^H_{opt}$ is not updated as was the case for the other BLAST schemes. This is attributed to the fact that each DBLAST codeword is spread over all the spatial layers, and thus we cannot delete any of the rows of $\boldsymbol{H}$. The newly computed $\tilde{\boldsymbol{Y}}$ can be then used to detect $\boldsymbol{d}_1, \ldots, \boldsymbol{d}_4$, similar to how $\boldsymbol{a}_1, \ldots, \boldsymbol{a}_4$ were detected. The same process repeats exactly in the same way in detecting the rest of the codewords.

A summary of the MMSE-IC algorithm for the DBLAST scheme is given in Table 6.4.

6.2.2 Examples

Now we examine the bit error rate performance of the DBLAST scheme using the ZF and MMSE detectors with and without interference cancellation. The channel codes used for all layers were identical, with each being a rate 1/2 convolutional code with generators $(171,133)_{octal}$. The results are shown in Figure 6.11. We observe from the figure that, consistent with what we have seen for the other BLAST schemes, the MMSE detector outperforms the ZF detector. When interference cancellation is employed, additional performance improvements can be obtained. To examine the impact of error propagation on the overall performance, we plot in Figure 6.12 the bit error rate performance of DBLAST using the ZF-IC and MMSE-IC detectors when there is no error propagation, i.e., it is assumed that previously detected symbols are correct. It is clear from the figure that error propagation has little impact on the performance.

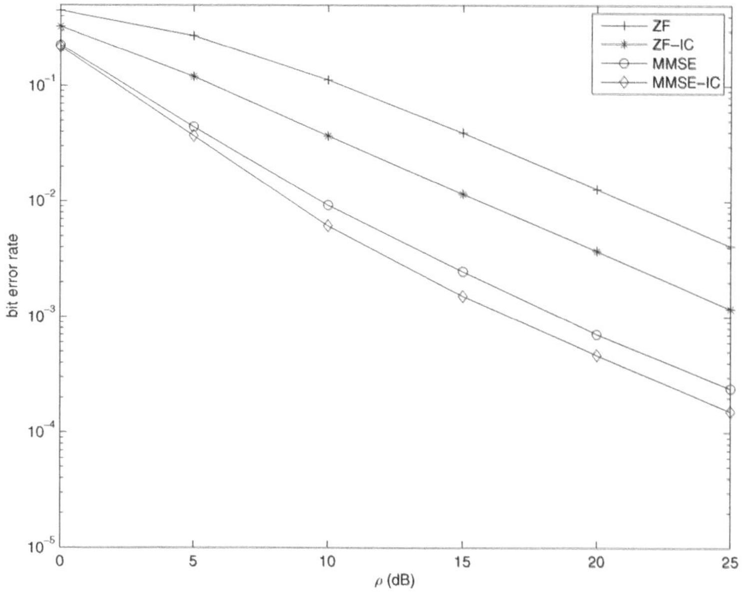

Figure 6.11 DBLAST bit error rate performance with various detection criteria for $N_t = N_r = 4$ and uncoded BPSK.

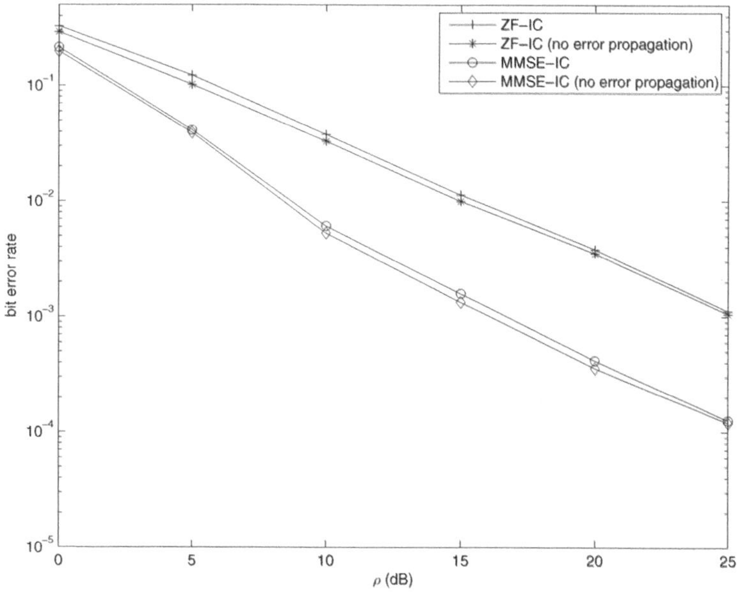

Figure 6.12 DBLAST bit error rate performance with various detection criteria for $N_t = N_r = 4$ and uncoded BPSK.

6.3 Multilayered Space-Time Codes

6.3.1 Encoder Structure

The approach used by the BLAST architectures is in general purely signal processing in the sense that there is no space-time coding done at the transmitter to achieve transmit diversity. These architectures employ, as discussed in the previous sections, interference cancellation and suppression techniques at the receiver to detect the transmitted streams by removing or reducing the effects of inter-layer interference. As mentioned earlier, as an alternative, the multilayered space-time coding scheme (MLSTC), introduced by Tarokh et al. (1999), gives a somewhat better compromise between spatial diversity and spatial rate. This scheme combines both array processing at the receiver and coding techniques at the transmitter in an effort to improve the system performance by enhancing its diversity order while increasing the data rate by using the concept of data layering. In particular, the transmit antennas are partitioned into subgroups, and each subgroup is used to transmit an independent signal after being coded using a space-time coding scheme such as a space-time trellis or block code. These space-time codes are called component space-time codes.

The MLSTC encoder accepts as its input a block of K information bits at a time. These bits are divided into q subgroups, namely $k_1, k_2, \ldots, k_q$, such that $\sum_{i=1}^{q} k_i = K$. The available transmit antennas are also divided into q subgroups, denoted by $n_1, n_2, \ldots, n_q$ where $\sum_{i=1}^{q} n_i = N_t$. Each k_i, $1 \leq i \leq q$ is applied to a space-time encoder, and the resulting code is denoted by C_i (see Figure 6.13). Codewords belonging to code C_i are then transmitted from their respective subgroup of antennas, namely n_i. Consequently, the number of layers is q. There is no restriction on the number of antennas per subgroup. In the special case when $q = N_t$, the MLSTC scheme obviously reduces to a BLAST scheme.

In the following sections, we shall consider a detection approach that can be used for MLSTCs, based on the ZF criterion.

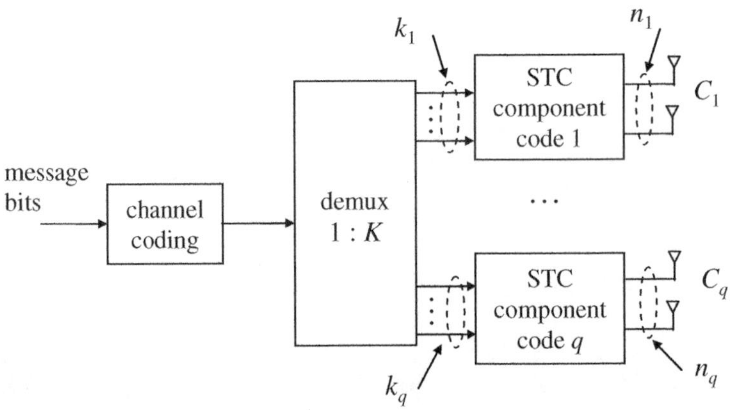

Figure 6.13 MLSTC encoder structure.

6.3.2 Group Interference Cancellation Detection

The combined transmitted codes $\mathcal{C}_i$, $1 \leq i \leq q$ can be viewed as a product code, denoted by $\mathcal{C} = \mathcal{C}_1 \times \mathcal{C}_2 \times \cdots \times \mathcal{C}_q$. Such a product code can be decoded jointly, but the associated complexity is too high since the number of trellis states grows exponentially in the number of transmit antennas and K. A seemingly much lower complexity decoding approach is to apply the concept of group interference cancellation whereby each code $\mathcal{C}_i$ is decoded separately while the rest of the codes are viewed as interference, and thus can be suppressed and/or cancelled. This approach was inspired by the decoding methods applied to the BLAST schemes which combine interference cancellation and suppression (as discussed in the previous sections).

Let $\boldsymbol{X}$ be the transmitted signal, which can be expressed in vector form as

$$\boldsymbol{X} = \left[\begin{array}{cccc} \boldsymbol{X}_{\mathcal{C}_1} & \boldsymbol{X}_{\mathcal{C}_2} & \cdots & \boldsymbol{X}_{\mathcal{C}_q} \end{array} \right], \tag{6.31}$$

$$= \left[\begin{array}{ccccccc} x_1(1) & \cdots & x_{n_1}(1) & \cdots & x_{n_1+n_2}(1) & \cdots & x_{N_t}(1) \\ x_1(2) & \cdots & x_{n_1}(2) & \cdots & x_{n_1+n_2}(2) & \cdots & x_{N_t}(2) \\ \vdots & \cdots & \vdots & \cdots & \vdots & \cdots & \vdots \\ x_1(N) & \cdots & x_{n_1}(N) & \cdots & x_{n_1+n_2}(N) & \cdots & x_{N_t}(N) \end{array} \right]. \tag{6.32}$$

Note that the first n_1 columns of $\boldsymbol{X}$ in (6.32) correspond to $\mathcal{C}_1$, columns $n_1 + 1, \ldots, n_2$ correspond to $\mathcal{C}_2$, and so on. The received signal $\boldsymbol{Y}$ can then be expressed as

$$\boldsymbol{Y} = \sqrt{\rho}\boldsymbol{X}\boldsymbol{H} + \boldsymbol{N} \tag{6.33}$$

where $\boldsymbol{H}$ is the channel matrix (of size $N_t \times N_r$), and $\boldsymbol{N}$ is the noise matrix (of size $N \times N_r$). The entries of $\boldsymbol{H}$ and $\boldsymbol{N}$ are independent and $\mathcal{CN}(0, 1)$ distributed.

Suppose we are interested in decoding $\mathcal{C}_1$. The objective here is to suppress the interference coming from all other layers. To proceed with this, we first partition the channel matrix $\boldsymbol{H}$ as

$$\boldsymbol{H} = \left[\begin{array}{c} \boldsymbol{H}_{\mathcal{C}_1} \\ \boldsymbol{H}_{\mathcal{C}-\mathcal{C}_1} \end{array} \right], \tag{6.34}$$

where $\boldsymbol{H}_{\mathcal{C}_1}$ is defined as

$$\boldsymbol{H}_{\mathcal{C}_1} = \left[\begin{array}{cccc} h_{1,1} & h_{1,2} & \cdots & h_{1,N_r} \\ h_{2,1} & h_{2,2} & \cdots & h_{2,N_r} \\ \vdots & \vdots & \ddots & \vdots \\ h_{n_1,1} & h_{n_1,2} & \cdots & h_{n_1,N_r} \end{array} \right], \tag{6.35}$$

and

$$\boldsymbol{H}_{\mathcal{C}-\mathcal{C}_1} = \left[\begin{array}{cccc} h_{n_1+1,1} & h_{n_1+1,2} & \cdots & h_{n_1+1,N_r} \\ h_{n_1+2,1} & h_{n_1+2,2} & \cdots & h_{n_1+21,N_r} \\ \vdots & \vdots & \ddots & \vdots \\ h_{N_t,1} & h_{N_t,2} & \cdots & h_{N_t,N_r} \end{array} \right]. \tag{6.36}$$

The size of $\boldsymbol{H}_{\mathcal{C}_1}$ is $n_1 \times N_r$ and the size of $\boldsymbol{H}_{\mathcal{C}-\mathcal{C}_1}$ is $(N_t - n_1) \times N_r$.

Assume that $N_r \geq N_t - n_1 + 1$. This assumption is needed to be able to suppress the interference coming from the other layers, similar to what was assumed in detecting the BLAST schemes that involve interference suppression. Note that $\text{rank}\left[\boldsymbol{H}_{C-C_1}\right] \leq N_t - n_1$. Let Ω_{C-C_1} denote the matrix whose columns span the null space $\mathcal{N}$ of $\boldsymbol{H}_{C-C_1}$. The size of Ω_{C-C_1} is $N_r \times (N_r - N_t + n_1)$ and it has the property that

$$\Omega_{C-C_1}^H \Omega_{C-C_1} = \boldsymbol{I}_{N_r - N_t + n_1}.$$

Also, since Ω_{C-C_1} spans the null space of $\boldsymbol{H}_{C-C_1}$, it has the property that

$$\boldsymbol{H}_{C-C_1} \Omega_{C-C_1} = \boldsymbol{0},$$

where $\boldsymbol{0}$ is the $(N_t - n_1) \times (N_r - N_t + n_1)$ all-zero matrix. Armed with this result, by right multiplying both sides of (6.33) by Ω_{C-C_1}, with $\boldsymbol{H}$ defined by (6.34), we arrive at

$$\begin{aligned} \tilde{\boldsymbol{Y}} &= \sqrt{\rho}\,(\boldsymbol{X}\boldsymbol{H} + \boldsymbol{N})\,\Omega_{C-C_1}, \\ &= \sqrt{\rho}\boldsymbol{X}_{C_1}\boldsymbol{H}_{C_1}\Omega_{C-C_1} + \tilde{\boldsymbol{N}}, \end{aligned} \tag{6.37}$$

where $\tilde{\boldsymbol{N}} \triangleq \boldsymbol{N}\Omega_{C-C_1}$.

The second line of (6.37) suggests that the interference coming from antennas $n_1 + 1, n_1 + 2 \ldots, N_t$, i.e., the other codes, has been suppressed, as desired. After C_1 is detected and the interference caused by it is cancelled, then C_2 is detected. This is accomplished by updating $\boldsymbol{Y}$ defined by (6.33) by subtracting the interference coming from C_1, denoted by $\sqrt{\rho}\hat{\boldsymbol{X}}_{C_1}\boldsymbol{H}_{C_1}$ where $\hat{\boldsymbol{X}}_{C_1}$ is a matrix that represents the estimate of $\boldsymbol{X}_{C_1}$. Once C_2 is detected and its contribution to interference is cancelled, C_3 is detected the same way. This process continues until the last code C_q is detected.

In decoding C_1, if the interference coming from the other codes was completely suppressed, then this code achieves a diversity order of up to $n_1 \cdot (n_1 + N_r - N_t)$. When C_2 is detected, assuming that the interference coming from C_1 is perfectly cancelled and the interference coming from the other codes is completely suppressed, then this code achieves a diversity order that is as high as $n_2 \cdot (n_2 + n_1 + N_r - N_t)$. In general, code C_i achieves a diversity order of up to $n_i \cdot (n_i + \cdots + n_2 + n_1 + N_r - N_t)$. Note that code C_q achieves a maximum diversity of $n_q \cdot N_r$. These results hold for quasi-static fading channels. When the channel is block or fast fading, the detection method will not change, but space-time code designs suitable for these channel models should be used instead (see Tarokh et al. (1998)), and the available diversity orders are much larger.

As in the standard BLAST schemes, for all layers to achieve a somewhat similar performance, which is often desired, the respective transmit powers will have to be adjusted accordingly. One may also order the received sequences according to their powers such that the best sequence is detected first, similar to what was suggested for the BLAST schemes. One may also go further by assigning a different number of antennas per subgroup to compensate for the discrepancy in performance.

6.3.3 Example

In this example, we consider a MIMO system with $N_t = N_r = 4$. We set $q = 2$, i.e., there are two antenna groups with two transmit antennas per group. The Alamouti scheme is employed for each transmit antenna group, and BPSK modulation is used. The codeword length is set to $N = 100$. The underlying channel is assumed to be quasi-static fading and the subchannels are independent. As such, the channel matrix, $\boldsymbol{H}$, can be expressed as

$$\boldsymbol{H} = \begin{bmatrix} \boldsymbol{H}_{C_1} \\ \boldsymbol{H}_{C_2} \end{bmatrix},$$

which is equivalent to

$$\boldsymbol{H} = \begin{bmatrix} h_{1,1} & h_{1,2} & h_{1,3} & h_{1,4} \\ h_{2,1} & h_{2,2} & h_{2,3} & h_{2,4} \\ h_{3,1} & h_{3,2} & h_{3,3} & h_{3,4} \\ h_{4,1} & h_{4,2} & h_{4,3} & h_{4,4} \end{bmatrix}.$$

At the receiver, the two groups are detected one at a time. When the first group is detected, the second group is nulled out. Nulling the second group involves generating the corresponding nulling matrix, denoted by Ω_{C_2}, and then multiplying it by the received vector $\boldsymbol{Y}$. Note that Ω_{C_2} has the properties that

$$\Omega_{C_2}^H \Omega_{C_2} = \boldsymbol{I}_2,$$

and

$$\boldsymbol{H}_{C_2} \Omega_{C_2} = \boldsymbol{0}_2.$$

When $\Omega_{C_2}$is multiplied with $\boldsymbol{Y}$, the resulting matrix, denoted by $\tilde{\boldsymbol{Y}}_{C_1}$, is essentially the received signal corresponding to the first group, i.e.,

$$\begin{aligned} \tilde{\boldsymbol{Y}}_{C_1} &= \sqrt{\rho}\boldsymbol{X} \begin{bmatrix} \boldsymbol{H}_{C_1} \\ \boldsymbol{H}_{C_2} \end{bmatrix} \Omega_{C_2} + \boldsymbol{N}\Omega_{C_2} \\ &= \sqrt{\rho}\boldsymbol{X}_{C_1}\boldsymbol{H}_{C_1}\Omega_{C_2} + \boldsymbol{N}\Omega_{C_2}. \end{aligned}$$

Then $\tilde{\boldsymbol{Y}}_{C_1}$ is passed to the STBC decoder corresponding to the first group. Once detected, its interference contribution is calculated and subtracted from the received signal as

$$\tilde{\boldsymbol{Y}}_{C_2} = \boldsymbol{Y} - \sqrt{\rho}\hat{\boldsymbol{X}}_{C_1}\boldsymbol{H}_{C_1}.$$

The result of the subtraction yields the received signal corresponding to the second group. If the decisions $\hat{\boldsymbol{X}}_{C_1}$ are all correct, the signal $\tilde{\boldsymbol{Y}}_{C_2}$ is free of interference.

We plot in Figure 6.14 the bit error rate performance of the above-mentioned system. In the same figure, we give the overall bit error rate performance. We also give the bit error rate performance for each layer separately assuming that the interference is perfectly

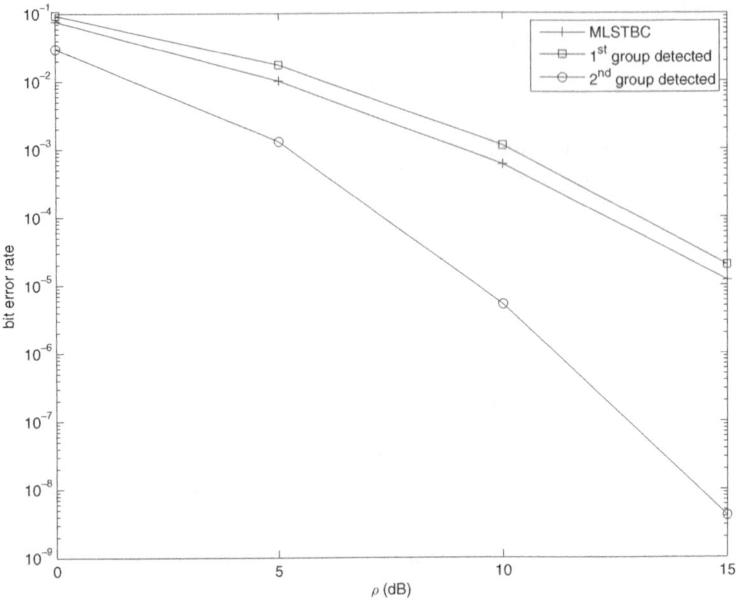

Figure 6.14 MLSTC bit error rate performance for a 4×4 MIMO system employing orthogonal STBCs as component codes with two transmit antennas per group.

cancelled from the second group. As per the foregoing discussion, the diversity order of the first detected group is equal to four, whereas it is eight for the second group. This is very clear from the figure where the slopes of the two curves corresponding to the two groups are different. We also observe from the figure that the overall performance is dominated by that of the first detected group.

6.4 Threaded Space-Time Codes

The approach used in threaded space-time coding combines ideas from standard space-time coding, e.g., space-time trellis or block coding, and the DBLAST scheme in an effort to maximize the spatial and temporal diversity available in the system while increasing the transmission rate. The flexibility in the trade-off between diversity and rate offered by threaded space-time codes (TSTCs) is higher than that offered by other layering schemes. In addition, at about the same transmission rate and complexity, the threaded space-time coding scheme is superior to multilayered space-time coding in terms of performance. For instance, the threaded space-time coding scheme comes within 3 dB from the outage capacity at 10% frame error rate, as opposed to being 6–9 dB for the multilayered space-time coding approach (El Gamal and Hammons (2001)).

The TSTC design involves three main steps. First, the layers are formed such that the spatial and temporal diversity can be maximized. This is done independently of the space-time codes used at the transmitter and of the signal processing used at the receiver.

Second, algebraic codes based on the codes developed by Hammons and El Gamal (2000) are designed in a way that allows for capturing the spatial and temporal diversity available in the system. Third, iterative signal processing algorithms depending on the algebraic code design and the layering approach are used at the receiver to optimize the performance. In the following subsections, we address these steps in the same order.

6.4.1 Layering Approach

A thread is a layer that extends over the full spatial span N_t and the full temporal span N, where N denotes the codeword length at the output of each space-time component encoder after modulation. Thus a layer can be viewed as an array of size $N \times N_t$. An element of this array may be identified by two indices, namely (a, k) where a represents the transmitting antenna and k represents the transmission time interval. Let L_i, $1 \leq i \leq N_t$ denote the ith layer. A layer is formed by distributing the codeword at the output of each component space-time encoder (of length k) among the N_t antennas along the diagonal layers such that

$$L_i = \left\{ \left((k + i - 1)_{N_t}, k \right) : 0 \leq k < N \right\}, \tag{6.38}$$

where $(x)_{N_t}$ denotes x modulo N_t. Accordingly, consecutive symbols from the same codeword are transmitted from different transmit antennas at different time intervals. As an example, consider Figure 6.15 in which there are four transmit antennas, and hence four layers. In the figure, each codeword is of length $N = 16$ and has a different pattern. It is obvious that each codeword spans the entire spatial domain in an effort to maximize the (possible) spatial diversity. Also, each codeword is fully interleaved before it is assigned to the transmit antennas in order to maximize the temporal diversity. Interleaving is incorporated into the space-time code design, as will be explained in the next subsection.

By comparing Figures 6.10 and 6.15, one can see that the threaded space-time coding scheme is a generalization of the DBLAST scheme and both schemes become equivalent when $N = N_t$. On the other hand, the threaded space-time coding scheme is slightly more bandwidth efficient since all antennas are active at all times as opposed to the DBLAST signaling in which some of the antennas remain idle at the beginning and end of the transmission periods. One might argue that, in a multi-user environment, some of the

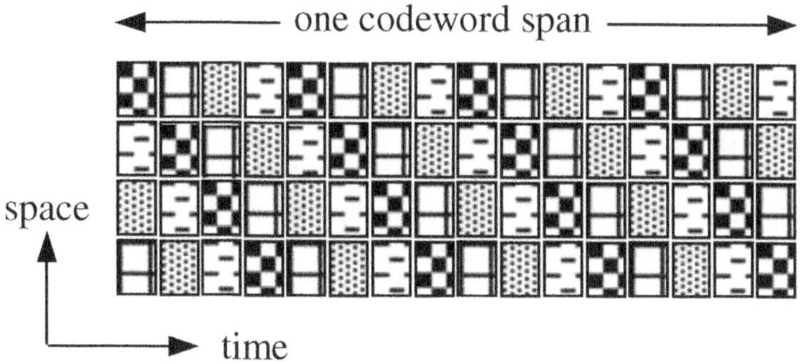

Figure 6.15 Layering in threaded space-time coding.

antennas in the threaded space-time coding scheme could be idle if there is no data to transmit from some of the users. But this problem can be alleviated by assigning codewords coming from other active users to fill up these empty time slots.

6.4.2 Threaded Space-Time Code Design

Recall that the number of layers equals the number of transmit antennas, N_t. Each layer L_i, $1 \leq i \leq N_t$ is encoded by a space-time component encoder, denoted by E_i, resulting in code $\mathcal{C}_i$. The codewords of $\mathcal{C}_i$ are essentially complex matrices of size $N \times N_t$. This set-up encompasses different modulation schemes, as well as binary and nonbinary codes. Let $\{\boldsymbol{d}^i_j : 1 \leq j \leq 2^m\}$ be the set of all m-tuple information bit sequences at the input of encoder E_i, and let $\left\{\boldsymbol{c}^i_j : 1 \leq j \leq 2^m\right\}$ be the set of all corresponding codewords in $\mathcal{C}_i$. A codeword $\boldsymbol{c}^i_j$ has the form

$$\boldsymbol{c}^i_j = \left[\begin{array}{cccc} \boldsymbol{d}^i_j\boldsymbol{M}_1 & \boldsymbol{d}^i_j\boldsymbol{M}_2 & \cdots & \boldsymbol{d}^i_j\boldsymbol{M}_{N_t} \end{array}\right], \tag{6.39}$$

where $\boldsymbol{M}_q$, $1 \leq q \leq N_t$ are binary matrices each of size $m \times bN/N_t$. The role of these matrices becomes clear later. Note that the length of $\boldsymbol{c}^i_j$ is bN, but it is implied here that each space-time component encoder is followed by a modulator that maps every consecutive b bits onto a symbol drawn from a constellation of size 2^b, hence the length of the transmitted codeword, after modulation, is effectively N. In (6.39), $\boldsymbol{d}^i_j\boldsymbol{M}_q$ corresponds to the portion of $\boldsymbol{c}^i_j$ that is assigned to antenna q. While the set of matrices $\boldsymbol{M}_q$ can be different for different codes $\mathcal{C}_i$, they can also be the same for all codes at no performance penalty. Therefore, to simplify the presentation, we shall assume henceforth that $\boldsymbol{M}_q$ are the same for all $\mathcal{C}_i$.

To incorporate interleaving into the code design, the individual portions $\boldsymbol{d}^i_j\boldsymbol{M}_q$ are interleaved before they are assigned to the transmit antennas so that each portion spans the entire time span N. An effective way of performing interleaving it to spread the components of $\boldsymbol{d}^i_j\boldsymbol{M}_q$ so that the time separation between such components is maximized. Since the length of each $\boldsymbol{d}^i_j\boldsymbol{M}_q$ is N/N_t, and the goal is to span these components across the time span N, the separation between consecutive symbols would be $N/N_t - 1$. A constraint that interleaving has to respect is that only one symbol per codeword is allowed to be transmitted in any given time slot. To satisfy this constraint, one may use periodic interleaving.

Periodic interleaving can be seen in the example shown in Figure 6.15. In the figure, referring to the substream transmitted from the first antenna, each of the four symbols corresponding to the same codeword (same pattern) are separated by three symbol intervals. The substream transmitted from the second antenna follows the same interleaving principle, except that it is shifted to the right by one symbol interval to avoid transmitting two symbols from the same codeword at the same time. The same rule applies to the rest of the substreams. This way of interleaving essentially converts a quasi-static fading channel into a block fading channel. Of course there are other ways of interleaving that do not violate the TST code design principles, but this approach is simple and achieves the design objectives.

Code $\mathcal{C}_i$ achieves a spatial diversity of dN_r in quasi-static fading if and only if d is the largest integer such that the matrices $\boldsymbol{M}_q$, $1 \leq q \leq N_t$ have the property that

$$\boldsymbol{M} = \left[\begin{array}{cccc} a_1\boldsymbol{M}_1 & a_2\boldsymbol{M}_2 & \cdots & a_{N_t}\boldsymbol{M}_{N_t} \end{array}\right]$$

is of rank m over the binary Galois field, $GF(2)$, for all $a_1, a_2, \ldots, a_{N_t} \in GF(2)$ where $\sum_{i=1}^{N_t} a_i = N_t - d + 1$. Such a code is said to achieve a d-level diversity. When all of the matrices $\boldsymbol{M}_q$ are of rank m, then the maximum diversity $N_t N_r$ is achieved. As for the spatial rate, the maximum rate that can be achieved by a d-level code with a signal constellation of size 2^b and channel code rate R_c is $R_c b(N_t - d + 1)$. This gives a clear trade-off between diversity and rate. For instance, the TST coding scheme would become equivalent to a BLAST scheme when $d = 1$, whereas when $d = N_t$, it becomes equivalent to an STTC or STBC.

We now discuss how the matrices $\boldsymbol{M}_q$ are designed. Let C denote a binary convolutional code of rate m/N_t where the convolutional encoder accepts m binary input sequences, denoted by $x_1(k), x_2(k), \ldots, x_m(k)$ and produces N_t coded binary sequences, denoted by $y_1(k), y_2(k), \ldots, y_{N_t}(k)$. These coded sequences are then multiplexed to form a codeword. The underlying assumption here is that the coded sequence length is N_t where there is a natural match between this and the number of available antennas. It is also assumed that binary modulation is employed. When 2^b-ary modulation schemes are considered, to fit the code design within the above framework, the code rate should be m/bN_t.

Define the generator matrix of C as

$$\boldsymbol{G}(D) = \begin{bmatrix} \boldsymbol{G}_{1,1}(D) & \boldsymbol{G}_{1,2}(D) & \cdots & \boldsymbol{G}_{1,N_t}(D) \\ \boldsymbol{G}_{2,1}(D) & \boldsymbol{G}_{2,2}(D) & \cdots & \boldsymbol{G}_{2,N_t}(D) \\ \vdots & \vdots & \ddots & \vdots \\ \boldsymbol{G}_{m,1}(D) & \boldsymbol{G}_{m,2}(D) & \cdots & \boldsymbol{G}_{m,N_t}(D) \end{bmatrix}, \tag{6.40}$$

where $\boldsymbol{G}_{i,j}(D)$ represents the D-transform of the impulse response relating the ith input $x_i(k)$ and the jth output $y_j(k)$. As such, the input and output of the convolutional encoder may be related to each other in the D-domain as

$$\boldsymbol{Y}(D) = \boldsymbol{X}(D)\boldsymbol{G}(D)$$

where

$$\boldsymbol{Y}(D) = \begin{bmatrix} \boldsymbol{Y}_1(D) & \boldsymbol{Y}_2(D) & \cdots & \boldsymbol{Y}_{N_t}(D) \end{bmatrix},$$

and

$$\boldsymbol{X}(D) = \begin{bmatrix} \boldsymbol{X}_1(D) & \boldsymbol{X}_2(D) & \cdots & \boldsymbol{X}_m(D) \end{bmatrix}.$$

The objective now is to make the connection between C and C_i. One way to make this connection is to assign the ith coded sequence, that is $\boldsymbol{Y}_i(D)$, to the ith transmit antenna. But we still need to verify what spatial diversity this scheme achieves. To this end, define the matrix

$$\boldsymbol{F}(a_1, a_2, \ldots, a_{N_t}) = \begin{bmatrix} a_1\boldsymbol{F}_1 & a_2\boldsymbol{F}_2 & \cdots & a_{N_t}\boldsymbol{F}_{N_t} \end{bmatrix}$$

for $a_1, a_2, \ldots, a_{N_t} \in GF(2)$, where $\boldsymbol{F}_m$ denotes the mth column of $\boldsymbol{G}(D)$ defined by (6.40). $\boldsymbol{G}(D)$ can then be expressed as

$$\boldsymbol{G}(D) = \begin{bmatrix} \boldsymbol{F}_1(D) & \boldsymbol{F}_2(D) & \cdots & \boldsymbol{F}_{N_t}(D) \end{bmatrix}.$$

Accordingly, the codeword defined by (6.39) has now the form

$$\boldsymbol{c}_j^i = \begin{bmatrix} \boldsymbol{d}_j^i\boldsymbol{F}_1 & \boldsymbol{d}_j^i\boldsymbol{F}_2 & \cdots & \boldsymbol{d}_j^i\boldsymbol{F}_{N_t} \end{bmatrix}. \tag{6.41}$$

As for the spatial diversity achieved by this code, let v be the smallest integer that has the property that, when $\sum_{i=1}^{N_t} a_i = v$, the matrix $\boldsymbol{F}(a_1, a_2, \ldots, a_{N_t})$ has full rank m over the field of binary polynomials $GF(2)[D]$. Then C_i achieves a d-level spatial diversity over the quasi-static fading channel when $d = N_t - v + 1$ and $v \geq m$. The above code design principles extend to convolutional codes with rate $1/n$ with $n < N_t$, but with a slight variation. The reader may refer to the paper by El Gamal and Hammons (2001) for details.

We now extend these code design rules from quasi-static fading to block fading channels. In quasi-static fading, with these design rules, from the perspective of any space-time component code, the channel is equivalent to a block fading channel that has N_t independently faded blocks per codeword. In a truly block fading channel, the subchannels between transmit and receive antenna pairs are block fading. Let B be the number of faded blocks per subchannel. Then this channel, from the perspective of any space-time component code, is equivalent to a quasi-static fading channel with $N_t B$ transmit antennas.

As such, a codeword $\boldsymbol{c}^i_j$ belonging to the code C_i, according to the new design rules, will now have the form

$$\boldsymbol{c}^i_j = \begin{bmatrix} \boldsymbol{d}^i_j \boldsymbol{M}_{1,1} & \boldsymbol{d}^i_j \boldsymbol{M}_{2,1} & \cdots & \boldsymbol{d}^i_j \boldsymbol{M}_{N_t,1} & \cdots & \boldsymbol{d}^i_j \boldsymbol{M}_{1,B} & \cdots & \boldsymbol{d}^i_j \boldsymbol{M}_{N_t,B} \end{bmatrix},$$

where $\boldsymbol{M}_{i,j}$, $1 \leq i \leq N_t$, $1 \leq b \leq B$ are binary matrices each of size $m \times bN/N_t B$. Code C_i achieves a d-level spatial diversity if and only if d is the largest integer such that matrices $\boldsymbol{M}_{i,j}$, $1 \leq i \leq N_t$, $1 \leq b \leq B$ have the property that for all $a_{1,1}, a_{2,1}, \ldots, a_{N_t,B} \in GF(2)$

$$\sum_{i=1}^{N_t} \sum_{b=1}^{B} a_{i,b} = N_t B - d + 1,$$

the matrix $\boldsymbol{M} = \begin{bmatrix} a_{1,1}\boldsymbol{M}_1 & a_{2,1}\boldsymbol{M}_{2,1} & \cdots & a_{N_t,B}\boldsymbol{M}_{N_t,B} \end{bmatrix}$ is of rank m over $GF(2)$.

6.4.3 Example

We borrow this example from the paper by El Gamal and Hammons (2001). Consider the rate-$1/2$, four-state convolutional code whose generator matrix is given by

$$\mathbf{G}(D) = \begin{bmatrix} 1 + D^2 & 1 + D + D^2 \end{bmatrix}$$

with two transmit antennas. The rank of $\mathbf{G}(D)$ is one, i.e., $v = 1$. Hence, this code is a $d = 2$ level diversity code.

When the number of transmit antennas increases to four, the above rate-$1/2$ convolutional code can be viewed as a rate-$2/4$ convolutional code with the generator matrix

$$\mathbf{G}(D) = \begin{bmatrix} 1 + D & 0 & 1 + D & 1 \\ 0 & 1 + D & D & 1 + D \end{bmatrix}.$$

By inspection, one can see that every pair of columns is linearly independent over $GF(2)[D]$. Therefore, the corresponding TSTC achieves a $d = 3$ level diversity.

When the number of transmit antennas is increased to six, the above rate-$1/2$ convolutional code can be viewed as a rate-$3/6$ convolutional code with the generator matrix

$$\mathbf{G}(D) = \begin{bmatrix} 1 & 0 & 1 & 1 & 1 & 1 \\ D & 1 & 0 & D & 1 & 1 \\ 0 & D & 1 & D & D & 1 \end{bmatrix}.$$

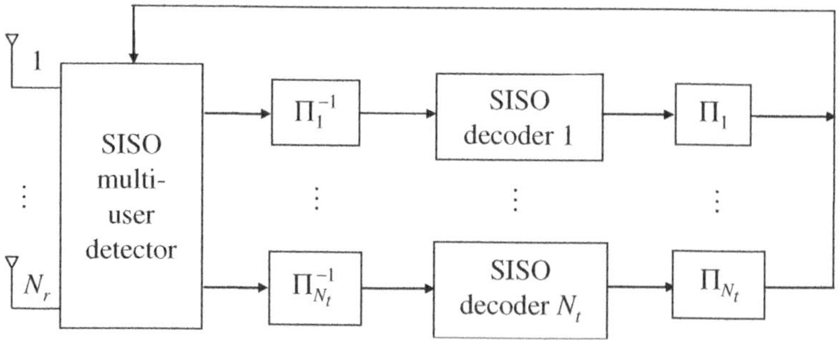

Figure 6.16 Iterative multiuser detector for threaded space-time coded MIMO systems.

Also by inspection, one can find that every set of three columns has full rank over $GF(2)[D]$. Therefore, the corresponding TSTC achieves a $d = 4$ level spatial diversity.

6.4.4 Detection of Threaded Space-Time Codes

The detection of TSTCs can be viewed as a joint multiuser detection and decoding problem, and thus the turbo decoding principle applies (Hagenauer (1997)). A block diagram of a generic decoder is shown in Figure 6.16. In the figure, the soft-input soft-output (SISO) multiuser detector module produces soft information on the N_t transmitted sequences. The soft information sequences are then demultiplexed into N_t parallel substreams, each corresponding to a transmitted codeword. The outer N_t SISO decoders in turn produce their own soft information about the transmitted sequences, and this soft information is fed back to the multiuser SISO detector to refine its estimates for the next iteration. This process repeats for a number of iterations before final decisions are made.

There are several algorithms that the SISO multiuser detector can employ, including the MAP algorithm (Reed et al. (1997), Moher (1998)); the iterative MMSE algorithm (El Gamal and Geraniotis (2000), Wang and Poor (1999)), and the soft interference canceler (Alexander et al. (1998)). The latter algorithm is an approximation of the iterative MMSE algorithm. As for the individual SISO decoders, they are matched to the individual space-time component encoders at the transmitter. These SISO decoders may use any suitable SISO algorithm, depending on the type of code employed.

6.5 Other Detection Algorithms for Spatial Multiplexing Systems

The detection schemes we have considered so far are based on the ZF and MMSE criteria. Several other detection schemes have been introduced in the literature. We briefly discuss some of these algorithms in the following sections.

6.5.1 Greedy Detection

The greedy detector introduced by AlRustamani and Vojcic (2002) was originally developed for multiuser systems as a suboptimal alternative detector. It was then extended by AlRustamani et al. (2002) to MIMO systems. In particular, it was extended to turbo coded MIMO systems, which are discussed in Chapter 7. This detector can also be extended to BLAST transmission schemes. In a coded MIMO system, the front end receiver is required to compute log-likelihood values for the transmitted bits. The associated computational complexity can be prohibitively high since it grows exponentially in the number of transmit antennas and the constellation size. With greedy detection, the receiver computes the log-likelihood values for the transmitted bits in a greedy fashion. This is achieved by obtaining only a subset of the possible transmitted vectors that have high conditional probabilities. The motivation behind this is that the vectors that have higher probabilities contribute more to the log-likelihoods, whereas the ones with lower probabilities have very low contributions, and thus they can be ignored. The consequence of this simplification is a significant reduction in the computational complexity as compared with that of the optimal detector, which comes as a trade-off for a slight degradation in performance.

6.5.2 Belief Propagation Detection

Belief propagation (BP) detection of BLAST schemes is proposed by Hu and Duman (2007). It is based on the observation that a MIMO channel can be represented as a factor graph, and consequently the belief propagation algorithm proposed by Pearl (1988) can be employed. A factor graph normally consists of two types of nodes: bit and factor nodes. A factor node is connected to a bit node if it is a function of this bit node. In a BLAST-coded system, the bit nodes are simply the transmitted bits, whereas the factor nodes are the received signals. In this setting, the number of bit nodes is bN_t where 2^b is the signal constellation size, and the number of factor nodes is N_r. There is no restriction on N_r, that is, it need not be greater than or equal to N_t as in the case of ZF detection.

The bit and factor nodes exchange messages iteratively. The bit and factor nodes are capable of generating extrinsic information that can be used as the *a priori* information by the receiving node in the next iteration. After a certain number of iterations, the BP detector produces soft information about the information bits, and thus it can be viewed as a SISO detector. As such, when the BLAST system is coded by an outer channel code, the BP detector is capable of working iteratively with the outer channel decoder where they exchange soft information on the coded bits for a certain number of iterations before final decisions are made.

Since the MIMO channel factor graph contains cycles, the BP detector is not optimal. However, its performance approaches that of the maximum likelihood detector. In terms of complexity, since the BP detector is based on the fully-connected factor graph, its complexity increases exponentially, similar to that of the maximum likelihood detector. To reduce this complexity, some approximations can be made to reduce the number of computations, particularly by reducing the number of graph edges considered in the calculations. The reduced complexity BP detector achieves a performance close to that of the full complexity BP detector with much lower levels of complexity.

6.5.3 Turbo-BLAST Detection

The turbo-BLAST scheme borrows ideas from the HBLAST and TSTC schemes in an effort to maintain a good performance for MIMO systems, especially when the number of transmit antennas is larger than the number of receive antennas and when the data frame length is relatively short. As was shown above, the detection of BLAST schemes assumes knowledge of the CSI before any interference cancellation and suppression takes place. When the data frame length is short, in order to maintain an acceptable bandwidth efficiency, the training sequence will be short as well. This consequently impacts the reliability of the CSI estimates, and hence the overall performance. To overcome this problem, the turbo-BLAST detector proposed by Sellathurai and Haykin (2002) attempts to perform the CSI estimation and interference cancellation simultaneously using the MMSE criterion. This joint estimation/interference cancellation is performed iteratively. At the first iteration, a short training sequence is used to obtain a rough estimate of the CSI. These estimates are used in the interference cancellation process, and the resulting decisions are used to re-estimate the CSI in the next iteration. This process repeats for a few iterations (four to five iterations have been shown to be sufficient) before final decisions are made. The performance of the turbo-BLAST detector has been shown by Sellathurai and Haykin (2002) to approach that achieved when the receiver has perfect knowledge of the CSI.

6.5.4 Reduced Complexity ZF/MMSE Detection

The sorted MMSE detector we have discussed for the BLAST schemes involves computing $\boldsymbol{W}_{opt}^{H}$ for every detected layer, finding the optimal detection order, and performing successive interference suppression and cancellation. It is implied here that the receiver has already done the CSI estimation via some training sequence. These steps are repeated for every detected layer. The computational complexity of this detector is relatively high where most of this complexity results from finding the MMSE nulling vectors, i.e., computing $\boldsymbol{W}_{opt}^{H}$, as pointed out by Hassibi (2000). A reduced complexity detector is introduced by Hassibi (2000) which is referred to as the square-root detector. This algorithm combines ideas from the ZF and MMSE detection criteria. Specifically, it involves augmenting the channel matrix $\boldsymbol{H}$ and then applying the QR decomposition to the augmented matrix. It is shown that the inverse of the resulting $\boldsymbol{R}$ matrix is related to the square root of $\boldsymbol{W}_{opt}^{H}$. Once the inverse of $\boldsymbol{R}$ is computed, the same matrix can be used in detecting all layers without recalculating it for every detected layer, as opposed to the standard MMSE detector, and thus the complexity is reduced. A simplified version of this square-root algorithm is introduced in Zhu et al. (2004) where further complexity reduction is achieved.

6.5.5 Sphere Decoding

For certain transmission schemes over MIMO channels, an alternative decoder structure is to employ sphere decoding (Viterbo and Boutros (1999); Fincke and Pohst (1985); Damen et al. (2000); Forney (1989)). Specifically, if the transmitted signals form a lattice, i.e., they can be written in the form $\boldsymbol{X} = \boldsymbol{MA}$ where $\boldsymbol{M}$ is the cubic lattice over the set of integers, one can employ sphere decoding. Some of the important transmission schemes described in this chapter, and in earlier chapters, such as the VBLAST and orthogonal STBCs, or more generally linear dispersion codes with BPSK or QAM constellations, fall

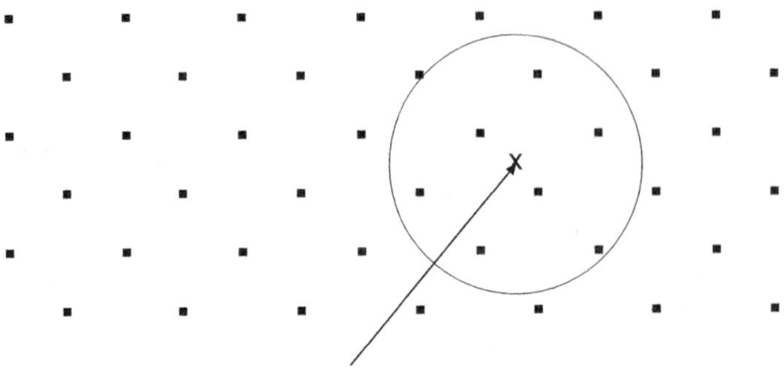

Figure 6.17 Illustration of sphere decoding.

in this category. Clearly, there are also limitations for sphere decoding. For example, it is not useful if space-time trellis coding is used. Also, it is not applicable for VBLAST or linear dispersion codes if 8-PSK is employed.

The sphere decoder is basically a bounded distance decoder. A sphere of a certain radius is selected around the received signal point in the high-dimensional space (of the transmitted signals), as illustrated in Figure 6.17. Then, all the codewords (lattice points) within this sphere are found. In the presence of AWGN, the ML codeword corresponding to the received signal is simply the closest lattice point in the Euclidean distance sense. If there are no lattice points in the sphere selected, then a decoder failure is declared. Or, we can simply increase the radius of the sphere and redo the search for the lattice points and continue this process until at least one point is found. To find the list of lattice points, the structure of the transmission (along with the known channel matrix) can be employed, and an efficient step-by-step algorithm can be derived (see for example the work by Viterbo and Boutros (1999)). We note that the worst-case complexity of sphere decoding is exponential. However, for the usual noise levels in communications, often the average complexity is only polynomial (Hassibi and Vikalo (2005)) which makes it a useful approach to find the ML codeword in a practical system.

The sphere decoding approach has received significant attention in recent years. For example, Damen and Beaulieu (2003); Zhao and Giannakis (2005) derive different versions of the sphere decoding algorithm. Vikalo and Hassibi (2002) apply the sphere decoding algorithm to MIMO channels with intersymbol interference. Vikalo et al. (2004) use sphere decoding with concatenated codes over MIMO channels. A soft-input soft-output version of the sphere decoding algorithm is developed in Boutros et al. (2003).

6.6 Diversity/Multiplexing Gain Trade-off

As it is now clear from our discussion on space-time codes and spatial multiplexing schemes, in general, there are two types of gains provided by MIMO systems, namely, the diversity and multiplexing gains. The former is defined as the negative (asymptotic) slope of the error rate curve as a function of the signal-to-noise ratio in the log-log scale, and the latter is defined as the ratio of the achievable rate by the specific MIMO scheme to the logarithm of the signal-to-noise ratio (again asymptotically) which is basically a measure of the capacity increase. Mathematically, the diversity gain is defined as (Zheng and Tse (2003))

$$d = \lim_{\rho\to\infty} -\frac{\log P_e(\rho)}{\log \rho}, \tag{6.42}$$

and the spatial multiplexing gain is defined as

$$r = \lim_{\rho\to\infty} \frac{R(\rho)}{\log \rho}, \tag{6.43}$$

where $R(\rho)$ denotes the rate of the transmission scheme as a function of the signal-to-noise ratio.

The maximum diversity gain that can be achieved by a MIMO system with N_t transmit and N_r receive antennas is $N_t N_r$, which is the total number of independent fading coefficients that one can average over. As we have seen in previous chapters, schemes such as STTCs and STBCs can achieve this maximum diversity gain. As for the multiplexing gain, it cannot exceed the number of degrees of freedom provided by the MIMO channel, which is $\min(N_t, N_r)$. It has been shown by Zheng and Tse (2003) that it is not possible to achieve both full diversity and full multiplexing gains with a specific system, and there is an inherent trade-off between these two forms of gains. Curves showing the diversity-multiplexing trade-off for several MIMO systems are given by Zheng and Tse (2003).

By examining the expression in (6.43), it is obvious that to achieve a nonzero multiplexing gain, the underlying scheme must support a data rate that increases with the signal-to-noise ratio. The reason is that the capacity increases with the signal-to-noise ratio, and thus, to achieve a non-zero fraction of this capacity asymptotically, the data rate must be increasing with the signal-to-noise ratio as well. This may be achieved by increasing the signal constellation size with the signal-to-noise ratio. Otherwise, the multiplexing gain will be zero, as is the case for fixed rate schemes.

The optimal multiplexing gain and diversity trade-off for any scheme in a Rayleigh fading channel is given by (Zheng and Tse (2003))

$$d(r) = (N_t - r)(N_r - r) \tag{6.44}$$

where $r = 0, 1, \ldots, \min(N_t, N_r)$. It is assumed here that the coherence time of the channel is greater than or equal to $N_t + N_r - 1$. It is clear from (6.44) that the maximum diversity gain is $N_t N_r$ while the multiplexing gain is zero; on the other hand, when the multiplexing gain is maximum, (i.e., $\min(N_t, N_r)$), the diversity gain is zero.

As a specific example for the Alamouti scheme, the diversity multiplexing trade-off curve is given by

$$d(r) = 2N_r (1 - r)^+, \tag{6.45}$$

where $d(r)$ is the achievable diversity gain as a function of the asymptotic rate r (where a^+ is equal to a if $a > 0$, and zero otherwise). This expression clearly shows that for a fixed

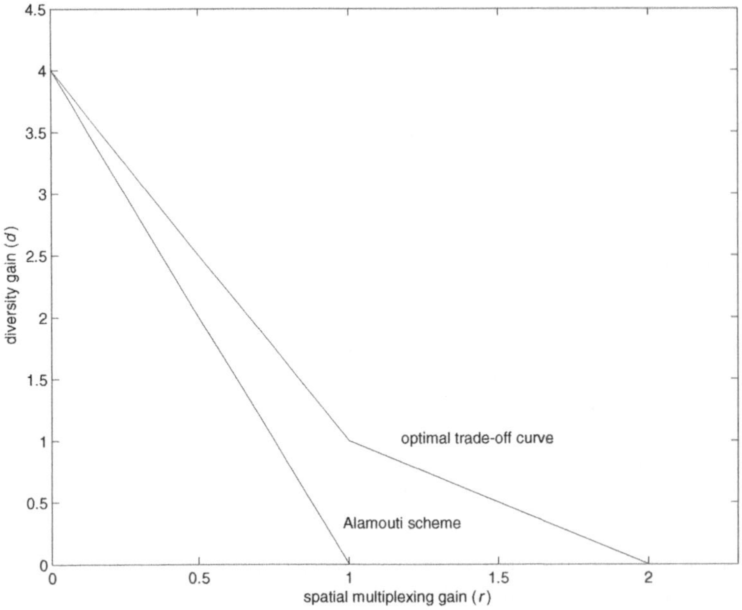

Figure 6.18 Diversity/multiplexing trade-off for the Alamouti scheme and the optimal trade-off curve ($N_r = 2$, coherence time ≥ 3).

rate transmission (i.e., for an asymptotic rate of $r = 0$), full diversity is obtained. However, if $r \neq 0$, full diversity is impossible to attain. In fact, if we would like to transmit at the maximum spatial gain, the diversity order reduces to zero.

In Figure 6.18, we plot the trade-off curve for the Alamouti scheme with two receive antennas. The coherence time is assumed to be greater than or equal to three symbol durations. We plot in the same figure the optimal trade-off curve. It is clear from the figure that the performance achieved by the Alamouti scheme is always suboptimal for any nonzero multiplexing gain. The optimal performance is achieved only when the multiplexing gain is zero. We remark, however, that the Alamouti scheme achieves the optimal trade-off performance when the number of receive antennas is one.

Another scheme for which it is interesting to study the multiplexing gain and diversity trade-off is the VBLAST scheme since it was developed with the intention of maximizing the transmission rate. Consider a square MIMO system where $N_t = N_r$. When the transmission rate is fixed across all N_t layers and detection is performed in a certain prescribed way independently of the channel realization, i.e., no sorting, the diversity/rate trade-off function for this scheme is given by

$$d(r) = \left(1 - \frac{r}{N_t}\right)^+. \tag{6.46}$$

Further improvements can be achieved by using sorting in the detection process. In this case, while the transmission rates are fixed for all layers, the trade-off can be upper

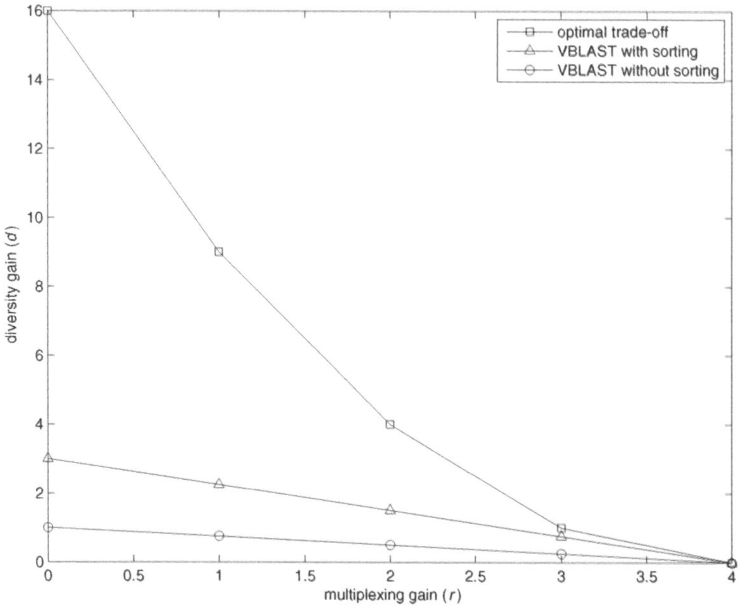

Figure 6.19 Diversity/multiplexing trade-off for the VBLAST scheme (with and without sorting in the detection process) and the optimal trade-off curve.

bounded as

$$d(r) \leq (N_t - 1)\left(1 - \frac{r}{N_t}\right). \tag{6.47}$$

Additional performance gains can also be achieved by assigning different rates for the different layers according to the channel condition. This concept is similar to that of waterfilling whereby higher rates are assigned to better channels.

The trade-off performance curves for the VBLAST scheme with and without sorting are plotted in Figure 6.19 along with the optimal trade-off curve. The MIMO system considered here is a 4×4 system. It is evident from the figure that the VBLAST scheme is optimal only when the diversity gain is zero at which point the maximum multiplexing gain is achieved. This should not be surprising because this scheme, as mentioned in previous sections, has been designed to achieve the maximum transmission rate possible. It is also interesting to observe the performance improvement achieved when sorting is used in the detection process.

Another observation that can be made from Figure 6.19 is that the performance of both VBLAST schemes (with and without sorting) is still far from the optimal one. This is attributed to the lack of coding across the transmitted substreams. For this scheme, all transmitted substreams are independent for different transmit antennas. In contrast with the Alamouti scheme and other orthogonal designs, for small values of r, the orthogonal designs are superior, whereas the VBLAST scheme becomes superior for larger values of r. Therefore to achieve further improvements, one could code across the different substreams. This is what is essentially done in the DBLAST architecture. As a matter fact, it is argued by Zheng and Tse (2003) that the DBLAST scheme with MMSE detection achieves the optimal trade-off performance, provided that a Gaussian random code with infinite block size is used. However, this is rather unrealistic from a practical point of view.

6.7 Chapter Summary and Further Reading

In this chapter, we discussed various layered space-time coding schemes. We also presented several detection approaches. It was shown that the BLAST schemes are suitable for applications where high data rates are required as they achieve the maximum transmission rate for a given number of transmit and receive antennas. In applications where a trade-off between diversity and rate is required, however, the multilayered and threaded space-time coding schemes provide better alternatives. We have also shown that any space-time coding or spatial multiplexing approach offers a trade-off between diversity and multiplexing gain.

Many other aspects of layered space-time codes have been considered in the literature. For example, Lozano and Papadias (2002) design layered space-time coding schemes for frequency selective fading channels where successive interference cancellation based on the MMSE criterion is employed at the receiver. Zhu and Murch (2003) consider layered space-time codes with decision feedback equalization for frequency selective fading channels. Combined orthogonal frequency division multiplexing (OFDM) and VBLAST coding are introduced by Lee et al. (2006) and Xin and Giannakis (2002). Wrapped space-time codes are proposed in Caire and Colavolpe (2003).

Coded layered space-time codes have also been considered. For example, Ariyavisitakul (2000) propose a turbo-coded layered space coding scheme with iterative decoding where coding is applied across the different layers. Matache and Wesel (2003) consider universal trellis codes designed such that they have a distance structure that is matched to the variation in the signal-to-noise ratio of the channel due to DBLAST. Li et al. (2006) consider LDPC coded layered space time coding.

Ma and Giannakis (2003) design a concatenated code comprising a linear complex-field outer code and a layered space-time code. It is shown that full diversity and full rate is possible. Autocoding and autocapacity were introduced by Hochwald et al. (2001). El Gamal (2002a) considers the design of layered space-time codes for autocoding where it is shown that these codes achieve the autocapacity with reasonable complexity. Generalized layered space-time codes are introduced by Tao and Cheng (2004) which combine ideas from BLAST schemes and space-time coding principles, in a somewhat similar manner with the threaded space-time codes.

Problems

6.1 As shown in Equation (6.5), the zero forcing detector involves expressing the channel matrix $\boldsymbol{H}$ using the QR factorization principle. Show that this factorization can be accomplished using the Gramm–Schmidt orthogonalization process.

6.2 Derive Equation (6.11).

6.3 In the MMSE-IC detection of the VBLAST scheme, the channel matrix $\boldsymbol{H}$ is deflated by removing one of its rows every time a new layer is detected (see the discussion following Equation (6.19)). Argue that $\boldsymbol{H}$ cannot be deflated in the MMSE-IC detection of the DBLAST scheme.

6.4 We have seen that it is possible to do sorting for the MMSE detection of the VBLAST scheme. Can we do sorting in the MMSE detection of the DBLAST scheme? Explain your answer.

6.5 Reproduce Figure 6.7.

6.6 Reproduce Figure 6.14.

6.7 Derive the diversity/multiplexing trade-off function for the Alamouti scheme given by Equation (6.45).

7

Concatenated Codes and Iterative Decoding

In this chapter, we study various code concatenation schemes for MIMO systems that have been proposed over the past decade or so. Since most of the existing concatenation schemes borrow ideas from turbo codes, either in terms of encoding, iterative decoding or both, we introduce turbo codes in detail. In particular, we give details about the encoding, iterative decoding and performance analysis of such codes over AWGN channels. We extend the use of these codes to fading channels and see how they can be combined with space-time codes to provide additional diversity and coding gains. Several concatenated coding schemes are described, some in great detail and others superficially. The chapter is concluded with a brief summary followed by a few suggestions for further reading on this subject.

Before proceeding to the introduction of turbo codes, it is instructive to give a brief summary of the development of the major concatenated coding approaches that have been introduced in the literature.

7.1 Development of Concatenated Codes

Concatenated codes were first introduced by Forney (1966), where he proposed a scheme that involves concatenating two single codes in a serial fashion. The inner code is a convolutional code and the outer code is a high-rate algebraic Reed–Solomon (RS) code which has a powerful error correction capability. The performance improvements achieved by this concatenated coding scheme were very promising and opened the door for further developments in this area.

Motivated by the work of Forney on concatenated codes and that of Ungerboeck on trellis coded modulation (TCM) (Ungerboeck (1982)), Deng and Costello (1989) proposed an alternative concatenation scheme, which involves using a TCM scheme as an inner code and an RS code as an outer code. The inner code comprises a convolutional code combined with a higher order modulation scheme such as 8-PSK. This concatenated code can be decoded in two different ways. First, the inner code is decoded using a soft-decision Viterbi algorithm (VA) that provides hard decisions to the outer RS decoder. Second, the VA is modified such that it provides, in addition to hard decisions, reliability information to the outer decoder in the form of symbol erasures. The RS decoder is then modified in such a way that it uses the reliability information as side information to make better decisions. The latter decoding technique achieves better performance than the former one.

Another concatenation scheme which involves serial concatenation of two convolutional codes separated by a pseudo-random interleaver was proposed by Hagenauer and Hoeher (1989). The inner code is decoded using a soft-output VA (SOVA) that provides to the outer Viterbi decoder soft information on the received sequence. The function of the interleaver is to separate the bursts of errors produced by the inner decoder in order to help the outer decoder to have a better error correction capability.

The discovery of turbo codes by Berrou et al. (1993) marks one of the most important breakthroughs in the history of coding theory, and changed the way concatenated codes are exploited to achieve further improvements. Turbo codes represent a different way of concatenating two simple codes to obtain an overall powerful code. A typical turbo code comprises two parallel concatenated convolutional codes separated by an interleaver, and is decoded using iterative decoding techniques. Since their invention, turbo codes have generated an abundance of literature, mainly because of their exceptional performance for very low signal-to-noise ratios. To give a simple example, for a rate $1/2$ turbo code with an interleaver of size 65,536 and memory-4 component codes, it was demonstrated by Berrou et al. (1993) that a bit error probability of 10^{-5} at a signal-to-noise ratio per bit of 0.7 dB over an AWGN channel is possible. For an AWGN channel, the capacity is equal to the transmission rate, i.e., $1/2$, when the signal-to-noise ratio is 0 dB. Therefore, if we consider 10^{-5} probability of error as the goal, the performance is only 0.7 dB away from the channel capacity, which is remarkable.

Inspired by turbo codes, a new concatenation scheme was proposed by Benedetto et al. (1998) which involves the serial concatenation of two convolutional codes separated by an interleaver and decoded using iterative decoding techniques. These codes were introduced as alternatives to turbo codes as, for some applications, they are less complex with performance comparable to or better than that of turbo codes. Another concatenated coding scheme called hybrid concatenated codes has been introduced by Divsalar and Pollara (2000). A hybrid concatenated code is a parallel concatenation of a serial concatenated code and a single convolutional code separated by two interleavers, one for the serial concatenated code and the other for the combined code.

Another major development in capacity achieving coding has been the resurrection of low-density parity check (LDPC) codes. These codes were first proposed by Gallager (1962). They were generalized by Tanner (1981) where a graph representation for these codes is also proposed. Since their introduction, LDPC codes did not receive much attention from the coding community until MacKay and others (MacKay and Neal (1995), MacKay (1999), Alon and Luby (1996)) showed their great potential. In general, LDPC codes are a class of block codes generated using large, sparse (low-density) generator matrices, resulting in a large minimum distance. They achieve a near-capacity performance, and they are decoded via the so-called message passing or belief propagation algorithm.

Most of the above concatenated coding schemes and their variants have been developed and optimized for AWGN channels. They have also been successfully applied to wireless communication systems. In fact, many coding schemes have been introduced in the literature which involve the concatenation of classical channel codes, such as turbo and convolutional codes, with space-time codes. Such code concatenation schemes have been shown to be very effective in terms of providing performance improvement, and they motivate the material in this chapter.

7.2 Concatenated Codes for AWGN Channels

In this section, we consider parallel and serial concatenated codes for AWGN channels. In particular, we introduce the encoder and iterative decoder structures for such codes, and present their performance analysis.

7.2.1 Encoder Structures

The idea in turbo coding is to concatenate two recursive systematic convolutional (RSC) codes in parallel via an interleaver, as shown in Figure 7.1. Each RSC encoder is normally described by two polynomials, namely $(g_1(D), g_2(D))$, where $g_1(D)$ is the feedforward polynomial and $g_2(D)$ is the feedback polynomial. We henceforth refer to this scheme as a parallel concatenated convolutional code (PCCC). The encoding process proceeds as follows. The information sequence at the input of the turbo encoder is divided into blocks of a certain length, and these blocks are encoded sequentially. The input to the first encoder is an information block, whereas the input of the second encoder is an interleaved version of this information block. The encoded sequence (codeword) corresponding to that information block is then the information block itself, the first parity block and the second parity block. Thus, the natural code rate of a turbo code is $1/3$. When higher code rates are desired, puncturing of parity bits may be used, as shown in the figure where the punctured bits are omitted, i.e., not transmitted.

To obtain a good performance, the component codes of the PCCC should be chosen as convolutional codes with feedback (hence, the term recursive), with the feedback polynomial being primitive (Benedetto et al. (1998)). On the other hand, these codes can be non-systematic. However, if the component codes are non-systematic, the overall code rate

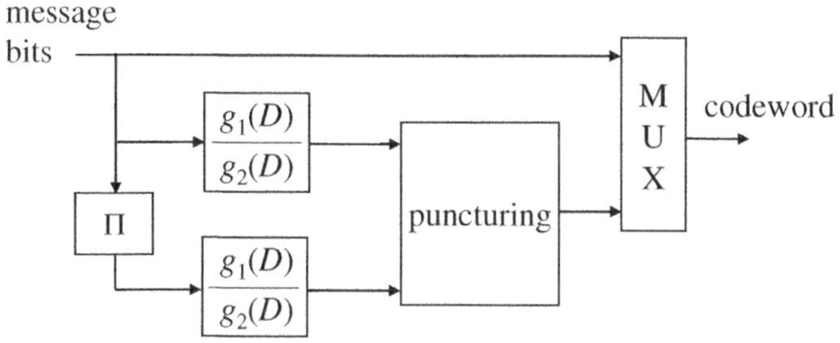

Figure 7.1 PCCC encoder structure.

will be smaller as the non-systematic information used in place of the systematic part should be transmitted twice, once for each component code. It is also a common practice to choose identical component encoders. One can argue that any recursive convolutional code has an equivalent feedback-free convolutional encoder, and thus there is no need to employ recursive component encoders. This is not true in the context of turbo codes, that is, although the codewords generated by the two encoders are the same, they correspond to different input sequences. Therefore, for our purposes, the recursive encoder and its feedback-free version are not equivalent. The other ingredient of the PCCC scheme, namely the interleaver, can be chosen pseudo-randomly to obtain a good performance. The performance of the PCCC improves when the interleaver length is increased, but that comes at the expense of increased delay. For more on the interleaver design, the reader may refer to Duman (2002).

There are some obvious generalizations of the standard turbo coding scheme. The most prominent generalization is the serial concatenated convolutional code (SCCC), which comprises of two serial concatenated convolutional codes separated by an interleaver. The SCCC encoder is shown in Figure 7.2. In the figure, the generator polynomials of the outer RSC encoder are $\left(g_1^o(D), g_2^o(D)\right)$, and the polynomials of the inner RSC encoder are $\left(g_1^i(D), g_2^i(D)\right)$. It is required that the inner RSC code be recursive in order to obtain an interleaver gain. As for the outer RSC code, although it need not be recursive, it may be convenient to make it recursive. The overall rate of the SCCC code is $R_c = R_o \cdot R_i$, where R_o is the rate of the outer code and R_i is the rate of the inner code. Often, the inner code is selected as a simple differential code with polynomials $g_1^i(D) = 1$, $g_2^i(D) = 1 \oplus D$, and thus $R_i = 1$.

Let us now give a brief explanation of why concatenated codes perform so well. This argument applies to PCCCs; a similar argument holds for SCCCs. PCCCs are linear block codes. Therefore, for the purposes of analysis, we can simply assume that the all-zero codeword is transmitted. The possible error sequences corresponding to this transmission are all the non-zero codewords. Consider a codeword with information weight one, that is, a codeword obtained by encoding a weight one information sequence. Since the component encoders are recursive, the parity sequences corresponding to this information sequence will not terminate until the end of the block is reached because it is the impulse response of an infinite impulse response (IIR) filter. With a good selection of the interleaver, if the single "1" occurs towards the end of the information block for one of the component encoders, it will likely occur towards the beginning of the input block for the other component encoder. Therefore, the codewords with information weight one, typically, have large parity weights, hence large total weights. The reader may refer to Ryan (2003) for more detailed discussion on this point.

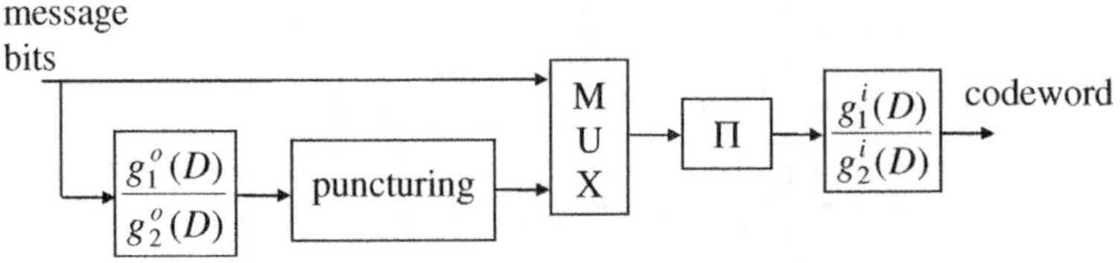

Figure 7.2 A typical SCCC encoder structure.

Furthermore, the interleaver "breaks down" sequences with information weight greater than one, that is, if the information block results in a lower weight parity sequence corresponding to one of the encoders, it will have, with a high probability, a larger weight parity sequence corresponding to the other encoder. Therefore, most of the codewords will have large Hamming weights and they are less likely to be decoded as the correct codeword when the all-zero codeword is transmitted, at least for AWGN channels. On the other hand, the interleaver cannot break down all the sequences, and therefore, there will be some codewords with low weights as well, hence the overall free distance of turbo codes is generally similar to or even smaller than that of single convolutional codes. Since the number of low-weight codewords is typically small, although the asymptotic performance of the code (for large signal-to-noise ratios) is limited by its relatively low free distance, its performance is very good for low signal-to-noise ratios. We also note that the most troublesome error sequences for a turbo code are the ones with information weight two, since those are the most difficult ones for the interleaver to "break down." The reader may refer to the paper by Perez et al. (1996) in which sparseness of the distance spectrum of turbo codes is demonstrated.

We finally remark here that terminating the trellis of a recursive convolutional code (that is, bringing the state of the encoder to the all-zero state) is not possible by appending a number of zeros at the end of the information sequence. Instead, a non-zero sequence (depending on the current state of the encoder) should be appended. For a turbo code, however, due to the presence of the interleaver and also the fact that the states of the two component encoders are in general different from each other, terminating the trellis of both encoders at the same time is not straightforward. What is normally done for PCCCs is that the top encoder starts and ends at the zero state, whereas the bottom encoder starts at the zero state but it does not end at the zero state. For SCCCs, usually the outer encoder is terminated but not the inner encoder. The penalty for not terminating the trellis of one of the encoders is a degradation in performance by a fraction of a dB (see Barbulescu and Pietrobon (1995); Blackert et al. (1995); Divsalar and Pollara (1995); Joerssen and Meyr (1994)).

7.2.2 Iterative Decoder Structures

Performing ML decoding of concatenated codes is very difficult mainly because of the presence of the interleaver. In general, one has to consider all the possible codewords (there are 2^N possibilities where N is the interleaver size), compute the cost of decoding each one and take the one with the lowest cost as the transmitted codeword. Even for short interleavers, this is a tedious task. As an alternative, suboptimal iterative decoding algorithms offer performance that is very close to that of the maximum likelihood at much reduced complexity. The iterative decoder is normally comprised of component decoders that work together iteratively and cooperatively. The number of component decoders should match the number of component encoders. When turbo codes were first introduced, it was proposed to use the *a posteriori* probability (APP) algorithm for each component decoder since this algorithm is optimal in the sense of minimizing the bit (or symbol) error rate (for each component code).

The APP algorithm was introduced in McAdam et al. (1972), and also independently in Bahl et al. (1974). An algorithm that is similar in concept was also introduced in Chang and Hancock (1966). The APP algorithm is also known as the maximum *a posteriori* probability (MAP) algorithm, or the BCJR algorithm. Right after the discovery of turbo codes, the APP algorithm was modified by Hagenauer et al. (1996) to suit the iterative decoding structure of concatenated codes. However, the original APP suffers from stability problems, especially at relatively high signal-to-noise ratios. To avoid such problems, it was modified to operate in the logarithm domain, where the modified version is usually referred to as the log-APP algorithm. These decoders, as well as any other decoder that accepts soft information at its input and produces soft information at its output, are referred to as SISO decoders.

In the following subsections, we introduce the APP algorithm in the probability domain and then convert it to operate in the logarithm domain. In this development, we assume a single code, that is, no iterative decoding. We then modify the log-APP algorithm to suit iterative decoding procedure, making it useful for concatenated codes including PCCCs and SCCCs.

The APP Algorithm

Consider a rate $1/2$ RSC code on an AWGN channel. Let $\boldsymbol{b} = [b_1, b_2, \ldots, b_N]$ denote an information sequence of length N at the input of the RSC encoder, and let $\boldsymbol{p} = [p_1, p_2, \ldots, p_N]$ be the corresponding parity sequence. The corresponding codeword, denoted by $\boldsymbol{x}$, is formed by multiplexing $\boldsymbol{b}$ and $\boldsymbol{p}$, that is, $\boldsymbol{x} = [x_1^b, x_1^p, \ldots, x_N^b, x_N^p]$. Since we are assuming BPSK modulation, $b_k, p_k \in \{\pm 1\}$. The noisy received sequence is $\boldsymbol{y} = \sqrt{\rho}\boldsymbol{x} + \boldsymbol{n}$ where $\boldsymbol{y} = [y_1^b, y_1^p, \ldots, y_N^b, y_N^p]$, and $\boldsymbol{n} = [n_1^b, n_1^p, \ldots, n_N^b, n_N^p]$ is a sequence of i.i.d. AWGN samples of length $2N$.

In the symbol-by-symbol APP decoder, the decoder decides $b_k = +1$ if $P(b_k = +1|\boldsymbol{y}) > P(b_k = -1|\boldsymbol{y})$, and it decides $b_k = -1$ otherwise. In the logarithm domain, the decision $\hat{b}_k$ is given by $\hat{b}_k = \text{sign}\,[L(b_k)]$ where $L(b_k)$ is the log *a posteriori* probability (LAPP) ratio defined as

$$L(b_k) \triangleq \log\left[\frac{P(b_k = +1|\boldsymbol{y})}{P(b_k = -1|\boldsymbol{y})}\right]. \tag{7.1}$$

When the inputs are equally likely, we may use Bayes' rule to express (7.1) as

$$L(b_k) = \log\left[\frac{p(\boldsymbol{y}|b_k = +1)}{p(\boldsymbol{y}|b_k = -1)}\right] + \log\left[\frac{P(b_k = +1)}{P(b_k = -1)}\right]. \tag{7.2}$$

The second term in (7.2) represents the *a priori* information of bit b_k, and it is zero since the inputs are assumed to be equally likely. As such, instead of using the APP algorithm, one may also use the ML decoding algorithm implemented via the VA, which is much less complex than the APP algorithm. For single codes over AWGN channels, both algorithms achieve almost identical performances, although the VA minimizes the sequence error probability and the APP algorithm minimizes the bit (or symbol) error rate. In iterative decoding, however, this *a priori* term is usually non-zero and it represents the soft (extrinsic) information that the component decoders pass to each other.

In our derivation of the APP algorithm, we make no assumptions regarding the probability distribution of the inputs. To this end, we express (7.1) as

$$L(b_k) = \log\left[\frac{P(b_k = +1, \boldsymbol{y})/P(\boldsymbol{y})}{P(b_k = -1, \boldsymbol{y})/P(\boldsymbol{y})}\right], \tag{7.3}$$

$$= \log\left[\frac{P(b_k = +1, \boldsymbol{y})}{P(b_k = -1, \boldsymbol{y})}\right]. \tag{7.4}$$

Computing (7.4) requires incorporating the trellis of the RSC encoder. For example, to evaluate the numerator, we need to consider all state transitions resulting from input $b_k = +1$; the same argument holds for the denominator. Now define T^+ to be the set of pairs (s', s) for the state transitions $(s_{k-1} = s', s_k = s)$ corresponding to $b_k = +1$, where s_k is the encoder state at time k; T^- is similarly defined. As such, (7.4) can be expressed as

$$L(b_k) = \log \frac{\sum_{T^+} p(s_{k-1} = s', s_k = s, \mathbf{y})}{\sum_{T^-} p(s_{k-1} = s', s_k = s, \mathbf{y})}. \tag{7.5}$$

Clearly $p(s_{k-1} = s', s_k = s, \mathbf{y})$ can be expressed as (Bahl et al. (1974))

$$p(s_{k-1} = s', s_k = s, \mathbf{y}) = \tilde{\alpha}_{k-1}(s') \cdot \tilde{\gamma}_k(s', s) \cdot \tilde{\beta}_k(s), \tag{7.6}$$

where

$$\tilde{\alpha}_k(s) \triangleq P(s_k = s, \mathbf{y}_1^k), \tag{7.7}$$

$$\tilde{\gamma}_k(s', s) \triangleq P(s_k = s, y_k | s_{k-1} = s'), \tag{7.8}$$

$$\tilde{\beta}_k(s) \triangleq P(\mathbf{y}_{k+1}^N | s_k = s), \tag{7.9}$$

where $\mathbf{y}_v^w \triangleq [y_v, y_{v+1}, \ldots, y_w]$. It is important to note that $\mathbf{y}_1^k$ in (7.7) corresponds to $(\mathbf{b}_1^k, \mathbf{p}_1^k)$, y_k in (7.8) corresponds to the pair (b_k, p_k), and $\mathbf{y}_{k+1}^N$ in (7.9) corresponds to the pair of sequences $(\mathbf{b}_{k+1}^N, \mathbf{p}_{k+1}^N)$.

To understand how the APP algorithm works, it may be helpful to relate it to something we already know, which is the VA. The APP algorithm can be simply viewed as two VAs: forward VA and backward VA. In this context, the quantities $\tilde{\alpha}_{k-1}(s')$ and $\tilde{\beta}_k(s)$ represent, respectively, the forward recursion used by the forward algorithm and the backward recursion used by the backward algorithm. As for the quantity $\tilde{\gamma}_k(s', s)$, it represents the state transition between $s' \rightarrow s$, which serves as a bridge between the forward and backward algorithms. In the context of the VA, $\tilde{\gamma}_k(s', s)$ may be called the branch metric corresponding to state transition $s' \rightarrow s$.

In general, the forward recursion $\tilde{\alpha}_k(s)$, known as the cumulative metric, can be expressed as

$$\tilde{\alpha}_k(s) = \sum_{s'} \tilde{\gamma}_k(s', s) \cdot \tilde{\alpha}_{k-1}(s'), \tag{7.10}$$

where the index of the sum runs over all valid state transitions $s' \rightarrow s$. The state transition

$$\tilde{\gamma}_k(s', s) = P(b_k) \cdot p(y_k | b_k), \tag{7.11}$$

can be simplified, for AWGN channels, as

$$\tilde{\gamma}_k(s', s) = \begin{cases} \frac{P(b_k)}{\pi} \exp\left(-\left(y_{2k-1} - \sqrt{\rho} b_k\right)^2 - \left(y_{2k} - \sqrt{\rho} p_k\right)^2\right) & \text{for valid } s' \rightarrow s \\ 0, & \text{for invalid } s' \rightarrow s. \end{cases} \tag{7.12}$$

The backward recursion, $\tilde{\beta}_k(s)$, can be expressed as

$$\tilde{\beta}_{k-1}(s') = \sum_{s} \tilde{\beta}_k(s) \cdot \tilde{\gamma}_k(s', s). \tag{7.13}$$

The forward and backward recursions are initialized as

$$\tilde{\alpha}_0(s) = \begin{cases} 1, & s = 0 \\ 0, & s \neq 0, \end{cases} \tag{7.14}$$

and

$$\tilde{\beta}_N(s) = \begin{cases} 1, & s = 0 \\ 0, & s \neq 0, \end{cases} \tag{7.15}$$

respectively. In (7.14), it is assumed that the RSC encoder starts at the zero state, whereas in (7.15) it is assumed that the RSC encoder ends at the zero state. The latter is achieved by appending the information sequence at the input of the encoder with a particular bit pattern to force the encoder to end at the zero state.

Once $\tilde{\alpha}_k(s)$, $\tilde{\gamma}_k(s', s)$ and $\tilde{\beta}_k(s)$ are computed for all k, one can easily compute $L(b_k)$ defined by (7.5) to obtain hard decisions on b_k using $\hat{b}_k = \text{sign}\,[L(b_k)]$.

The log-APP Algorithm

So far we have presented the APP algorithm in the probability domain. We now convert it to the logarithm domain. To this end, the log-domain forward recursion becomes

$$\begin{aligned} \alpha_k(s) &= \log(\tilde{\alpha}_k(s)) \\ &= \log\left(\sum_{s'} \exp\left(\gamma_k(s', s) + \alpha_{k-1}(s')\right)\right) \end{aligned} \tag{7.16}$$

where

$$\begin{aligned} \gamma_k(s', s) &= \log\left(\tilde{\gamma}_k(s', s)\right) \\ &= \log\left(\frac{P(b_k)}{\pi}\right) - \left(\left(y_k^b - \sqrt{\rho} b_k\right)^2 + \left(y_k^p - \sqrt{\rho} p_k\right)^2\right). \end{aligned} \tag{7.17}$$

The first term can be dropped because it does not affect the calculations afterwards since it is a constant and it appears in all valid state transitions.

Similarly, the log-domain backward recursion is given as

$$\begin{aligned} \beta_k(s) &= \log\left(\tilde{\beta}_k(s)\right) \\ &= \log\left(\sum_{s} \exp\left(\beta_k(s) + \gamma_k(s', s)\right)\right). \end{aligned}$$

As for the initial conditions for these new log-domain recursions, they are modified as

$$\alpha_0(s) = \begin{cases} 0, & s = 0 \\ -\infty, & s \neq 0 \end{cases}$$

and

$$\beta_N(s) = \begin{cases} 0, & s = 0 \\ -\infty, & s \neq 0. \end{cases}$$

We now write $L(b_k)$ defined by (7.5) using the log-domain expressions as

$$\begin{aligned} L(b_k) &= \log \frac{\sum_{T^+} \tilde{\alpha}_{k-1}(s') \cdot \tilde{\gamma}_k(s', s) \cdot \tilde{\beta}_k(s)}{\sum_{T^-} \tilde{\alpha}_{k-1}(s') \cdot \tilde{\gamma}_k(s', s) \cdot \tilde{\beta}_k(s)}, \\ &= \log\left(\sum_{T^+} \exp\left(\alpha_{k-1}(s') + \gamma_k(s', s) + \beta_k(s)\right)\right) \\ &\quad - \log\left(\sum_{T^-} \exp\left(\alpha_{k-1}(s') + \gamma_k(s', s) + \beta_k(s)\right)\right). \end{aligned} \tag{7.18}$$

The expression in (7.18) can be simplified further via the following result

$$\max(v, w) = \log\left(\frac{e^v + e^w}{1 + e^{-|v-w|}}\right),$$

which can be applied to more than two inputs by grouping them in pairs. For instance, $\max(u, v, w) = \max(u, \max(v, w))$.

Now define

$$\max{}^*(v, w) \triangleq \log\left(e^v + e^w\right).$$

Consequently,

$$\max{}^*(v, w) = \max(v, w) + \log\left(1 + e^{-|v-w|}\right). \tag{7.19}$$

The second term in (7.19) is normally referred to as *correction term*. In practice, to reduce the complexity arising from computing (7.19), the correction term is approximated by using a lookup table. The lookup table is generated by quantizing the range of values of the function $\log\left(1 + e^{-|v-w|}\right)$, i.e., $[0, 0.693]$, into a certain (finite) number of levels. It has been shown by Robertson et al. (1995) that using a lookup table of size eight should be sufficient to get a good approximation. The resulting performance degradation due to this approximation is about a few tenths of a dB.

Using (7.19), we can express the log-domain recursions as

$$\alpha_k(s) = \max_{s'}{}^*\left(\gamma_k(s', s) + \alpha_{k-1}(s')\right), \tag{7.20}$$

$$\beta_{k-1}(s') = \max_{s}{}^*\left(\gamma_k(s', s) + \beta_k(s)\right), \tag{7.21}$$

and

$$\begin{aligned} L(b_k) &= \max_{T^+}{}^*\left(\alpha_{k-1}(s') + \gamma_k(s', s) + \beta_k(s)\right) \\ &\quad - \max_{T^-}{}^*\left(\alpha_{k-1}(s') + \gamma_k(s', s) + \beta_k(s)\right). \end{aligned} \tag{7.22}$$

Similar to the case in the previous section, once $L(b_k)$ is computed via (7.22) for $k = 1, 2, \ldots, N$, the decisions on b_k are obtained using $\hat{b}_k = \text{sign}\,[L(b_k)]$.

PCCC Iterative Decoder Structure

In this section, we describe a suitable iterative decoder structure for the PCCC system using the log-APP algorithm as a building block. In this case, we need two log-APP decoders that are matched to the two RSC component encoders. The two decoders work together iteratively. Detailed block diagrams of the PCCC encoder and iterative decoder are shown in Figures 7.3 and 7.4, respectively. In the figures, to simplify the notation, we refer to the top RSC encoder by E_1 and to the bottom one by E_2. Similarly, we refer to the decoder matched to E_1 by D_1, and to the decoder matched to E_2 by D_2.

Let $\boldsymbol{b} = [b_1, b_2, \ldots, b_N]$ denote the information sequence of length N at the input of E_1 and let $\boldsymbol{b}' = [b'_1, b'_2, \ldots, b'_N]$ be the input to E_2 which is an interleaved version of $\boldsymbol{b}$. Also, let $\boldsymbol{p} = [p_1, p_2, \ldots, p_N]$ be the parity sequence at the output of E_1 and $\boldsymbol{q} = [q_1, q_2, \ldots, q_N]$ be the parity sequence at the output of E_2. The codeword corresponding to $\boldsymbol{b}$, denoted by $\boldsymbol{c}$, is formed by multiplexing $\boldsymbol{b}$, $\boldsymbol{p}$, and $\boldsymbol{q}$, which will be in the form $\boldsymbol{x} = [x_1^b, x_1^p, x_1^q, , \ldots, x_N^b, x_N^p, x_N^q]$. We denote by $\boldsymbol{y}$ the received noisy sequence which is simply $\boldsymbol{y} = \sqrt{\rho}\boldsymbol{x} + \boldsymbol{n}$ where $\boldsymbol{n}$ is a sequence of i.i.d. AWGN samples of length $3N$. Thus, the kth element of $\boldsymbol{y}$ is defined as $y_k \triangleq [y_k^b, y_k^p, y_k^q]$. The assumption here is that the overall code rate is 1/3, i.e., no puncturing is performed. When puncturing is used, the punctured bits are not transmitted and the corresponding received signals (or log-likelihoods) are replaced by zeros in the decoding process.

When the inputs are equally likely, the first term in (7.17) disappears when $\gamma_k(s', s)$ is calculated in (7.18) because the same term appears in both terms of (7.18). However, in iterative decoding, this is no longer true because this is essentially the extrinsic information that decoders exchange between each other back and forth. Accordingly, $L(b_k)$ is modified to incorporate this change as

$$L(b_k) = L^e(b_k) + \max_{T^+} {}^* \left(\alpha_{k-1}(s') + 2\sqrt{\rho} b_k y_k^b + 2\sqrt{\rho} p_k y_k^p + \beta_k(s)\right)$$
$$- \max_{T^-} {}^* \left(\alpha_{k-1}(s') + 2\sqrt{\rho} b_k y_k^b + 2\sqrt{\rho} p_k y_k^p + \beta_k(s)\right),$$

where

$$L^e(b_k) = \log \left[\frac{P(b_k = +1)}{P(b_k = -1)}\right].$$

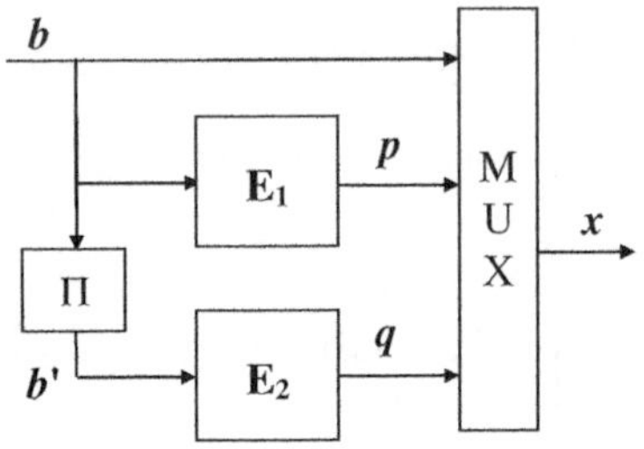

Figure 7.3 Rate 1/3 PCCC encoder structure.

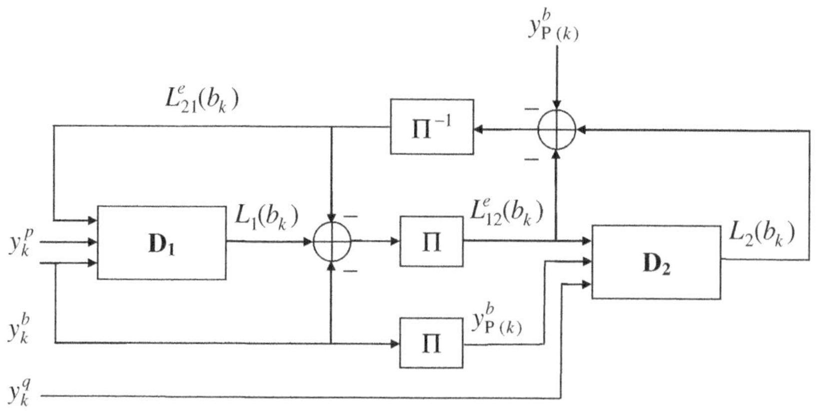

Figure 7.4 PCCC iterative decoder structure.

Since $b_k = +1$ in the term corresponding to T^+ and it is -1 for the term corresponding to T^-, $L(b_k)$ can be further simplified as

$$L(b_k) = 4\sqrt{\rho}y_k^b + L^e(b_k) + \max_{T^+} {}^* \left(\alpha_{k-1}(s') + 2\sqrt{\rho}p_k y_k^p + \beta_k(s)\right) - \max_{T^-} {}^* \left(\alpha_{k-1}(s') + 2\sqrt{\rho}p_k y_k^p + \beta_k(s)\right). \quad (7.23)$$

The first term in (7.23) is the channel value, the second term is the extrinsic information on bit b_k received from the companion decoder, and the last two terms combined correspond to the extrinsic information on bit b_k to be passed to the companion decoder. Note that the last two terms are not functions of the channel value or $L^e(b_k)$. Therefore, this information is used by the other decoder as *a priori* information.

From the foregoing discussion, as shown in Figure 7.4, we can write the output of D_1 as

$$L_1(b_k) = 4\sqrt{\rho}y_k^b + L_{21}^e(b_k) + L_{12}^e(b_k) \quad (7.24)$$

where $L_{21}^e(b_k)$ is extrinsic information received from D_2, and $L_{12}^e(b_k)$ is extrinsic information to be passed to D_2. The output of D_2 is similarly defined.

We now give an outline for the PCCC iterative decoder.

Initialization

- $\alpha_0^{(1)}(s) = 0$ for $s = 0$, and $\alpha_0^{(1)}(s) = -\infty$ for $s \neq 0$ (for D_1)
- $\beta_N^{(1)}(s) = 0$ for $s = 0$, and $\beta_N^{(1)}(s) = -\infty$ for $s \neq 0$ (for D_1)
- $\alpha_0^{(2)}(s) = 0$ for $s = 0$, and $\alpha_0^{(2)}(s) = -\infty$ for $s \neq 0$ (for D_2)
- $\beta_N^{(2)}(s) = \alpha_N^{(2)}(s)$ for $\forall\, s$ (for D_2)
- $L_{21}^e(b_k) = 0$ for $\forall\, b_k : k = 1, 2, \ldots, N$

The ith iteration: D$_1$

for $k = 1 : N$

obtain $y_k = [y_k^b, y_k^p]$ and $L_{21}^e\left(b_{\Pi^{-1}(k)}\right)$.

compute $\gamma_k\left(s', s\right)$ for all valid transitions $s' \rightarrow s$ using

$$\gamma_k\left(s', s\right) = b_k L_{21}^e(b_{\Pi^{-1}(k)})/2 + 2\sqrt{\rho} b_k y_k^b + 2\sqrt{\rho} p_k y_k^p$$

compute $\alpha_k^{(1)}(s)$ from (7.20) for all s.

end

for $k = N : -1 : 2$

compute $\beta_{k-1}^{(1)}(s)$ from equation (7.21) for all s.

end

for $k = 1 : N$

set numer $= -\infty$ and denom $= -\infty$

for all transitions $s' \rightarrow s$

if $\left(b_k\left(s', s\right) = +1\right)$

$$\text{numer} = \max{}^*\left[\text{numer}, \alpha_{k-1}^{(1)}(s) + \gamma_k\left(s, s'\right) + \beta_k^{(1)}\left(s'\right)\right]$$

else if $\left(b_k\left(s', s\right) = -1\right)$

$$\text{denom} = \max{}^*\left[\text{denom}, \alpha_{k-1}^{(1)}(s) + \gamma_k\left(s, s'\right) + \beta_k^{(1)}\left(s'\right)\right]$$

end

$L_{12}^e(b_k) = \text{numer} - \text{denom} - L_{21}^e\left(b_{\Pi^{-1}(k)}\right) - 4\sqrt{\rho} y_k^b$

end

The ith iteration: D$_2$

for $k = 1 : N$

obtain $y_k = [y_{\Pi(k)}^b, y_k^q] = [y_k^{b'}, y_k^q]$ and $L_{12}^e\left(b_{\Pi(k)}\right)$.

compute $\gamma_k\left(s', s\right)$ for all valid transitions $s' \rightarrow s$ using

$$\gamma_k\left(s', s\right) = b_k' L_{12}^e(b_{\Pi(k)})/2 + 2\sqrt{\rho} b_k' y_{\Pi(k)}^b + 2\sqrt{\rho} q_k y_k^p$$

compute $\alpha_k^2(s)$ from (7.20) for all s.

end

for $k = N : -1 : 2$

compute $\beta_{k-1}^{(2)}(s)$ from equation (7.21) for all s.

end

for $k = 1 : N$

set numer $= -\infty$ and denom $= -\infty$

for all transitions $s' \rightarrow s$

if $\left(b_k\left(s', s\right) = +1\right)$

$$\text{numer} = \max{}^*\left[\text{numer}, \alpha_{k-1}^{(2)}(s) + \gamma_k\left(s, s'\right) + \beta_k^{(2)}\left(s'\right)\right]$$

else if $\left(b_k\left(s', s\right) = -1\right)$

$$\text{denom} = \max{}^*\left[\text{denom}, \alpha_{k-1}^{(2)}(s) + \gamma_k\left(s, s'\right) + \beta_k^{(2)}\left(s'\right)\right]$$

end

$L_{21}^e\left(b_k'\right) = \text{numer} - \text{denom} - L_{12}^e\left(b_{\Pi(k)}\right) - 4\sqrt{\rho} y_{\Pi(k)}^b$

end

After the last iteration: making final decisions

for $k = 1 : N$

compute $L_1(b_k)$ using

$$L_1(b_k) = 4\sqrt{\rho}y_k^b + L_{12}^e(b_k) + L_{21}^e(b_{\Pi^{-1}(k)})$$

make final decisions on b_k using

$$\hat{b}_k = \text{sign}(L_1(b_k))$$

end.

SCCC Iterative Decoder Structure

As mentioned before, the SCCC iterative decoder comprises two log-APP decoders matched to the two RSC encoders. Detailed diagrams of the SCCC encoder and decoder are shown in Figures 7.5 and 7.6, respectively. As shown in the figures, the outer and inner RSC encoders are denoted by E_1 and E_2, respectively. The corresponding outer and inner log-APP decoders are denoted by D_1 and D_2, respectively.

For illustration purposes, both RSC encoders are assumed to be of rate 1/2, resulting in an overall rate of 1/4. To achieve higher code rates, puncturing may be used where the punctured bits are not transmitted and are replaced by zeros in the decoding process. Let the input to E_1 be $\boldsymbol{b} = [b_1, b_2, \ldots, b_N]$ and the corresponding output be $\boldsymbol{c} = [c_1^b, c_1^p, \ldots, c_N^b, c_N^p]$. The input to E_2 is denoted by $\mathbf{u}$, which is an interleaved version of $\boldsymbol{c}$, and whose corresponding output is $\boldsymbol{x} = [x_1^u, x_1^q, \ldots, x_{2N}^u, x_{2N}^q]$. Let $\boldsymbol{y} = \sqrt{\rho}\boldsymbol{x} + \boldsymbol{n}$ be the received noisy sequence which is in the form $\boldsymbol{y} = [y_1^u, y_1^q, \ldots, y_{2N}^u, y_{2N}^q]$. Note that $\boldsymbol{n}$ is a sequence of i.i.d. AWGN samples of length $4N$. Since the inner and outer encoders might have different numbers of states, we let M_1 denote the set of states corresponding to E_1 and M_2 be the set of states corresponding to E_2.

As shown in Figure 7.6, D_2 is identical to any of the decoders discussed for the PCCC case (shown in Figure 7.4) where it receives three inputs at any given time, which are the channel value y_k^u, the corresponding parity y_k^q and the extrinsic information received from D_1, that is $L_{12}^e(u_k)$. In turn, this decoder produces LLR values on the bits u_k. As for D_1, it is slightly different in the sense that it receives all its inputs from D_2, i.e., it does not have access to the channel values. To account for this difference, the state transition metric is calculated as

$$\gamma_k(s', s) = b_k L_{21}^e(b_k)/2 + p_k L_{21}^e(p_k)/2.$$

In addition, unlike in the PCCC case, D_1 must produce LLR values for both of the information bits b_k and their parity bits p_k so they can be used by D_2 as extrinsic information in the next iteration. Computing the LLRs for b_k is similar to what is done in the D_2 case. As for computing the LLRs for p_k, it can be accomplished by following the same procedure

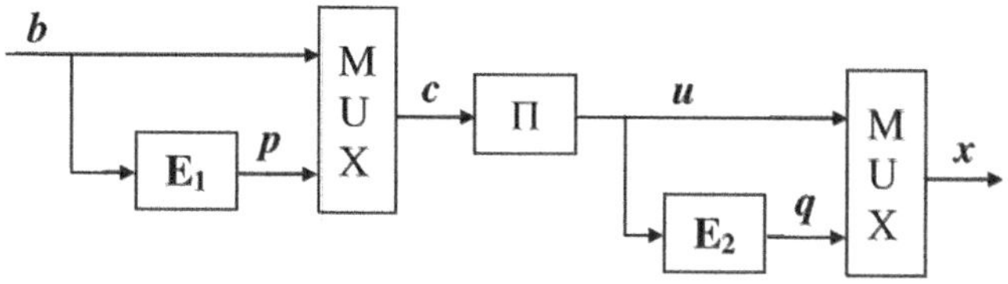

Figure 7.5 Rate 1/4 SCCC encoder structure.

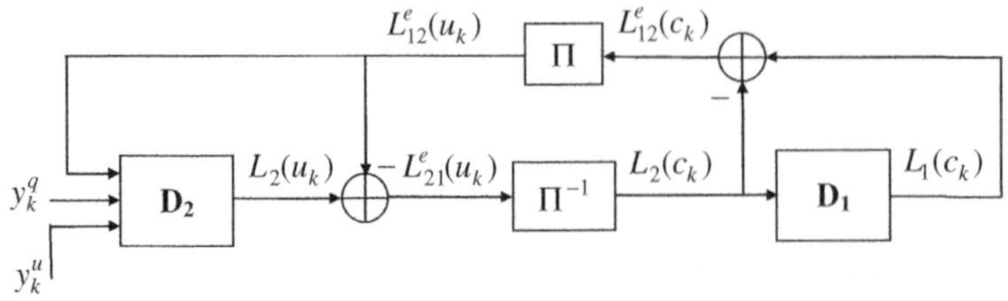

Figure 7.6 SCCC iterative decoder structure.

used to compute the LLRs for b_k. This is easy to do because the code is systematic, and thus the branch labels have both b_k and p_k appearing explicitly in them. With that being said, the LLRs for p_k can be computed as

$$L(p_k) = \max_{P^+} * \left(\alpha_{k-1}(s') + \gamma_k(s', s) + \beta_k(s)\right) - \max_{P^-} * \left(\alpha_{k-1}(s') + \gamma_k(s', s) + \beta_k(s)\right) \tag{7.25}$$

where P^+ is the set of state transitions $(s' \to s)$ that correspond to $p_k = +1$, and P^- is similarly defined.

We now give an outline of the SCCC iterative decoder.

Initialization

- $\alpha_0^{(1)}(s) = 0$ for $s = 0$, and $\alpha_0^{(1)}(s) = -\infty$ for $s \neq 0$ (for D_1)
- $\beta_N^{(1)}(s) = 0$ for $s = 0$, and $\beta_N^{(1)}(s) = -\infty$ for $s \neq 0$ (for D_1)
- $\alpha_0^{(2)}(s) = 0$ for $s = 0$, and $\alpha_0^{(2)}(s) = -\infty$ for $s \neq 0$ (for D_2)
- $\beta_{2N}^{(2)}(s) = \alpha_{2N}^{(2)}(s)$ for $\forall\, s \in M_2$ (for D_2)
- $L_{12}^e(u_k) = 0$ for $\forall\, u_k : k = 1, 2, \ldots, 2N$

The *i*th iteration: $\mathbf{D}_2$

for $k = 1 : 2N$
    obtain $y_k = [y_k^u, y_k^q]$ and $L_{12}^e(u_k)$.
    compute $\gamma_k(s', s)$ for all valid transitions $s' \to s$ using
        $\gamma_k(s', s) = u_k L_{12}^e(u_k)/2 + 2\sqrt{\rho} u_k y_k^u + 2\sqrt{\rho} q_k y_k^q$
    compute $\alpha_k^{(2)}(s)$ from (7.20) for all $s \in M_2$.
end
for $k = 2N : -1 : 2$
    compute $\beta_{k-1}^{(2)}(s)$ from equation (7.21) for all $s \in M_2$.
end
for $k = 1 : 2N$
    set numer $= -\infty$ and denom $= -\infty$
    for all transitions $s' \to s$
        if $(u_k(s', s) = +1)$
            numer $= \max^*\left[\text{numer}, \alpha_{k-1}^{(2)}(s) + \gamma_k(s, s') + \beta_k^{(2)}(s')\right]$

else if $\left(u_k\left(s', s\right) = -1\right)$

$$\text{denom} = \max{}^*\left[\text{denom}, \alpha_{k-1}^{(2)}(s) + \gamma_k\left(s, s'\right) + \beta_k^{(2)}\left(s'\right)\right]$$

end

$L_{21}^e(u_k) = \text{numer} - \text{denom} - L_{12}^e(u_k)$

end

The $(i+1)$th iteration: $\mathbf{D}_1$

for $k = 1 : N$

obtain $L_{21}^e(u_k) = L_2(c_k)$ (see Figure 7.6)

compute $\gamma_k\left(s', s\right)$ for all valid transitions $s' \to s$ using

$\gamma_k\left(s', s\right) = b_k L_{21}^e(b_k)/2 + p_k L_{21}^e(p_k)/2$

compute $\alpha_k^{(1)}(s)$ from (7.20) for all $s \in M_1$.

end

for $k = N : -1 : 2$

compute $\beta_{k-1}^{(1)}(s)$ from equation (7.21) for all $s \in M_1$.

end

for $k = 1 : N$

// *we first compute* $L_{12}^e(b_k)$

set numer $= -\infty$ and denom $= -\infty$

for all transitions $s' \to s$

if $\left(b_k\left(s', s\right) = +1\right)$

$$\text{numer} = \max{}^*\left[\text{numer}, \alpha_{k-1}^{(1)}(s) + \gamma_k\left(s, s'\right) + \beta_k^{(1)}\left(s'\right)\right]$$

else if $\left(b_k\left(s', s\right) = -1\right)$

$$\text{denom} = \max{}^*\left[\text{denom}, \alpha_{k-1}^{(1)}(s) + \gamma_k\left(s, s'\right) + \beta_k^{(1)}\left(s'\right)\right]$$

end

$L_{12}^e(b_k) = \text{numer} - \text{denom} - L_{21}^e(b_k)$

// *we then compute* $L_{12}^e(p_k)$

set numer $= -\infty$ and denom $= -\infty$

for all transitions $s' \to s$

if $\left(p_k\left(s', s\right) = +1\right)$

$$\text{numer} = \max{}^*\left[\text{numer}, \alpha_{k-1}^{(1)}(s) + \gamma_k\left(s, s'\right) + \beta_k^{(1)}\left(s'\right)\right]$$

else if $\left(p_k\left(s', s\right) = -1\right)$

$$\text{denom} = \max{}^*\left[\text{denom}, \alpha_{k-1}^{(1)}(s) + \gamma_k\left(s, s'\right) + \beta_k^{(1)}\left(s'\right)\right]$$

end

$L_{12}^e(p_k) = \text{numer} - \text{denom} - L_{21}^e(p_k)$

end

After the last iteration:

for $k = 1 : N$

set numer $= -\infty$ and denom $= -\infty$

for all transitions $s' \to s$

if $\left(b_k\left(s', s\right) = +1\right)$

$$\text{numer} = \max{}^*\left[\text{numer}, \alpha_{k-1}^{(1)}(s) + \gamma_k\left(s, s'\right) + \beta_k^{(1)}\left(s'\right)\right]$$

else if $\left(b_k\left(s',s\right)=-1\right)$

$$\text{denom} = \max{}^*\left[\text{denom}, \alpha_{k-1}^{(1)}(s) + \gamma_k\left(s,s'\right) + \beta_k^{(1)}\left(s'\right)\right]$$

end

$L_1(b_k) = \text{numer} - \text{denom}$

make final decisions on b_k using

$\hat{b}_k = \text{sign}(L_1(b_k))$

end.

7.2.3 The SOVA Decoder

In this section, we present the SOVA algorithm as an alternative to the APP algorithm. This is motivated by the fact that the complexity of the SOVA algorithm is roughly half that of the APP algorithm. This advantage comes at the expense of a degradation in the performance, as compared with the APP algorithm, by about one dB over AWGN channels and about two dBs over fading channels. However, as we will show later in this section, most (and sometimes all) of this degradation can be recovered by applying simple modifications to the SOVA algorithm without introducing much additional complexity. In fact, these modifications can be applied to the iterative decoding algorithms using the APPs as component decoders in certain cases to achieve better performance.

In our derivation of the SOVA algorithm, we consider the PCCC scheme whose encoder and decoder are shown in Figures 7.3 and 7.4, respectively. We also use the same notation introduced in that section. Extension of the SOVA algorithm to the SCCC case is straightforward and thus it is not considered here.

In the standard Viterbi algorithm, the cumulative metric $m_k(s)$ for state s at time k is updated according to

$$m_s(k) = \max\left\{\lambda(s',s) + m_{k-1}(s'),\ \lambda(s'',s) + m_{k-1}(s'')\right\} \tag{7.26}$$

where $\lambda(s',s)$ is the branch metric for the transition from state s' to state s, and the other quantities are obviously defined. In iterative decoding, the branch metric, when incorporating the *a priori* information received from a companion decoder, is given by

$$\lambda(s',s) = \frac{1}{2}L^e(b_k)b_k + 2\sqrt{\rho}y_k^b b_k + 2\sqrt{\rho}y_k^p p_k \tag{7.27}$$

where $L^e(b_k)$ is the extrinsic information that could be either $L_{21}^e(b_k)$ or $L_{12}^e(b_k)$ depending on which decoder is being considered. The branch metric $\lambda(s'',s)$ is similarly defined. Now define the difference metric for state s at time k as

$$\Delta_k = \left|\left(m_{k-1}(s') + \lambda(s',s)\right) - \left(m_{k-1}(s'') + \lambda(s'',s)\right)\right|. \tag{7.28}$$

It can be shown that Δ_k can be well approximated by

$$\Delta_k \approx \log\frac{P(\text{correct})}{1-P(\text{correct})}, \tag{7.29}$$

where $P(\text{correct})$ is the probability that the path decision of the survivor at time k was correct. Therefore, Δ_k represents the reliability that the path ending at state s at time k was correct.

To obtain the soft output for bit b_k, we first obtain the hard decision $\hat{b}_k$ after a delay δ (i.e., at time $k+\delta$) , where δ is the *decoding depth.* At time $k+\delta$, we select the surviving path that ends at the state that has the highest metric, and the selected path is considered to be the ML path. We trace back the ML path to obtain the hard decision $\hat{b}_k$. Along the ML path, there are $\delta+1$ nonsurviving paths that have been discarded, and each nonsurviving path has a certain difference metric Δ_j where $k \leq j \leq k+\delta$. Clearly, this is because along the ML path there are $\delta+1$ states, and each state has a difference metric that was calculated using (7.28).

Now define

$$\Delta_k^* = \min\{\Delta_k,\ \Delta_{k+1},\ \ldots\ ,\ \Delta_{k+\delta}\}, \tag{7.30}$$

where the minimum is taken *only* over the nonsurviving paths within the time window $[k\ ,\ k+\delta]$ that would have led to a different decision $\hat{b}_k$. We note that Δ_k^* represents the reliability of the hard decision $\hat{b}_k$, and the reliability increases with increasing δ. Given $\hat{b}_k$ and Δ_k^*, the soft output for bit b_k is approximated by

$$L(b_k) \approx \hat{b}_k \cdot \Delta_k^*. \tag{7.31}$$

To obtain the best performance possible, $\hat{b}_k$ must be as reliable as possible. Since the two RSC encoders start at the zero state and the top encoder is terminated at the zero state, this setup should be exploited in the decoding process as follows. The SOVA decoders proceed decoding the usual way, i.e., choosing a survivor and calculating the metric as well as difference metric for each state for all $k \in \{1,\ 2,\ \ldots\ ,\ N\}$. We emphasize here that no hard decisions are made until $k = N$. When $k = N$, we select for D_1 the surviving path that ends at *state zero* as the ML path. For D_2, we select the surviving path that ends at the state that has the highest metric. We then trace back the ML path to obtain the hard decisions $\hat{b}_k$, and the corresponding Δ_k, for $k = 1, ..., N$. As for Δ_k^*, it can be obtained as described above. We stress the fact that obtaining hard decisions following the conventional way degrades the performance significantly, especially for high code rates, as shown by Ghrayeb and Ryan (1999) and Ghrayeb (2003).

From Figure 7.7, and expressions (7.27) and (7.28) it is easy to show that Δ_k^* has the following structure

$$\begin{aligned}\Delta_k^* = &\left(M_{j<k}^{(1)} - M_{j<k}^{(2)}\right) + \left(M_{k<j<k+i}^{(1)} - M_{k<j<k+i}^{(2)}\right)\\ &+2\sqrt{\rho}\, y_k^p\,(p_{k1} - p_{k2}) + L^e(b_k)\hat{b}_k + 4\sqrt{\rho}\, y_k^b \hat{b}_k,\end{aligned} \tag{7.32}$$

where p_{k1}, and p_{k2} are the parity bits corresponding to the state transitions at time k along the ML and nonsurviving paths, respectively, i is the time index at which $\Delta_k^* = \Delta_{k+i}$, and $M_{j<k}^{(1)}$ is the cumulative metric at time $k-1$ along the ML path. The other terms are similarly defined.

In light of the above arguments, we can rewrite (7.31) as

$$\begin{aligned}L(b_k) \approx\ &4\sqrt{\rho}\, y_k^b + L^e(b_k) + \left(M_{j<k}^{(1)} - M_{j<k}^{(2)}\right)\cdot \hat{b}_k\\ &+ \left(M_{k<j<k+i}^{(1)} - M_{k<j<k+i}^{(2)}\right)\cdot \hat{b}_k\\ &+2\sqrt{\rho}\, y_k^p\,(p_{k1} - p_{k2})\cdot \hat{b}_k,\end{aligned} \tag{7.33}$$

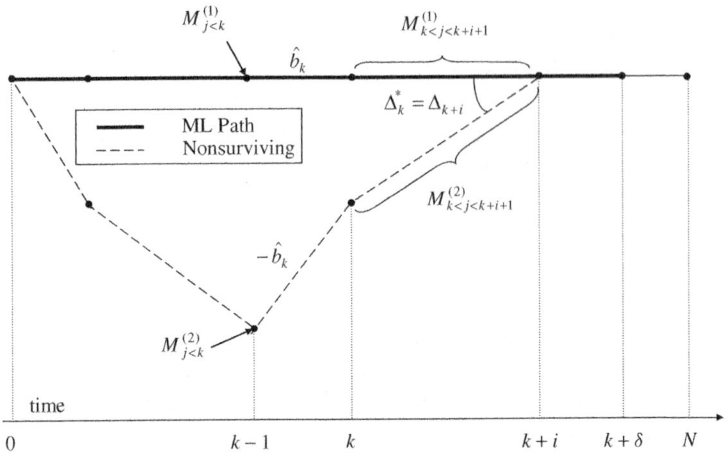

Figure 7.7 Illustration of obtaining the SOVA output for bit b_k.

which has a structure that is analogous to the structure of (7.23). By examining (7.33), we can see that the first two terms are known to the companion decoder, whereas the last three terms represent the extrinsic information to be passed to this companion decoder. To simplify the notation, we may express (7.33) as

$$L(b_k) = 4\sqrt{\rho}y_k^b + L^e(b_k) + L^a(b_k), \tag{7.34}$$

where $L^a(u_k)$ represents the last three terms, which is essentially the *a priori* information of bit b_k to be passed to a companion decoder.

In the conventional SOVA algorithm, it is normally assumed that the term $L^a(b_k)$ in (7.34) is weakly correlated with the other two terms, and thus it can be obtained as

$$L^a(b_k) = L(b_k) - L^i(b_k), \tag{7.35}$$

where $L^i(b_k) \triangleq 4\sqrt{\rho}y_k^b + L^e(b_k)$. However, it was demonstrated by Huang and Ghrayeb (2006) that the correlation between $L^i(b_k)$ and $L^a(b_k)$ can be high. We believe that this is essentially the reason behind the degradation in the SOVA performance. The consequence of this correlation is that the extrinsic information estimated by the SOVA is larger than its true value, which in turn results in misleading the companion decoder in the next iteration.

In light of the above discussion, it is clear that the output of the conventional SOVA is not really $L(b_k)$ as defined by (7.2), and consequently its extrinsic information is not $L^a(b_k)$ as defined by (7.35). Now, let us denote the actual SOVA output by $V(b_k)$ and the corresponding extrinsic information by $V^a(b_k)$, where the latter can be expressed as

$$V^a(b_k) \triangleq V(b_k) - L^i(b_k), \tag{7.36}$$

with $L_i(b_k)$ representing the intrinsic information (input to the SOVA.) The objective here is to modify $V(b_k)$ and $V_e(b_k)$ in order to obtain the true $L^i(b_k)$ and $L^a(b_k)$, which are supposed to be uncorrelated.

It has been shown by Hagenauer and Hoeher (1989) that $L^a(b_k)$ follows a Gaussian distribution, and so does $L^i(b_k)$ simply because it represents the extrinsic information passed from D_2 and $2y_k^b$ is the output of the channel. Therefore, the variables $L^i(b_k)$ and $V^a(b_k)$ are correlated Gaussian random variables with means m_i, m_a, and variances σ_i^2, σ_a^2, respectively. Using Bayes' rule, $L(b_k)$ can be computed as

$$\begin{aligned} L(b_k) &= \log \frac{P(b_k = +1 \mid V^a(b_k), L^i(b_k))}{P(b_k = -1 \mid V^a(b_k), L^i(b_k))}, \\ &= \mu_1 V^a(b_k) + \mu_2 L^i(b_k). \end{aligned} \tag{7.37}$$

where

$$\mu_1 = \frac{\left(\frac{2m_a}{\sigma_a^2} - \upsilon \frac{2m_i}{\sigma_a \sigma_i}\right)}{(1 - \upsilon^2)},$$

$$\mu_2 = \frac{\left(\frac{2m_i}{\sigma_i^2} - \upsilon \frac{2m_a}{\sigma_a \sigma_i}\right)}{(1 - \upsilon^2)},$$

and υ is the correlation between $V^a(b_k)$ and $L^i(b_k)$, which is computed as

$$\upsilon = \frac{E\left[(V^a(b_k) - m_a)(L^i(b_k) - m_i)\right]}{\sigma_i \sigma_a}.$$

Substituting (7.36) into (7.37) yields

$$L(b_k) = \mu_1 \left(V(b_k) - L^i(b_k)\right) + \mu_1 L^i(b_k), \tag{7.38}$$

and, consequently, the extrinsic information, $L^a(b_k)$, to be passed to the other SOVA decoder can be expressed as

$$\begin{aligned} L^a(b_k) &= L(b_k) - L^i(b_k) \\ &= (\mu_1 - \mu_2 + 1)\left[\frac{\mu_1}{\mu_1 - \mu_2 + 1} V(b_k) - L^i(b_k)\right], \end{aligned} \tag{7.39}$$

which can be implemented as shown in Figure 7.8 where c and d in the figure correspond to $\mu_1 - \mu_2 + 1$ and $\frac{\mu_1}{\mu_1 - \mu_2 + 1}$, respectively. This expression suggests that the immediate output

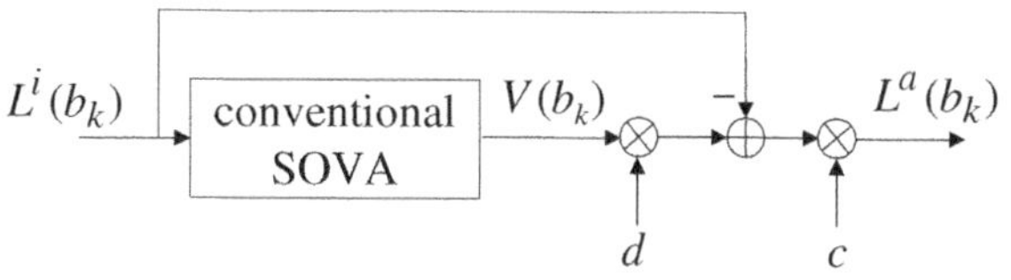

Figure 7.8 Structure of the MSOVA.

of the SOVA should be scaled by d before the intrinsic information is subtracted, and then the difference should be scaled by c. We refer to the SOVA with these modifications as the modified SOVA (MSOVA).

Since the values of c and d depend on the correlation between the extrinsic and intrinsic information, they are affected by the code structure and the puncturing mechanism. As such, when the constituent encoders are different (as in the SCCC case), these values should be computed separately for each component decoder. However, if the encoders are identical, such as the case in a PCCC, these values would be the same for both component decoders.

To obtain the values of c and d, in general, we need to compute the means and variances of $V^a(b_k)$ and $L^i(b_k)$, and the correlation coefficient between them for every received data frame and every decoding iteration. However, this method has two drawbacks. First, these computations increase the complexity of the decoder due to the additional processing delay. Second, the extrinsic information $V^a(b_k)$ may not be strictly Gaussian, especially when the data frame is relatively short, e.g., $N = 512$. As an alternative, one may perform a computer search to find the pair (c, d) that would give the best performance. Once the code generator polynomials and puncturing mechanism are determined, there is a fixed pair (c, d) that can be used for any received data frame and every decoding iteration. Moreover, this fixed pair of values works well for both channel models. We list in Table 7.1 the values of c and d that were found through Monte Carlo simulations for various PCCCs and SCCCs, and for various code rates.

We remark that the changes applied to the SOVA algorithm, as outlined above, extend in a straightforward manner to the APP algorithm (for both PCCCs and SCCCs). The modified APP (MAPP) will also have the same structure as that of the MSOVA. Extending these modifications to the APP algorithm was actually inspired after it was observed that the MSOVA substantially outperforms the APP algorithm in the SCCC system. This result should not be surprising though because iterative decoding of concatenated codes is not optimal in general, regardless of the decoder components employed. For such codes, the optimal decoder would be an APP decoder matched to the combined trellis of the constituent encoders and the interleaver. However, this decoder is very complex to realize, and thus an iterative decoder is employed instead.

Table 7.1 Values of c and d for various PCCCs and SCCCs for the MSOVA algorithm.

PCCC code		(c, d)	
$(7, 5)_{oct}, R_c = 1/2$		$(0.9, 0.8)$	
$(31, 33)_{oct}, R_c = 4/5$		$(0.8, 0.7)$	
SCCC code		(c, d)	
Outer	**Inner**	**Outer**	**Inner**
$(7, 5)_{oct},\ R_o = 1/2$	$\frac{1}{1+D},\ R_i = 1$	$(0.7, 0.8)$	$(0.8, 1.0)$
$(7, 5)_{oct},\ R_o = 8/9$	$\frac{1}{1+D},\ R_i = 1$	$(0.9, 0.8)$	$(1.0, 0.9)$

7.2.4 Performance with Maximum Likelihood Decoding

The challenge in finding performance bounds for concatenated codes is due to the presence of the interleaver, and particularly so since it is random. To overcome this hurdle, one may use one of the following two approaches, both of which invoke the union bound technique. In the first approach, the interleaver is assumed to be uniform, that is, it can be any one of the possible deterministic interleavers with equal probability. By this, the union bound becomes interleaver-independent since it is averaged with respect to all possible interleavers, as was done by Benedetto and Montorsi (1997). The second approach is based on deriving the union bound under the assumption that the interleaver is fixed. This involves finding, for a fixed code and interleaver, the most likely error events that impact the performance at high signal-to-noise ratios. In this subsection, we discuss the latter approach because it may be more useful from a practical point of view.

In general, turbo and turbo-like codes exhibit two different performance behaviors, depending on the range of signal-to-noise ratios. For relatively low signal-to-noise ratios, a slight increase in the signal-to-noise ratio results in a dramatic improvement in performance, leading to a sharp dip in the performance curve. This behavior is attributed to interleaving, as we shall explain later. The rest of the performance curve is referred to as the *floor* region, which covers the range of medium to high signal-to-noise ratios. In the floor region, the curve follows the Gaussian Q-function.

The optimal decoder for a concatenated code is the maximum likelihood sequence detector (MLSD). However, the MLSD is very complex to realize because of the complexity associated with the interleaver, as noted before. Therefore, as an alternative, the decoding of concatenated codes is achieved using a sub-optimal iterative decoding algorithm. Simulations have shown that the performance of PCCCs and SCCCs is very closely approximated by their ML performance. Additionally, since these codes are linear (the all-zero codeword can be used as a reference), their ML performance is very easy to derive compared with the performance of the iterative decoder. Therefore, we will derive the ML performance of these concatenated codes over AWGN channels. We begin with the PCCC case and later extend our derivation to the SCCC case.

Parallel Concatenated Codes

The ML analysis of concatenated codes involves invoking the union bound technique. Let us assume that the all-zero codeword was transmitted. Let the data block size be N. Assuming antipodal signaling, the number of possible codewords is then equal to 2^N. The pairwise error probability (i.e., the decoder picking codeword, $\boldsymbol{x}_i$, over the all-zero codeword, $\boldsymbol{x}_0$) is given by $P(\boldsymbol{x}_0 \rightarrow \boldsymbol{x}_i) = \Pr\left(\text{choosing } \boldsymbol{x}_i \text{ for } i \in \{1, 2, \ldots, 2^N - 1\} \mid \text{given that } \boldsymbol{x}_0 \text{ is transmitted}\right)$. Let d_i denote the Hamming distance between the two codewords. When the signalling scheme is BPSK with equiprobable signals, the pairwise error probability is therefore equal to

$$P(\boldsymbol{x}_0 \rightarrow \boldsymbol{x}_i) = Q\left(\sqrt{2d_i\rho}\right), \tag{7.40}$$

where ρ represents the signal-to-noise ratio at the receiver. Assuming that all possible information sequences are equally likely to be transmitted and ML soft decision decoding is employed, by invoking the union bound argument, the error probability can be upper bounded by

$$P_e \leq \sum_{i=1}^{2^N-1} Q\left(\sqrt{2d_i\rho}\right), \tag{7.41}$$

where the sum runs over all non-zero codewords. The probability of bit error P_b can then be upper bounded by

$$\begin{aligned} P_b &\leq \sum_{i=1}^{2^N-1} \frac{w_i}{N} Q\left(\sqrt{2d_i\rho}\right) \\ &= \sum_{w=1}^{N} \sum_{v=1}^{\binom{N}{w}} \frac{w}{N} Q\left(\sqrt{2d_{wv}\rho}\right). \end{aligned} \tag{7.42}$$

where d_{wv} is the total weight of the vth codeword with information weight w. The last equation reorganizes the summation firstly as a sum over all non-zero weight information inputs (since the all-zero information word would produce the all-zero codeword) and secondly, over the $\binom{N}{w}$ different codewords produced by the corresponding data words. d_{wv} is the weight of the vth codeword produced by a weight-w data word. Note that when we refer to the weight of a codeword we essentially mean the number of non-zero elements of that codeword.

When operating at medium to high signal-to-noise ratios, which is typical in real-life applications, only the first few terms in (7.42) will contribute to the sum. This is because d_{wv} typically increases as w increases. Thus, the last line of (7.42) can be approximated as

$$P_b \approx \sum_{w=2}^{t} \frac{wn_w}{N} Q\left(\sqrt{2d_{w,\min}^{PCCC}\rho}\right) \tag{7.43}$$

where n_w is the number of weight w input information sequences that result in the lowest weight codeword, $d_{w,\min}^{PCCC}$. Thus, (7.43) suggests that to find an upper bound for P_b, it suffices to find the minimum distance $d_{w,\min}^{PCCC}$ and its multiplicity n_w for all weight-t or lower input patterns. Typically, $1 < t \leq 4$. Note that weight-1 input patterns are excluded due to the recursive nature of the constituent encoders as such error patterns are likely to result in very high-weight codewords.

Serial Concatenated Codes

The derivation of the ML performance of the SCCC scheme is similar to that of the PCCC scheme. The reason is that the component encoders making up the SCCC encoder are also linear. Therefore, bit error rate in the floor region can be approximated as

$$P_b \approx \sum_{w=2}^{t} \frac{wn_w}{N} Q\left(\sqrt{2d_{w,\min}^{SCCC}\rho}\right). \tag{7.44}$$

Although (7.43) and (7.44) may appear to be the same, the parameters n_w and $d_{w,\min}$ in both equations can be different for PCCC and SCCC systems even if the same polynomials were used in both cases. This is attributed to the different structures these codes have.

Normally, for low code rates, e.g., 1/2, the SCCC scheme outperforms the PCCC scheme because $d_{w,\min}$ for the former scheme is typically higher, as was shown by Benedetto et al. (1998). Even the multiplicity n_w for SCCCs is lower than their PCCC counterparts.

To summarize, to find the performance parameters in (7.43) and (7.44), one needs to encode all weight-w input patterns and observe the corresponding outputs. The output with the minimum Hamming weight determines the minimum Hamming distance of the overall code. One can also obtain from the same experiment the corresponding multiplicity by counting the number of times the minimum distance occurs.

We conclude this section with the remark that other tighter bounds can also be used instead of the union bound (see Duman and Salehi (1998); Sason and Shamai (2000)).

7.2.5 Examples

In this section, we present several examples to demonstrate the performance of the PCCC and SCCC systems over AWGN channels with BPSK modulation. In Figure 7.9, we plot the bit error rate performance of the PCCC system for three and 15 decoder iterations. Code rates of the form $\frac{k_0}{k_0+1}$, where $k_0 = 4, 16$ and 64 were achieved by saving the second bit in every $2k_0$-bit parity block of each RSC encoder output and puncturing the rest. The RSC encoders are identical and employ the generator polynomials $(31, 33)_{octal}$. All simulations were done with a randomly-generated interleaver of size 4096. The iterative decoder employs two identical log-APP algorithms. We observe from the figure that a substantial improvement in the bit error rate relative to the uncoded case is achieved using

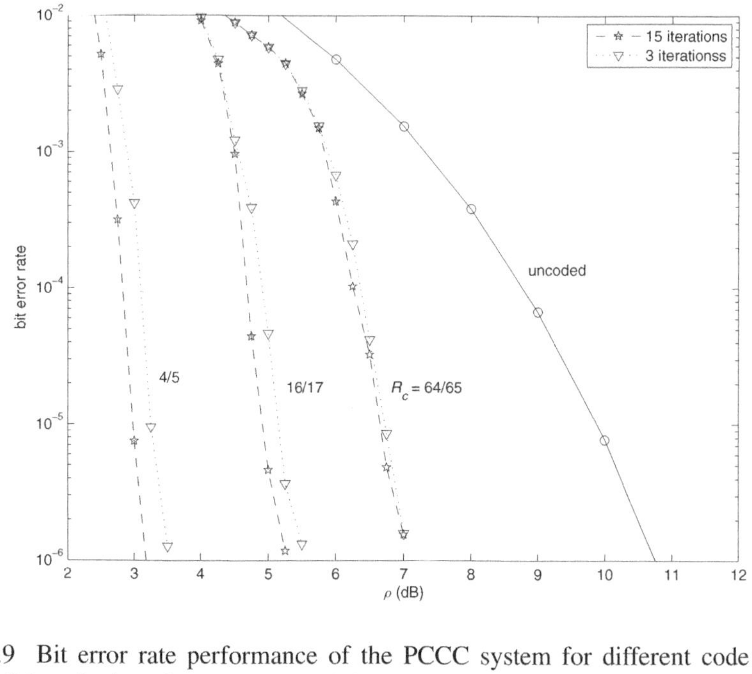

Figure 7.9 Bit error rate performance of the PCCC system for different code rates and number of decoder iterations over an AWGN channel. The interleaver length used is $N = 4096$. The iterative decoder employs two identical log-APP algorithms.

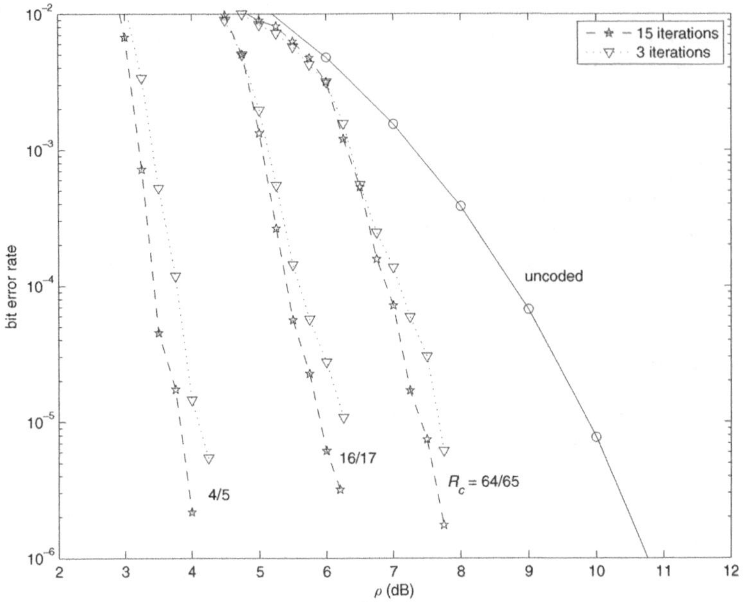

Figure 7.10 Bit error rate performance of the PCCC system for different code rates and different number of decoder iterations over an AWGN channel. The interleaver length used is $N = 4096$. The iterative decoder employs two identical SOVA algorithms. A decoding depth of $\delta = 50$ is used.

these codes: over 7 dB for the 4/5 rate, and over 3 dB for the 64/65 rate at a bit error rate of 10^{-5}. We also observe that the performance degrades by about 0.2 dB or less by using three iterations instead of 15.

The same PCCC system mentioned above was simulated with the two log-APP algorithms replaced by identical SOVA algorithms. The bit error rate results are plotted in Figure 7.10 (for three and 15 decoder iterations). A decoding depth of $\delta = 50$ is used in all simulations. We observe from the figure that the bit error rate performance suffers a degradation of about 0.7 dB by employing the SOVA relative to the log-APP algorithm. We also observe that there is a performance loss of close to 0.4 dB by using three iterations instead of 15 at a bit error rate of 10^{-5}. This indicates that the SOVA is more sensitive to the number of iterations than the APP algorithm.

We plot in Figure 7.11 the bit error rate performance of the SCCC system for eight decoder iterations. The outer convolutional code is of rate 1/2 and it employs the generator polynomials $(5, 7)_{octal}$. The inner code is of rate 1 and it has the transfer function $1/1 + D$. A pseudo-random interleaver of size $N = 512$ is used. The SCCC iterative decoder employs two MAPPs, two APPs, two conventional SOVAs, or two MSOVAs. As shown in the figure, the MSOVA improves the performance by about 1.2 dB relative to the SOVA at a bit error

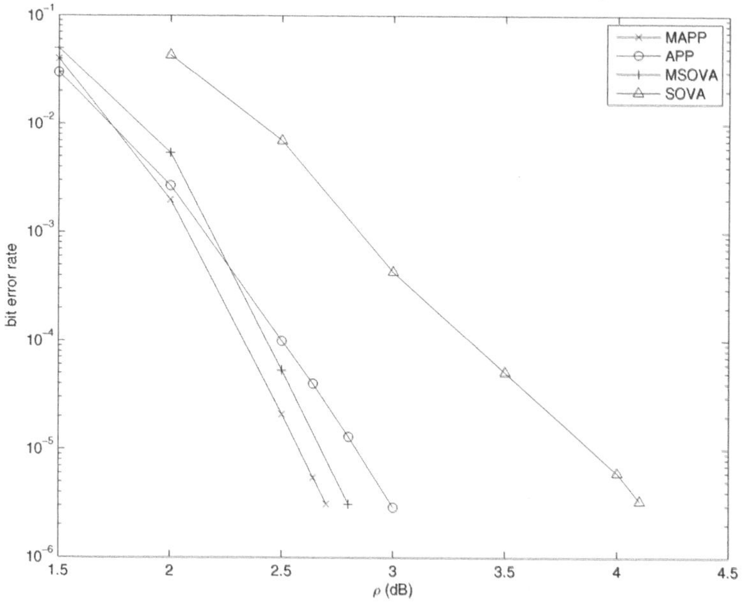

Figure 7.11 Bit error rate performance comparison of the SOVA, MSOVA, APP and MAPP decoders in AWGN for the SCCC scheme using a convolutional code with generator polynomials $(5, 7)_{octal}$, a differential encoder for the inner code, overall rate $1/2$, and $N = 512$ with eight iterations.

rate of 10^{-5}. Moreover, the MSOVA is superior to the APP by about 0.2 dB. It is also observed from the figure that the improvement provided by the MAPP, relative to the APP, is about 0.3 dB at a bit error rate of 10^{-5}. The values of c and d used for the MAPP were $(1.0, 1.1)$ and $(0.7, 0.8)$ for the inner and outer decoders, respectively, and they were $(0.7, 0.8)$ and $(0.8, 1.0)$ for the outer and inner MSOVAs, respectively.

In Figure 7.12, we present the bit error performance simulation results over an AWGN channel for the PCCC system with generator polynomials $(31, 33)_{octal}$ (16-state code), code rate $4/5$, data frame length $N = 512$, and eight decoder iterations. The PCCC iterative decoder uses two MAPPs, two APPs, two conventional SOVAs, or two MSOVAs. The values of c and d used were $(0.8, 0.7)$ for both decoders. We observe from the figure that the MSOVA achieves a performance improvement of about 0.8 dB relative to the SOVA, and is only about 0.3 dB away from the MAPP, all at bit error rate 10^{-5}. It is also interesting to observe that the MAPP is superior to the APP by about 0.3 dB at a bit error rate of 10^{-5}. We also plot on the same figure the ML bound for this code using (7.43) with the following parameters: $n_2 = 9$, $d_{2,\min} = 5$; and $n_3 = 1$, $d_{3,\min} = 3$. We can see the good agreement between the ML bound and the simulation results, which confirms the fact that iterative decoding achieves performance close to that of ML decoding.

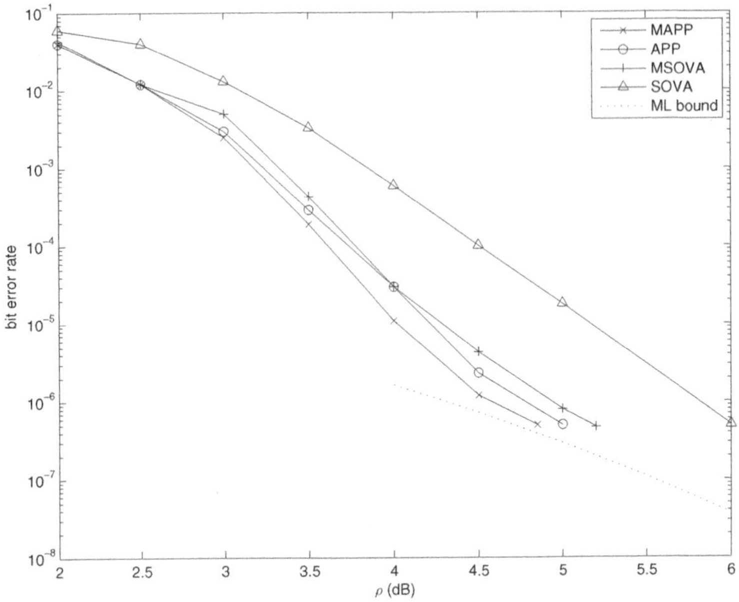

Figure 7.12 Bit error rate performance comparison of the SOVA, MSOVA, APP and MAPP decoders in AWGN for the PCCC scheme using 16-state RSC encoders, overall rate 4/5, $N = 512$, and eight iterations.

7.3 Concatenated Codes for MIMO Channels

In this and subsequent sections, we discuss how one can combine channel coding and space-time coding for MIMO channels in an effort to achieve further performance improvements. We note that part of this topic was discussed in the context of single-input single-output channels in Section 2.4.

One of the major advantages of using channel coding in conjunction with space-time coding is to achieve time diversity, especially for block fading channels. As mentioned before, time diversity is achieved by transmitting the signal components carrying the same information in multiple time intervals, provided that these time intervals are mutually separated by at least the coherence time of the channel. For example, if we consider a repetition code whereby each bit or symbol is transmitted several times, the diversity that can be achieved equals the length of the code, provided that the signal components fade independently and soft decision decoding is used. However, this is a naive way of achieving diversity because of the significant waste of the available bandwidth. To use the available bandwidth more efficiently, while achieving additional diversity, one may use a more sophisticated coding scheme such as convolutional code or TCM schemes in conjunction with interleaving, as discussed in Chapter 2.

For a given coded MIMO system employing N_t transmit and N_r receive antennas with i.i.d. channel fading coefficients (in time and space), the maximum diversity that can be achieved is $N_t N_r d^H_{\min}$ where $d^H_{\min}$ denotes the minimum Hamming distance of the outer code, provided that coherent soft decision decoding is employed. This is a significant enhancement in the diversity order. For example, if we consider a simple rate 1/2 convolutional code with $d^H_{\min} = 5$, which is a relatively weak code, the diversity of the coded MIMO system increases five times as compared with an uncoded MIMO system. On the other hand, if hard decision decoding is used, the achievable diversity is reduced to $\frac{1}{2} N_t N_r d^H_{\min}$. Similar results hold for nonbinary outer codes or TCM codes. The only difference here is that $d^H_{\min}$ is defined for these codes as the symbol-wise Hamming distance, and interleaving should be done on a symbol-by-symbol basis.

When the coherence time of the channel is relatively long, it may not be practical to employ a very long interleaver because of the associated delay. In cases like this, the outer channel code is still useful in the sense that it provides coding gains, and in some cases, in conjunction with limited interleaving, it may provide some diversity gain. For example, when the channel is quasi-static fading where interleaving is not useful, the outer channel code provides only coding gain, whereas when the channel is block fading, one may still use interleaving to achieve some increase in diversity. The level of diversity gain depends on the interleaving depth and the coherence time of the channel.

The advantages of employing an outer channel code obviously come at the expense of reducing the bandwidth efficiency, and this may become challenging in situations when the available bandwidth is scarce. To overcome this problem one may use higher order modulation schemes such as M-PSK and M-QAM to achieve a good use of the total available bandwidth. The results mentioned above concerning diversity and coding gains still hold for these modulation schemes.

Several code concatenation schemes have been developed for MIMO systems. In the following subsections, we discuss a few of them briefly.

7.3.1 Concatenated Space-Time Turbo Coding Scheme

Two concatenation schemes have been introduced by Liu et al. (2001a), one derives its structure from parallel concatenated codes as shown in Figure 7.13, whereas the other one derives its structure from serial concatenated codes as shown in Figure 7.14. As shown in the figure, the parallel concatenated code is designed for two transmit antennas. Puncturing is used to allow for flexible code rates. The channel interleaving and multiplexing are intended to take advantage of both time and space diversity. For this scheme to achieve full-rate, the number of bits fed at a time into each RSC encoder and the overall code rate are dictated by the modulation scheme used. For example, if QPSK is employed, one may use a rate 2/4 turbo encoder to achieve 2 bits/symbol. Although there is no guarantee that this code will achieve full diversity, if the interleavers (that of the channel and turbo encoder) are selected randomly, full diversity will be achieved with high probability. The decoder for this concatenation scheme is iterative and is based on belief propagation on Bayesian networks (McEliece et al. (1998)).

It is required for the serial concatenated code that the convolutional codes be identical, recursive and of rate one. The latter requirement is to ensure full-rate. It is also required that the numerator of the transfer function be one to avoid catastrophic codes. As for the interleavers, they must be identical. The decoder for this code resembles that of a

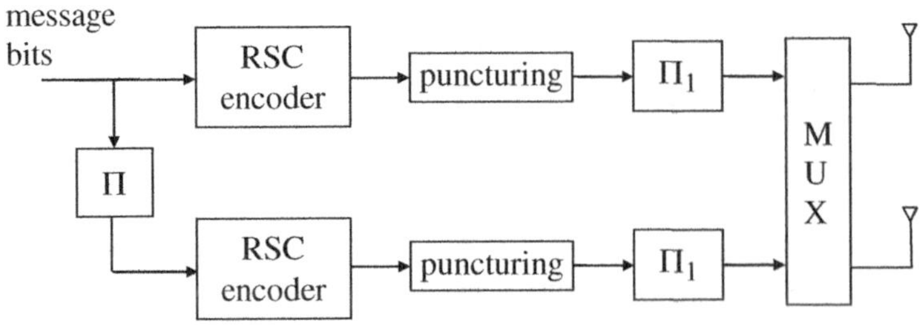

Figure 7.13 Parallel concatenated space-time turbo code encoder (for two transmit antennas).

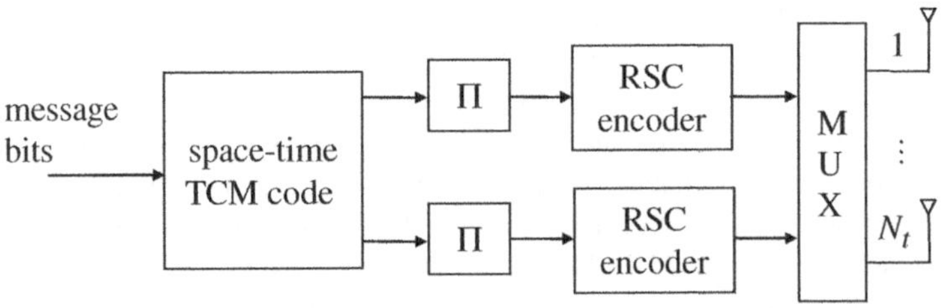

Figure 7.14 Encoder structure for the serial concatenated space-time turbo code.

standard serial concatenated code except that the inner decoder jointly decodes the recursive convolutional codes to ensure that all available diversity available in the channel is captured. In a standard serial concatenated code, the recursive convolutional codes are decoded separately.

7.3.2 Turbo Space-Time Trellis Coding Scheme

This concatenation scheme which borrows ideas from turbo codes and STTCs has been introduced by Gulati and Narayanan (2003). The encoder structure is depicted in Figure 7.15. The outer code is a binary code, whereas the inner code is a recursive STTC. The recursive nature of the inner code enhances the overall minimum Hamming distance of the combined code, thereby improving the available diversity. It is easy to design the inner code to achieve full diversity. This stems from the fact that any non-recursive space-time code that achieves full diversity, i.e. full rank, can be transformed easily into a recursive, full rank code. However, it does not achieve full-rate because of the presence of the outer code. This concatenation scheme is decoded iteratively where the front end receiver generates log-likelihoods for the mN_t bits which are deinterleaved and passed to the outer decoder. The outer decoder computes its own extrinsic information and passes them back to the

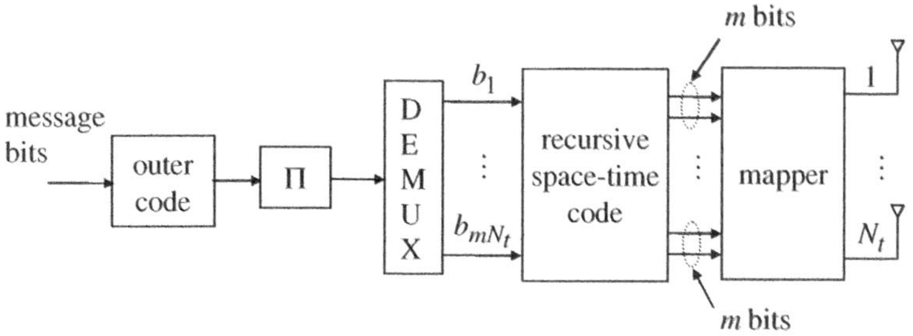

Figure 7.15 Encoder structure for the serial concatenation of an outer channel code and an inner recursive space-time code.

inner detector to refine its log-likelihoods for the next iteration. This repeats for a number of iterations before final decisions are made.

7.3.3 Turbo Space-Time Coding Scheme

This scheme was introduced by Cui and Haimovich (2001) and is referred to as turbo space-time coded modulation (turbo-STCM). The encoder structure is depicted in Figure 7.16. As shown in the figure, the code consists of two systematic (recursive) STTCs concatenated in a parallel fashion. One of the two antennas is permanently connected to the systematic part of the output of the top encoder and the second antenna is connected to the parity of the two STTC encoders. To achieve full-rate, puncturing must be used. In this case, puncturing is achieved by keeping one parity symbol for each systematic symbol and discarding the other parity symbol. Therefore, this scheme is guaranteed to achieve full-rate; however, it may achieve full diversity but this is not guaranteed because of puncturing.

This concatenation scheme is similar to a binary turbo code with its constituent convolutional codes replaced by recursive STTCs. However, there are some differences between the binary turbo codes and turbo-STCM, including the fact that the constituent STTCs need

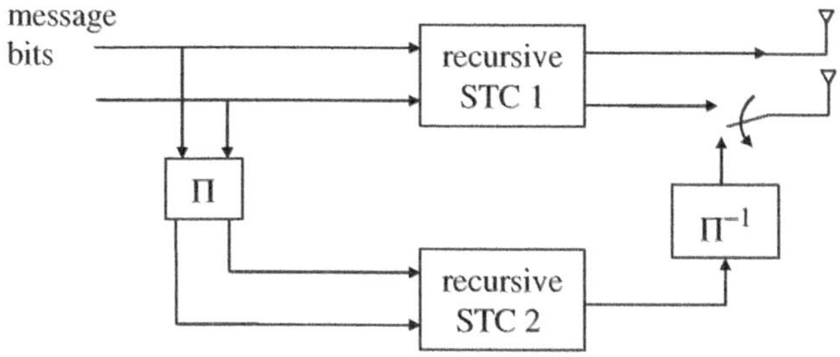

Figure 7.16 Encoder structure for the turbo-TCM scheme (Π denotes a symbol interleaver).

to be systematic at the symbol level rather than the bit level. Consequently, the interleaver operates on symbols. The symbol deinterleaver at the output of the lower STTC encoder is used to ensure that the systematic data at the output is identical to that of the upper encoder. The decoder for this scheme resembles that of an iterative turbo decoder. The only difference here is that the APP decoders are matched to the nonbinary trellis of the constituent STTC encoders.

In the next sections, we describe in detail two code concatenation schemes developed for MIMO systems, namely, turbo-coded modulation and combined channel coding and space-time block coding.

7.4 Turbo-Coded Modulation for MIMO Channels

The concatenation scheme we consider here involves the serial concatenation of a turbo code and a mapper with an interleaver separating them, which was introduced by Stefanov and Duman (2001c). In this scheme, the mapper is flexible in the sense that it can be an STTC, STBC or BLAST-like scheme. We refer to this scheme as *turbo-coded modulation* (TuCM). We remark that the concept of applying turbo-coded modulation to single-input single-output AWGN and fading channels was considered before (e.g., Duman (1998)).

7.4.1 Encoder Structure

A block diagram of the TuCM encoder is shown in Figure 7.17. As shown in the figure, the primitive data is first encoded by a turbo encoder. The turbo-coded sequence is interleaved so that bursts of errors resulting from deep fades are distributed, which makes it easier to deal with such errors. The interleaved sequence is demultiplexed into a number of parallel substreams. In this case we assume there are mN_t substreams. In general, the number of substreams depends on the type of modulation and mapping used. These mN_t streams are fed into a mapper. The function of the mapper is to map the incoming bits to a particular signal constellation. Specifically, to simultaneously transmit N_t symbols from the available N_t transmit antennas, the mapper maps every set of m coded bits to one symbol from a signal constellation of size 2^m.

It is assumed here that the underlying channel is block fading and coding is done across a number of consecutive differently faded blocks so as to achieve time diversity. Since block fading is assumed, the interleaver size is chosen in such a way that it is large enough to

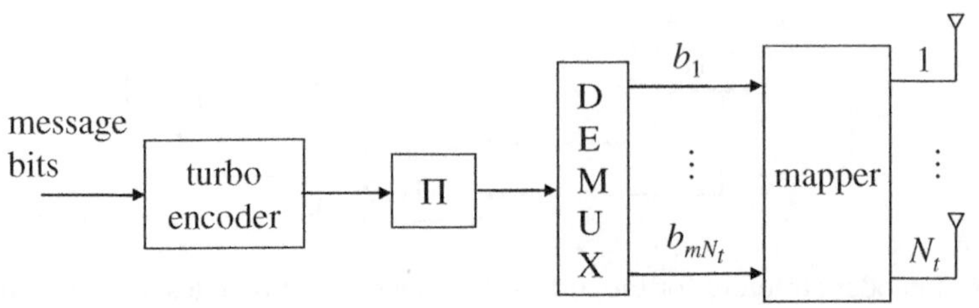

Figure 7.17 Encoder structure for the turbo-coded modulation scheme.

span a number of differently faded blocks to break the correlation between adjacent bits in an effort to achieve additional (time) diversity.

Now define $\boldsymbol{b}$ to be a vector of coded bits of length mN_t, which is the input to the mapper at any given time, i.e.,

$$\boldsymbol{b} \triangleq \left[b_1, \ldots, b_m, \ldots, b_{m+1}, \ldots, b_{mN_t}\right].$$

Let the corresponding vector of N_t symbols be $\boldsymbol{x}$, i.e.,

$$\boldsymbol{x} \triangleq \left[x_1, x_2, \ldots, x_{N_t}\right].$$

The components of $\boldsymbol{x}$ are then transmitted simultaneously from the available N_t transmit antennas. Let $y_j(k)$ denote the signal received by antenna j at time k, which can be expressed as

$$y_j(k) = \sqrt{\rho} \sum_{i=1}^{N_t} x_i(k) h_{i,j}(k) + n_j(k),$$

where $h_{i,j}(k)$ denotes the fading coefficient between the ith transmit and jth receive antenna at time k, and $n_j(k)$ is the complex Gaussian noise sample at time k corresponding to receive antenna j. Both the fading coefficients and AWGN samples are assumed to be independent and $\mathcal{CN}(0, 1)$ distributed. For notational convenience, we drop the time index k. With this, the signals received by all receive antennas can be expressed in a compact form as

$$\boldsymbol{y} = \sqrt{\rho}\boldsymbol{x}\boldsymbol{H} + \boldsymbol{n},$$

where $\boldsymbol{H}$ is the $N_t \times N_r$ channel matrix, $\boldsymbol{y} = \left[y_1, y_2, \ldots, y_{N_r}\right]$ and $\boldsymbol{n} = \left[n_1, n_2, \ldots, n_{N_r}\right]$.

7.4.2 Decoder Structure

The decoding of this concatenation scheme is done iteratively, as shown in Figure 7.18. In the figure, the log-likelihood computation module (or, the channel detector) uses the received signal $\boldsymbol{y}$ to compute the log-likelihood ratios for the bits comprising $\boldsymbol{b}$. These log-likelihoods are deinterleaved and passed to the turbo decoder for further processing. The turbo decoder consists of two SISO decoders that work iteratively between themselves, where each SISO decoder is matched to one of the component encoders of the turbo code. The channel detector and turbo decoder can work iteratively or noniteratively, depending on the performance/complexity requirements. In the former case, the turbo decoder feeds back its extrinsic information to the channel detector after the first iteration so that refined log-likelihoods are computed. In the latter case, the log-likelihoods are computed only once and the turbo decoder runs for a few times after which final decisions are made. We shall consider both scenarios next.

The log-likelihood for the kth bit in $\boldsymbol{b}$ is given by

$$L(b_k) = \log \frac{P\left(b_k = 1 | \boldsymbol{y}\right)}{P\left(b_k = 0 | \boldsymbol{y}\right)}, \tag{7.45}$$

$$= \log \frac{P\left(b_k = 1, \boldsymbol{y}\right)}{P\left(b_k = 0, \boldsymbol{y}\right)}. \tag{7.46}$$

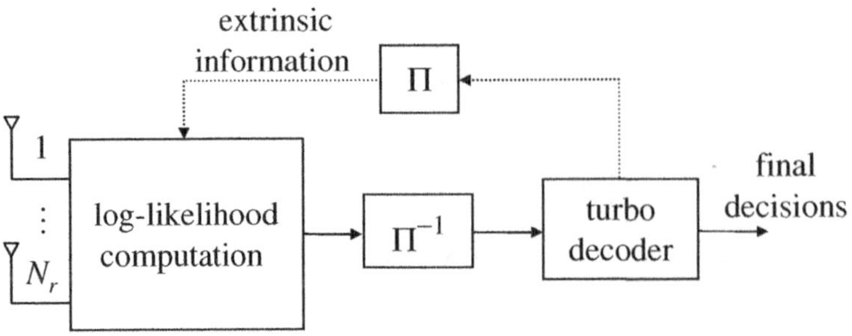

Figure 7.18 Iterative decoder structure for the turbo-coded modulation scheme.

Since the mapping between $\boldsymbol{b}$ and $\boldsymbol{x}$ is one-to-one, we can write (7.46) as

$$L(b_k) = \log \frac{\sum_{\boldsymbol{x}:\boldsymbol{x}=f(\boldsymbol{b}),b_k=1} P\left(\boldsymbol{y}, \boldsymbol{x} \text{ is transmitted}\right)}{\sum_{\boldsymbol{x}:\boldsymbol{x}=f(\boldsymbol{b}),b_k=0} P\left(\boldsymbol{y}, \boldsymbol{x} \text{ is transmitted}\right)}, \tag{7.47}$$

where $f(\cdot)$ is the mapping from $\boldsymbol{b}$ to $\boldsymbol{x}$. Assuming that the transmitted signals are equally likely, the joint p.d.f.s in (7.47) can be expressed as

$$L(b_k) = \log \frac{\sum_{\boldsymbol{x}:\boldsymbol{x}=f(\boldsymbol{b}),b_k=1} p\left(\boldsymbol{y}|\boldsymbol{x}\right)}{\sum_{\boldsymbol{x}:\boldsymbol{x}=f(\boldsymbol{b}),b_k=0} p\left(\boldsymbol{y}|\boldsymbol{x}\right)}, \tag{7.48}$$

where $p\left(\boldsymbol{y}|\boldsymbol{x}\right)$ is given by

$$p(\boldsymbol{y}|\boldsymbol{x}) = \prod_{j=1}^{N_r} p\left(y_j|\boldsymbol{x}\right), \tag{7.49}$$

$$= \prod_{j=1}^{N_r} \frac{1}{\pi} \exp\left(-\left|y_j - \sqrt{\rho}\sum_{i=1}^{N_t} x_i h_{i,j}\right|^2\right), \tag{7.50}$$

which assumes that the components of $\boldsymbol{y}$, given $\boldsymbol{x}$, are conditionally independent. By using the definition of $p\left(\boldsymbol{y}|\boldsymbol{x}\right)$ and the approximation

$$\log \sum_{\boldsymbol{x}:\boldsymbol{x}=f(\boldsymbol{b}),b_k=b} \prod_{j=1}^{N_r} p\left(y_j|\boldsymbol{x}\right) \approx \max_{\boldsymbol{x}:\boldsymbol{x}=f(\boldsymbol{b}),b_k=b} \log \prod_{j=1}^{N_r} p\left(y_j|\boldsymbol{x}\right), \tag{7.51}$$

where $b \in \{0, 1\}$, the expression in (7.48), with simple algebra, can be further simplified to

$$L(b_k) = \max_{\boldsymbol{x}:\boldsymbol{x}=f(\boldsymbol{b}),b_k=1} \log \prod_{j=1}^{N_r} p\left(y_j|\boldsymbol{x}\right) - \max_{\boldsymbol{x}:\boldsymbol{x}=f(\boldsymbol{b}),b_k=0} \log \prod_{j=1}^{N_r} p\left(y_j|\boldsymbol{x}\right). \tag{7.52}$$

Once the log-likelihoods in (7.52) are computed, they are deinterleaved and passed to the turbo decoder, which are used as if they are the log-likelihoods of an observation sequence from a BPSK modulated signal over an AWGN channel. The turbo decoder runs for several iterations of its own and at the end produces its own extrinsic information about the received bits. When the channel detector and turbo decoder work iteratively, the log-likelihoods in (7.48) are updated after the first iteration to utilize the extrinsic information produced by the turbo decoder. This extrinsic information is used by the channel detector as the *a priori* information on the transmitted sequence. As such, the updated log-likelihoods are computed using

$$L(b_k) = \log \frac{\sum_{\boldsymbol{x}:\boldsymbol{x}=f(\boldsymbol{b}),b_k=1} \left(\prod_{j=1}^{N_r} p\left(y_j|\boldsymbol{x}\right)\right) P\left(\boldsymbol{x}\right)}{\sum_{\boldsymbol{x}:\boldsymbol{x}=f(\boldsymbol{b}),b_k=0} \left(\prod_{j=1}^{N_r} p\left(y_j|\boldsymbol{x}\right)\right) P\left(\boldsymbol{x}\right)}. \tag{7.53}$$

Under the assumption that the components of $\boldsymbol{x}$ are mutually independent, we can write $P\left(\boldsymbol{x}\right) = \prod_{i=1}^{N_t} P(x_i)$. Clearly, because of interleaving, we may assume that the probabilities of the bits that correspond to a symbol are independent. Therefore, we have $P\left(\boldsymbol{x}\right) = P\left(\boldsymbol{b}\right) = \prod_{i=1}^{mN_t} P\left(b_i\right)$. Using this result, we can write (7.53) as

$$L(b_k) = \log \frac{\sum_{\boldsymbol{x}:\boldsymbol{x}=f(\boldsymbol{b}),b_k=1} \left(\prod_{j=1}^{N_r} p\left(y_j|\boldsymbol{x}\right)\right) \prod_{l=1,l\neq k}^{mN_t} P(b_l)}{\sum_{\boldsymbol{x}:\boldsymbol{x}=f(\boldsymbol{b}),b_k=0} \left(\prod_{j=1}^{N_r} p\left(y_j|\boldsymbol{x}\right)\right) \prod_{l=1,l\neq k}^{mN_t} P(b_l)}, \tag{7.54}$$

which is the log-likelihood for the bit b_k. The probability of each bit being 1 or 0 can be expressed in terms of its log-likelihood value as

$$P(b_l = 1) = \frac{e^{L(b_l)}}{1 + e^{L(b_l)}} \quad \text{and} \quad P(b_l = 0) = \frac{1}{1 + e^{L(b_l)}}.$$

To summarize, when the channel detector and turbo decoder work cooperatively and iteratively, in the beginning, the channel detector computes the sequence likelihoods according to (7.48) where the transmitted bits are assumed to be equally likely. The computed likelihoods are passed to the turbo decoder, which produces its own extrinsic information about the transmitted sequence. This information is fed back to the channel detector to compute new likelihoods according to (7.54). The updated likelihoods are then passed to the turbo decoder, and so on. This process, which we refer to as iterative demodulation/decoding, repeats for a certain number of times before final decisions are made. As for the performance analysis of this TuCM scheme, it is somewhat similar to that of STTCs, which was discussed in Section 5.5. For more on this, the reader may refer to Stefanov and Duman (2003a).

We finally remark that the TuCM scheme is not limited to turbo codes. Other codes, including convolutional codes, TCM code, etc., can also be used. If it is desired to have the channel detector and channel decoder work cooperatively and in an iterative fashion, then the channel decoder must be able to produce soft information at its output so that it can be used by the channel detector as *a priori* information in the next iteration. For example, when the outer code is a convolutional code, the corresponding decoder could be of APP or SOVA type so that iterative demodulation/decoding can be applied.

7.4.3 Examples

Several examples are considered in this section to demonstrate the performance of the TuCM scheme and to compare it with that of the STTC scheme. Note that interleaving is employed for the STTCs as well to provide time diversity in addition to the spatial diversity. In Figure 7.19, we plot the bit error rate performance for both schemes for information block lengths $N = 1300$ and 5200. We consider a MIMO system with $N_t = 2$ and $N_r = 1$. The channel is assumed to be fixed over 130 transmissions per transmit antenna. In all simulations, an S-random interleaver is used. We observe from the figure that the TuCM scheme achieves significant performance improvements over the STTC scheme. For example, the performance improvement is about 3.25 dB at a bit error rate of 10^{-5} for $N = 1300$, and this improvement increases to about 6 dB for $N = 5200$.

In Figure 7.20, we demonstrate the improvements achieved with iterative demodulation/decoding over iterative decoding only. In this case, we consider a MIMO system with $N_t = 2$ and $N_r = 1, 2$ with $N = 260$. As shown in the figure, at a frame error rate of 0.1,

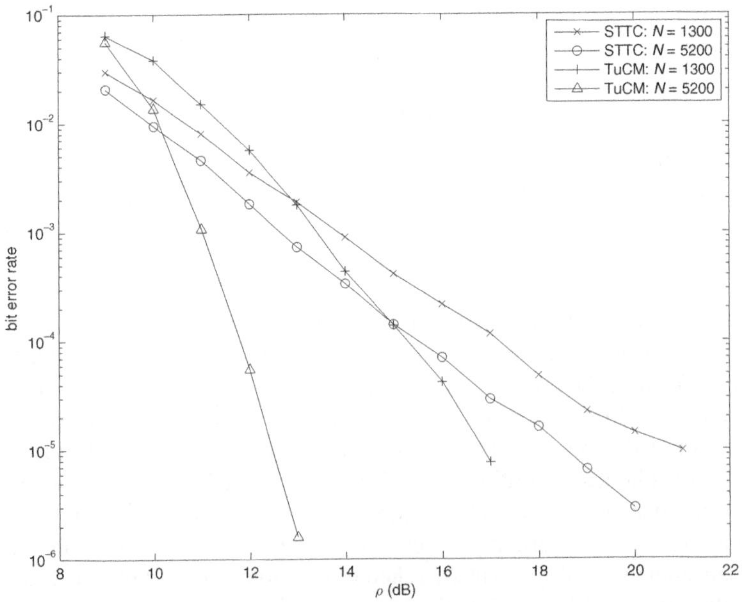

Figure 7.19 Bit error rate performance comparison between the TuCM and STTC schemes for $N_t = 2$, $N_r = 1$ with block lengths 1300 and 5200.

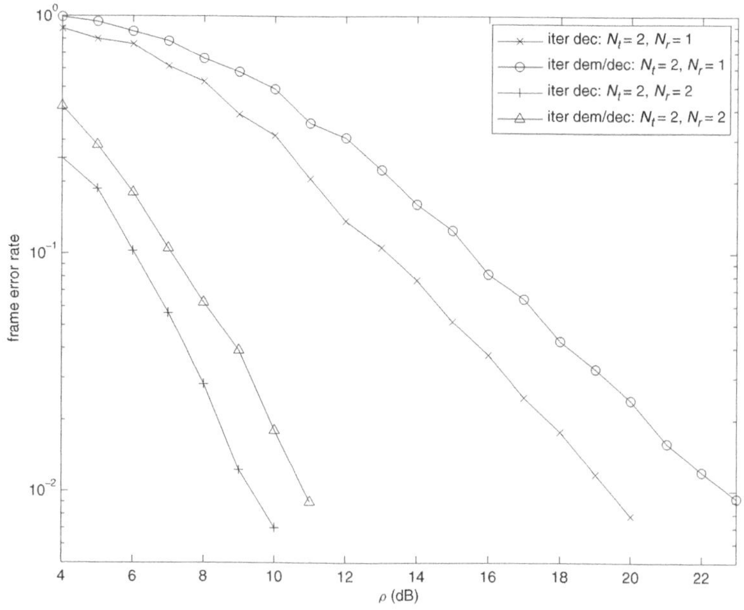

Figure 7.20 Frame error rate performance comparison between iterative demodulation/decoding and iterative decoding only.

the performance improvement due to combining iterative demodulation and decoding is about 2 dB and 1 dB for the case with one and two receive antennas, respectively.

7.5 Concatenated Space-Time Block Coding

The TuCM scheme, as mentioned previously, encompasses a wide variety of concatenation schemes since there is no restriction on its mapper. Among these schemes is the concatenation of an outer channel code and an inner orthogonal STBC, where in this case the STBC simply replaces the mapper as shown in Figure 7.17. This scheme was originally proposed by Bauch (1999). It is shown by Liew and Hanzo (2002) that the coded STBC scheme gives the best performance–complexity trade-off among other concatenation schemes when the outer code is a convolutional code or a turbo code. Additionally, as mentioned in Chapter 1, it has been included as an option in several standards for future wireless communication systems.

The coded STBC scheme owes its popularity to its simplicity and flexibility. For example, the complexity of the STBC decoder is much less than that of the TuCM scheme and it is more stable. The reason is that the STBC decoder exploits the orthogonality structure of the STBC to decouple the received signals. In addition, it is easier to obtain performance bounds for the coded STBC scheme for various outer channel codes. Motivated by these reasons, we discuss this scheme in detail in this section.

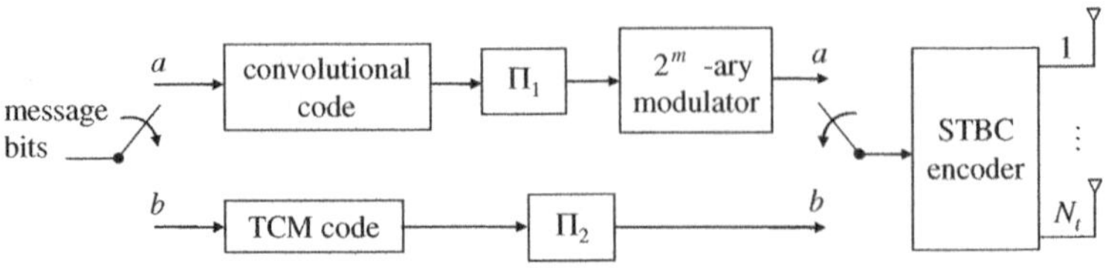

Figure 7.21 Encoder structure for the coded STBC scheme. (Π_1 denotes a bit interleaver whereas Π_2 denotes a symbol interleaver.)

7.5.1 Encoder Structure

The encoder of the coded STBC scheme is shown in Figure 7.21. In the figure, the primitive data stream is first encoded by an outer channel encoder. There is no restriction on the type of channel code employed, but the focus here is on convolutional codes, turbo codes and TCM codes because they are widely used. In addition, we only consider binary codes. The coded sequence is interleaved and demultiplexed. Each set of m bits at the output of the demultiplexer is mapped onto a symbol taken from a 2^m-ary signal constellation. The output of the modulator is then fed into the STBC encoder, which groups every N_t consecutive symbols and transmits them from the available N_t antennas according to the STBC encoding principles, assuming that the underlying STBC is full rate. As such, the transmission rate achieved is $R = mR_c$ bits per channel use, where R_c is the rate of the outer channel code.

7.5.2 Decoder Structure

The decoder structure for the coded STBC scheme is depicted in Figure 7.22. In the figure, when the switch is in position 'a', the resulting decoder corresponds to the convolutional (or turbo) coded system. In this case, the output of the STBC decoder is fed into the log-likelihood computation module, which computes the log-likelihoods for the bits comprising the corresponding symbols. These log-likelihoods are then deinterleaved and passed to the channel decoder.

The type of channel decoder employed is dictated by the channel encoder employed at the transmitter. For example, when a convolutional code is used, the corresponding decoder would be the Viterbi decoder. Whereas, when a turbo code is used, the corresponding

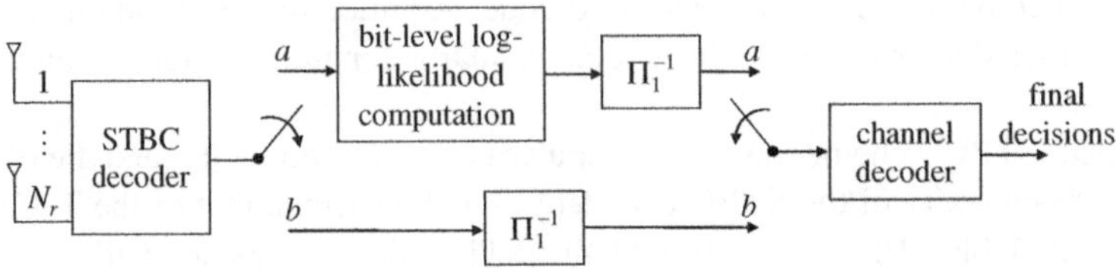

Figure 7.22 Decoder structure for the coded STBC scheme.

decoder would consist of two log-APP or SOVA decoders working together iteratively. When the switch is in position 'b', the resulting decoder corresponds to the TCM coded system. In this case, the sequence of the decoupled symbols at the output of the STBC decoder is deinterleaved and passed to the channel decoder, which is the Viterbi decoder.

We now give details as to how one can obtain the bit log-likelihoods that need to be passed to the outer channel decoder (see Figure 7.22). Consider a MIMO system with N_t transmit and N_r receive antennas. Let $x(k)$ denote the kth symbol at the input of the STBC encoder (see Figure 7.21). Assuming that the channel gains remain fixed over N_t consecutive symbol intervals and they are perfectly known at the receiver, the output of the STBC decoder corresponding to $x(k)$ (see Figure 7.22) can be expressed as

$$y(k) = \sqrt{\rho}x(k)\left(\sum_{j=1}^{N_r}\sum_{i=1}^{N_t}\left|h_{i,j}(k)\right|^2\right) + \tilde{n}(k) \tag{7.55}$$

where $\tilde{n}(k) \triangleq \left(\sum_{j=1}^{N_r}\sum_{i=1}^{N_t}\left|h_{i,j}(k)\right|^2\right)n(k)$ with $n(k) \sim \mathcal{CN}(0, 1)$. It is assumed in (7.55) that the underlying STBC is orthogonal such that decoupling of the received symbols can be done easily, as was explained in Chapter 4. To make the notation simpler, we temporarily drop the time index k. Note that y, conditioned on x and $\boldsymbol{H}$, is complex Gaussian distributed with mean $\mu_y \triangleq \sqrt{\rho}x\left(\sum_{j=1}^{N_r}\sum_{i=1}^{N_t}\left|h_{i,j}\right|^2\right)$ and variance $\sigma_y^2 \triangleq \sum_{j=1}^{N_r}\sum_{i=1}^{N_t}\left|h_{i,j}\right|^2$. Therefore, the conditional p.d.f. of y is given as

$$p(y|x) = \frac{1}{\pi\sigma_y^2}\exp\left(-\frac{\left|y-\mu_y\right|^2}{\sigma_y^2}\right). \tag{7.56}$$

Now we need to compute the log-likelihoods of the bits comprising x. Let $\boldsymbol{b} = [b_1, b_2, \ldots, b_m]$ be the m-tuple corresponding to x. Obviously it is implied here that the signal constellation is of size 2^m. Similar to what we did in the previous section, the log-likelihood for the kth bit of $\boldsymbol{b}$ is given by

$$L(b_k) = \log\frac{\displaystyle\sum_{x:x=f(\boldsymbol{b}),b_k=1} p(y|x)}{\displaystyle\sum_{x:x=f(\boldsymbol{b}),b_k=0} p(y|x)}, \tag{7.57}$$

where it is assumed here that the mapping between $\boldsymbol{b}$ and x is one-to-one, the components of $\boldsymbol{b}$ are independent and the transmitted signals are equally likely. By plugging (7.56) into (7.57), one can easily compute the log-likelihoods of the bits comprising $\boldsymbol{b}$. Once these log-likelihoods are computed, they are deinterleaved and passed to the outer decoder, which are used as if they are the log-likelihoods of an observation sequence from a BPSK modulated signal over an AWGN channel.

7.5.3 Performance Analysis

In this section, we derive performance bounds on the bit error rate of the coded STBC scheme. We consider two cases: fully-interleaved fading channels and slow fading channels. In the former case, which is also referred to ideal interleaving, the symbols comprising the sequence at the input of the channel decoder are assumed to experience independent fades. In the latter case, the channel is assumed to be quasi-static fading whereby the entire sequence at the input of the channel decoder experiences the same fade. These two cases represent the two extremes. That is, in ideal interleaving, the time diversity provided by the outer channel code translates into additional diversity and thus the maximum diversity possible is achieved. On the other hand, in slow fading, the presence of the outer channel code provides only coding gain but it does not provide any additional diversity. Therefore, when the channel is block fading (somewhere between fast and slow), the presence of the outer channel code provides both coding and diversity gains, with more diversity if the channel is closer to fast fading and more coding gain if the channel is closer to slow fading.

Fast Fading: Fully-Interleaved Channels

Let $\boldsymbol{x}_1$ and $\boldsymbol{x}_2$ denote the transmitted codeword (at the input of the STBC encoder) and the erroneously decoded codeword, respectively. Denote by $d(\boldsymbol{x}_1, \boldsymbol{x}_2)$ the symbol-wise Hamming distance between $\boldsymbol{x}_1$ and $\boldsymbol{x}_2$. Assuming ML soft-decision decoding, and that the CSI is perfectly known at the receiver, the conditional pairwise error probability that the receiver will select $\boldsymbol{x}_2$ over $\boldsymbol{x}_1$ conditioned on the channel gains is given by

$$P\left(\boldsymbol{x}_1 \rightarrow \boldsymbol{x}_2|\boldsymbol{H}\right) = Q\left(\sqrt{\frac{\rho}{2N_t} \sum_{k=1}^{d(\boldsymbol{x}_1,\boldsymbol{x}_2)} \sum_{i=1}^{N_t} \sum_{j=1}^{N_r} \left|h_{i,j}(k)\right|^2 \left|x_1(k) - x_2(k)\right|^2}\right) \tag{7.58}$$

where $|x_1(k) - x_2(k)|^2$ is the normalized squared Euclidean distance between the signal on the correct path and that on the error path at time index k, and ρ is the total transmitted power. Note that the expression in (7.58) is not limited to a particular outer code.

Now instead of using the Chernoff bound to find an upper bound on (7.58), we use Craig's formula for $Q(x)$ given by

$$Q(x) = \frac{1}{\pi} \int_0^{\pi/2} \exp\left(-\frac{x^2}{2\sin^2\theta}\right) d\theta.$$

The reason is that the Chernoff bound is quite loose especially for TCM codes since these codes generally have low symbol-wise Hamming distance as compared with other codes such as convolutional codes.

Consequently, the conditional pairwise error probability given in (7.58) can be represented as

$$\begin{aligned} P\left(\boldsymbol{x}_1 \rightarrow \boldsymbol{x}_2|\boldsymbol{H}\right) &= \frac{1}{\pi} \int_0^{\pi/2} \exp\left(-\frac{1}{\sin^2\theta} \sum_{k=1}^{d(\boldsymbol{x}_1,\boldsymbol{x}_2)} \sum_{i=1}^{N_t} \sum_{j=1}^{N_r} \left|h_{i,j}(k)\right|^2 \delta_k^2\right) d\theta, \\ &= \frac{1}{\pi} \int_0^{\pi/2} \prod_{k=1}^{d(\boldsymbol{x}_1,\boldsymbol{x}_2)} \exp\left(-\frac{\delta_k^2}{\sin^2\theta} \sum_{i=1}^{N_t} \sum_{j=1}^{N_r} \left|h_{i,j}(k)\right|^2\right) d\theta, \end{aligned} \tag{7.59}$$

where $\delta_k^2 \triangleq \frac{\rho}{4N_t} |x_1(k) - x_2(k)|^2$.

Now let us define Y_k as

$$Y_k = \sum_{j=1}^{N_r}\sum_{i=1}^{N_t} \left|h_{i,j}(k)\right|^2, \quad \text{for } k = 1, 2, \ldots, d(\boldsymbol{x}_1, \boldsymbol{x}_2). \tag{7.60}$$

Clearly, the random variables Y_k are independent and chi-square distributed, each with $2N_tN_r$ degrees of freedom and a p.d.f. given by

$$p_Y(y) = \frac{1}{(N_tN_r - 1)!} y^{(N_tN_r-1)} e^{-y}, \quad y > 0. \tag{7.61}$$

In order to evaluate the average pairwise error probability, we average (7.59) with respect to the distribution of Y_k as

$$\begin{aligned} P(\boldsymbol{x}_1 \to \boldsymbol{x}_2) &= \frac{1}{\pi}\int_0^{\pi/2} \prod_{k=1}^{d(\boldsymbol{x}_1,\boldsymbol{x}_2)} \int_0^{\infty} \exp\left(-\frac{\delta_k^2 y_k}{\sin^2\theta}\right) p_Y(y_k)\, dy_k d\theta \\ &= \frac{1}{\pi}\int_0^{\pi/2} \prod_{k=1}^{d(\boldsymbol{x}_1,\boldsymbol{x}_2)} \left(1 + \frac{\delta_k^2}{\sin^2\theta}\right)^{-N_tN_r} d\theta. \end{aligned} \tag{7.62}$$

At high signal-to-noise ratios, (7.62) can be approximated as (Zeng and Ghrayeb (2006))

$$\begin{aligned} P(\boldsymbol{x}_1 \to \boldsymbol{x}_2) &\approx \frac{1}{\pi}\int_0^{\pi/2} \prod_{k=1}^{d(\boldsymbol{x}_1,\boldsymbol{x}_2)} \left(\frac{\delta_k^2}{\sin^2\theta}\right)^{-N_tN_r} d\theta, \\ &= \binom{2N_tN_rd(\boldsymbol{x}_1, \boldsymbol{x}_2) - 1}{N_tN_rd(\boldsymbol{x}_1, \boldsymbol{x}_2)} \left(\prod_{k=1}^{d(\boldsymbol{x}_1,\boldsymbol{x}_2)} \left(|x_1(k) - x_2(k)|^2\right)^{-N_tN_r}\right) \\ &\quad \cdot \left(\frac{\rho}{N_t}\right)^{-N_tN_rd(\boldsymbol{x}_1,\boldsymbol{x}_2)}. \end{aligned} \tag{7.63}$$

The last line of (7.63) is obtained by changing the variable $t = \tan^{-1}\theta$, and then applying the definite integral

$$\frac{1}{\pi}\int_0^{\infty} \frac{1}{\left(x^2+1\right)^{L+1}} dx = \frac{1}{4^L}\binom{2L-1}{L}.$$

It is clear from (7.63) that the diversity order achieved is $N_tN_rd_{\min}$, where $d_{\min}$ represents the minimum distance of the outer code. For TCM codes, $d_{\min}$ represents the symbol-wise Hamming distance, whereas, for convolutional and turbo codes, it represents the minimum Hamming distance of the code.

For TCM codes, an upper bound on the bit error rate can then be found as

$$P_b \le \frac{1}{m\pi}\int_0^{\pi/2} \frac{\partial}{\partial I} T(D(\theta), I)|_{I=1}\, d\theta \tag{7.64}$$

where m is the number of bits per transmitted symbol, and $T(D(\theta), I)$ is the transfer function of the TCM code, which is obtained directly from the corresponding error state diagram. The branch labels of the error state diagram, denoted by $D(\theta)$, after being averaged with respect to the fading coefficients, are given by

$$D(\theta) = \left(1 + \frac{\delta_k^2}{\sin^2\theta}\right)^{-N_t N_r}. \tag{7.65}$$

The definite integral in (7.64) can be evaluated to any degree of accuracy by numerical integration methods (see for example the methods presented by Femenias and Furio (2000) and Tellambura (1996)).

We remark that when the outer code is a convolutional code with BPSK modulation, the pairwise error probability for this scheme can be obtained directly from (7.63) by setting $|x_1(k) - x_2(k)| = 2$. As for the average bit error rate for this scheme, it is upper bounded as

$$P_b \leq \frac{1}{n} \sum_{d=d_{\min}}^{\infty} \beta_d P(\boldsymbol{x}_1 \rightarrow \boldsymbol{x}_2), \tag{7.66}$$

where n is the number of new bits shifted into the shift register of the convolutional encoder each time, and β_d is the multiplicity corresponding to distance d, and represents the coefficients of the derivative of the convolutional code transfer function.

Quasi-Static Fading Channels

When the fading coefficients remain fixed for an entire frame, the conditional pairwise error probability defined by (7.58) is expressed as

$$P(\boldsymbol{x}_1 \rightarrow \boldsymbol{x}_2 | \boldsymbol{H}) = Q\left(\sqrt{\frac{\rho d^2}{2N_t} \sum_{i=1}^{N_t} \sum_{j=1}^{N_r} |h_{i,j}|^2}\right), \tag{7.67}$$

where

$$d^2 \triangleq \sum_{k=1}^{d(\boldsymbol{x}_1, \boldsymbol{x}_2)} |x_1(k) - x_2(k)|^2 \tag{7.68}$$

is the squared Euclidean distance of the outer code. Define Y as

$$Y \triangleq \sum_{i=1}^{N_t} \sum_{j=1}^{N_r} |h_{i,j}|^2, \tag{7.69}$$

which is obviously a chi-square distributed random variable with $2N_t N_r$ degrees of freedom and whose p.d.f. is given by (7.61). By using Craig's formula for $Q(x)$, we have

$$\begin{aligned} P(\boldsymbol{x}_1 \rightarrow \boldsymbol{x}_2) &= \int_0^{\infty} Q\left(\sqrt{\frac{\rho}{2N_t} d^2 y}\right) p_Y(y) dy, \\ &= \frac{1}{\pi} \int_0^{\pi/2} \left(1 + \frac{\rho d^2}{4N_t \sin^2\theta}\right)^{-N_t N_r} d\theta. \end{aligned} \tag{7.70}$$

At high signal-to-noise ratios, (7.70) can be approximated as

$$P(\boldsymbol{x}_1 \to \boldsymbol{x}_2) \approx \frac{1}{\pi}\int_0^{\pi/2}\left(\frac{1}{\sin^2\theta}\right)^{-N_tN_r} d\theta \left(\frac{d^2\rho}{4N_t}\right)^{-N_tN_r},$$

$$= \binom{2N_tN_r - 1}{N_tN_r}\left(\frac{d^2(\boldsymbol{x}_1, \boldsymbol{x}_2)\rho}{N_t}\right)^{-N_tN_r}. \tag{7.71}$$

The expression in (7.71) suggests that the diversity order achieved is N_tN_r, suggesting that employing an outer channel code does not provide any additional diversity when the channel is quasi-static fading, but it provides a coding gain, as expected.

Upper bounds on the bit error rate can be obtained by applying (7.64) and (7.66) for TCM and convolutional codes, respectively.

7.5.4 Examples

In this section, we examine the performance of the coded STBC scheme for various scenarios. We consider a convolutional code and a TCM code as the outer channel codes. In all cases, it is assumed, unless otherwise stated, that the channel fading coefficients remain fixed over a block of N_t consecutive symbols and change independently from one block to another. For the convolutional code, we use a rate 1/2 code with generator polynomials $(7, 5)_{octal}$, and employ BPSK modulation. The minimum distance for this code is $d_{\min} = 5$. Thus, the maximum diversity order for the ideally interleaved channel is $5N_tN_r$. To achieve ideal interleaving, the interleaving depth spans N_t consecutive codewords at the output of the convolutional encoder, which guarantees that consecutive coded bits in a codeword see independent fades. The bit error performance for this scheme is plotted in Figure 7.23 for $N_t = 2$ and $N_r = 1, 2, 3$. We also plot in the same figure the corresponding performance bounds for this scheme given by (7.66). It is clear from the figure that the maximum diversity is achieved, which is $10N_r$ in this case.

In Figure 7.24, we examine the bit error rate performance of the convolutional coded STBC system in block fading. In this case, the channel is not interleaved with the purpose of demonstrating the impact of interleaving (or lack of) on the performance. The length of the codeword at the output of the convolutional encoder is assumed to be $N = 20{,}000$. Four values of differently faded blocks are considered, namely, $L = 10$, 1000, 10,000 and 20,000. We observe from the figure that the convolutional code contributes more to the diversity when L is small and more to the coding gain as L increases. If interleaving was used, more diversity could be achieved.

For the TCM code, we use a rate 2/3, 4-state, 8-PSK TCM scheme presented by Wilson and Leung (1987). The corresponding symbol-wise Hamming distance for this code is two. Consequently, the maximum diversity order achieved is $2N_tN_r$. The bit error performance for this code is plotted in Figure 7.25 along with the corresponding upper bounds for the cases with $N_t = 2$ and $N_r = 1, 2, 3$. It is clear from the figure that the maximum diversity is achieved.

In Figure 7.26, we compare the bit error rate performance for the convolutionally coded and uncoded STBC systems with 8-PSK modulation over fully interleaved channels. The convolutional code is the rate 1/2, 4-state code mentioned above. (The acronym CC in

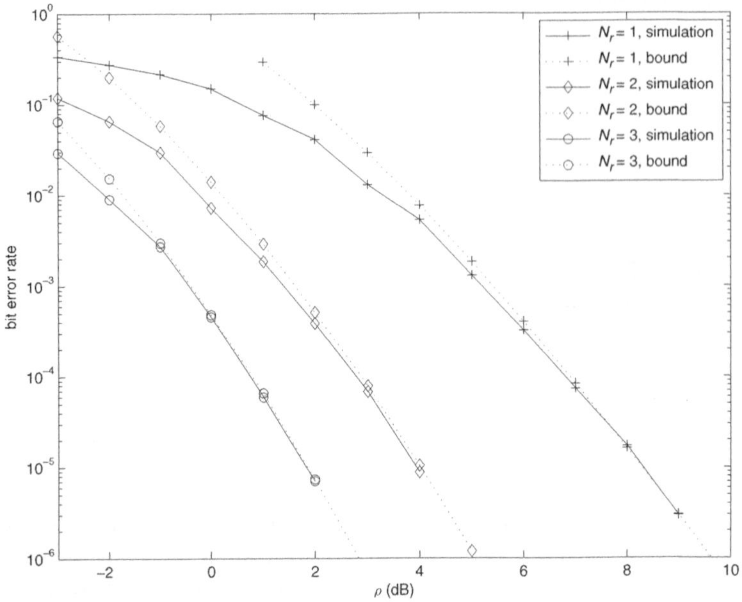

Figure 7.23 Bit error rate performance for the convolutional coded STBC scheme on an ideally interleaved channel with BPSK modulation for $N_t = 2$ and $N_r = 1, 2, 3$.

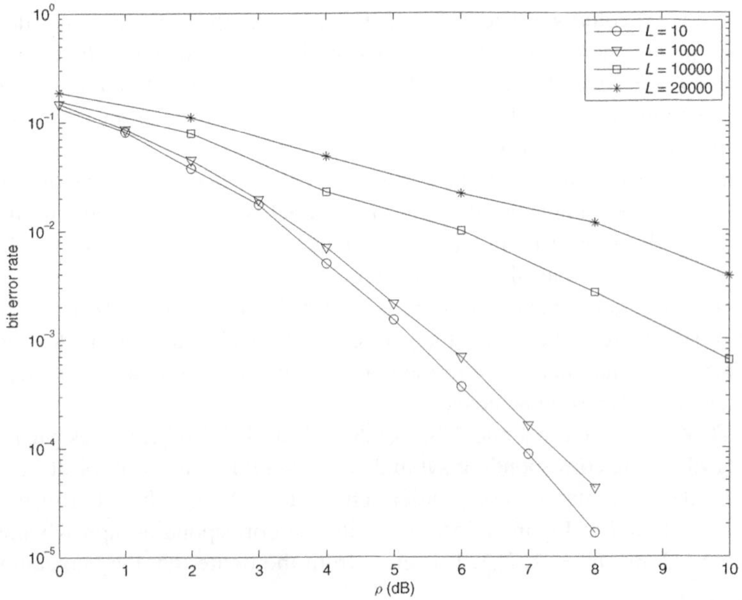

Figure 7.24 Bit error rate performance of the convolutional coded STBC over block fading channels without interleaving.

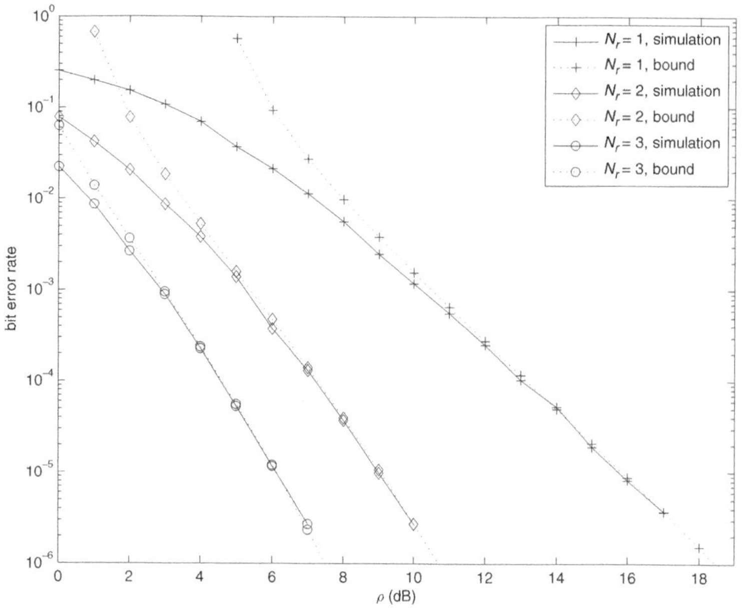

Figure 7.25 Bit error performance of a rate 2/3, 4-state TCM coded STBC system on an ideally interleaved channel for $N_t = 2$, $N_r = 1, 2, 3$.

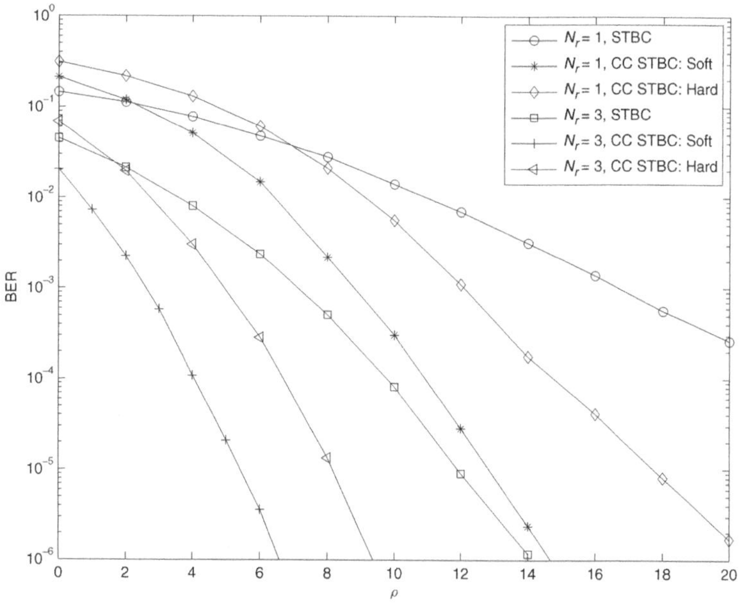

Figure 7.26 Bit error rate performance comparison between the convolutional coded STBC and uncoded STBC systems with 8-PSK modulation for $N_t = 2$ and $N_r = 1, 3$. Both soft and hard decision decoding are considered.

the figure means convolutional coding.) We consider a MIMO system with $N_t = 2$ and $N_r = 1, 3$. We also consider soft and hard decision decoding. The diversity improvement by employing a convolutional code over the uncoded case is clear from the figure. We also observe the degradation in diversity when hard decision decoding is used as opposed to soft decision decoding.

7.6 Chapter Summary and Further Reading

In this chapter, we discussed details of various code concatenation schemes involving channel coding and space-time codes. We gave a thorough treatment of parallel and serial concatenated codes over AWGN channels. We considered in detail several concatenated coding schemes designed for MIMO systems. We have seen that significant performance improvements can be achieved when channel coding is combined with space-time coding, particularly for block fading channels with interleaving. We note that this chapter may serve as a good starting point for learning concatenated coding and iterative decoding.

In the following, we point to some references for aspects of concatenated space-time codes that have not been considered in this chapter. Li et al. (2005b) propose a concatenation scheme, which comprises a turbo code and an STTC. It is referred to as assembled space-time turbo trellis code. Concatenated TCM and STBC is considered by Gong and Letaief (2002). The scheme considered by Jung and Cheun (2003) is simply a serial concatenation of a unitary precoder and the Alamouti scheme where it is shown that full diversity and full-rate are achieved. Tarasak and Bhargava (2004) describe a concatenation scheme comprising an outer TCM code and a differentially encoded STBC. A concatenation scheme consisting of multiple TCM codes and an STBC is considered by Siwamogsatham and Fitz (2005).

The scheme considered by Li et al. (2005a) is comprised of parallel concatenation of several rate $1/2$ two-state convolutional codes designed for an arbitrary number of transmit antennas in an effort to increase the bandwidth efficiency. Hou et al. (2005) and Wang et al. (2004b) consider the serial concatenation of an outer LDPC code and an inner space-time code. A serial concatenation of an outer standard space-time code and a rate-1 recursive inner code is proposed by Lin and Blum (2000). Concatenated code design for spatially correlated channels is presented by Siwamogsatham and Fitz (2002). Li and Xia (2006) propose an iterative demodulation/decoding method for convolutional space-time codes, and Reial and Wilson (2005) propose a low-complexity multistage iterative decoding algorithm for concatenated space-time codes.

Problems

7.1 Consider a rate $1/3$ PCCC whose encoder is shown in Figure 7.1 with generator polynomials $g_1(D) = 1 + D^2$ and $g_2(D) = 1 + D + D^2$. Assume that both RSC encoders are identical.

a) What is d_{free}? What are the input patterns that result in this d_{free}?

b) Why is it desirable to employ a primitive polynomial for $g_2(D)$?

c) What is the bit sequence that needs to be appended to the input information sequence to force the top RSC encoder to terminate at state zero?

7.2 Consider the SCCC scheme, whose encoder is shown in Figure 7.2. Assume that the input block length is N, the outer code rate is $2/3$ and the inner code rate is $1/2$. Both codes are systematic (and linear) and the interleaver is of length $3N/2$.

a) What is the overall code rate? Is the overall code linear?

b) Let $t_O(i, d)$ be the number of codewords of the outer code with information weight i and parity weight d, and $t_I(i, d)$ be the number of codewords of the inner code with information weight i and parity weight d. Assume that the received signal-to-noise ratio per code bit is ρ and the channel is AWGN with noise variance $1/2$. Assuming BPSK signaling, give the union bound on the bit error probability assuming that the interleaver is uniform.

7.3 Prove Equation (7.12).

7.4 Reproduce Figure 7.11.

7.5 Reproduce Figure 7.19.

7.6 Derive Equation (7.52).

7.7 Derive the pairwise error probability given by Equation (7.63) using the Chernoff bound instead of using Craig's formula.

7.8 The pairwise error probability given by Equation (7.71) assumes soft decision ML decoding. Derive the corresponding pairwise error probability with hard decision ML decoding.

7.9 Reproduce Figure 7.26 for the hard decision decoding case.

8

Unitary and Differential Space-Time Codes

The space-time codes we have considered so far assume coherent detection. That is, the receiver has perfect knowledge of the CSI a priori. In practical systems, the CSI estimation is accomplished by using pilot symbols transmitted along with the data which are known to the receiver. The underlying assumption for this scenario to be effective is that the channel remains fixed for a certain amount of time. The accuracy of the CSI estimation obviously improves with the number of pilot symbols used in the estimation process, giving rise to a clear trade-off between estimation accuracy and bandwidth efficiency. The subject of channel estimation and its impact on the performance is treated in Chapter 10. In applications where there is high mobility, the channel is likely to change frequently, and thus the CSI estimation based on pilot symbols placed periodically in a data frame may no longer be useful. The problem becomes even more difficult when the number of transmit antennas is relatively large as the channel may change before training is completed.

As an alternative, one may use noncoherent detection whereby the receiver does not need any knowledge of the CSI to perform data detection. To this end, two classes of space-time codes for noncoherent channels have been recently proposed, namely, unitary space-time codes (USTCs) and differential space-time codes (DSTCs).

In this chapter, we describe details of USTCs and DSTCs, including their encoding, decoding, code construction, and performance analysis. We begin with the channel capacity of noncoherent MIMO channels, and use this to motivate the use of USTCs and DSTCs as they have the potential of achieving this promised capacity. Furthermore, we consider the serial concatenation of USTCs and DSTCs with outer channel codes, including turbo codes. We demonstrate how such concatenation schemes can be decoded iteratively for improved performance. We conclude by giving a chapter summary and providing suggestions for further reading on the subject.

Coding for MIMO Communication Systems Tolga M. Duman and Ali Ghrayeb

8.1 Capacity of Noncoherent MIMO Channels

The capacity of coherent MIMO channels was discussed in great detail in Chapter 3 where it was shown that the capacity, for a rich scattering environment, increases by $\min(N_t, N_r)$ bits per channel use for every 3 dB increase in the signal-to-noise ratio. In this section, we consider the capacity of noncoherent MIMO channels and compare it with that of coherent MIMO channels. We also consider the signal structure that would achieve the resulting channel capacity.

Consider a MIMO system employing N_t transmit antennas and N_r receive antennas. The subchannels between these antennas are assumed to be independent and flat fading. The fading coefficients are assumed to remain constant over T consecutive symbol intervals. This can be a good model for time-division multiple access (TDMA), frequency hopping, and systems whose transmission is packet-based.

The signal received by antenna j at time k, denoted by $y_j(k)$, can be expressed as

$$y_j(k) = \sqrt{\rho}\sum_{i=1}^{N_t} x_i(k)h_{i,j} + n_j(k), \tag{8.1}$$

for $j = 1, 2, \ldots, N_r$; $k = 1, 2, \ldots, T$, where ρ denotes the average signal-to-noise ratio at each receive antenna, $h_{i,j}$ represents the fading coefficient between the ith transmit and jth receive antennas, $x_i(k)$ is the transmitted symbol from antenna i at time k, and $n_j(k)$ is the noise sample corresponding to the jth receive antenna at time k. The fading coefficients are assumed to be constant for $k = 1, 2, \ldots, T$. The fading coefficients and AWGN samples are assumed to be i.i.d. and complex Gaussian distributed with zero mean and variance 1/2 per dimension. It is also assumed that the signal constellation is scaled such that $E\left[|X|^2\right] \leq 1$. Consequently, the expected received signal-to-noise ratio at each receive antenna is ρ independently of N_t.

For the sake of making the notation simpler, we express the received signal over T symbol intervals, denoted by $\boldsymbol{Y}$, in a compact matrix form as

$$\boldsymbol{Y} = \sqrt{\rho}\boldsymbol{X}\boldsymbol{H} + \boldsymbol{N}, \tag{8.2}$$

where $\boldsymbol{H}$ is the channel matrix expressed as

$$\boldsymbol{H} = \begin{bmatrix} h_{1,1} & h_{1,2} & \cdots & h_{1,N_r} \\ h_{2,1} & h_{2,2} & \cdots & h_{2,N_r} \\ \vdots & \vdots & \ddots & \vdots \\ h_{N_t,1} & h_{N_t,2} & \cdots & h_{N_t,N_r} \end{bmatrix},$$

$\boldsymbol{X}$ is the transmitted signal expressed as

$$\boldsymbol{X} = \begin{bmatrix} x_1(1) & x_2(1) & \cdots & x_{N_t}(1) \\ x_1(2) & x_2(2) & \cdots & x_{N_t}(2) \\ \vdots & \vdots & \ddots & \vdots \\ x_1(T) & x_2(T) & \cdots & x_{N_t}(T) \end{bmatrix},$$

and N is the AWGN complex noise matrix expressed as

$$N = \begin{bmatrix} n_1(1) & n_2(1) & \cdots & n_{N_r}(1) \\ n_1(2) & n_2(2) & \cdots & n_{N_r}(2) \\ \vdots & \vdots & \ddots & \vdots \\ n_1(T) & n_2(T) & \cdots & n_{N_r}(T) \end{bmatrix}.$$

The columns of Y, conditioned on X, are independent and identically distributed (for every receive antenna). Let y_l denote the lth column of Y, which can be expressed as

$$y_l = \sqrt{\rho} X h_l + n_l,$$

where h_l and n_l denote the lth column of H and lth column of N, respectively. The corresponding $T \times T$ covariance matrix is given as

$$\begin{aligned} \Lambda &\triangleq E\left\{y_l y_l^H\right\}, \\ &= I_T + \rho X X^H, \end{aligned} \tag{8.3}$$

where I_T is the $T \times T$ identity matrix. As such, the probability density function of Y conditioned on X is given as

$$p(Y|X) = \frac{\exp\left(-\text{trace}\left\{\Lambda^{-1} Y Y^H\right\}\right)}{\pi^{T N_r} (\det \Lambda)^{N_r}}, \tag{8.4}$$

where $\det A$ denotes the determinant of A.

8.1.1 Channel Capacity

The mutual information, expressed in bits per channel use, is given in general as

$$I(Y; X) = H(Y) - H(Y|X), \tag{8.5}$$

where $H(Y)$ is the entropy of Y and $H(Y|X)$ is the entropy of Y given X. The joint p.d.f. of X, i.e., $p(X)$, should be such that (8.5) is maximized in order to obtain the capacity of this noncoherent channel, which we denote by C_{incoh}, i.e.,

$$C_{incoh} = \frac{1}{T} \sup_{p(X)} I(X; Y) \quad \text{(in bits per channel use)}. \tag{8.6}$$

The challenge in the optimization of (8.6) is the high dimensionality of the signals involved. An initial attempt to compute this capacity was made by Marzetta and Hochwald (1999) where the capacity of some simple scenarios is calculated, particularly for the $N_t = N_r = 1$ case. Then a solution to this problem was proposed by Zheng and Tse (2003), which involves transforming the optimization problem into a new coordinate system. With this transformation, the dimensionality of the problem can be reduced from $T N_t$ to $\min(N_t, N_r)$, which makes it a lot simpler to solve.

Define $\zeta \triangleq \min(N_t, N_r)$. When $T \geq \zeta + N_r$, the capacity can be approximated at high signal-to-noise ratios as

$$C_{incoh} \approx \zeta \left(1 - \frac{\zeta}{T}\right) \log_2 \rho \quad \text{bits per channel use,} \tag{8.7}$$

which suggests that as $T \to \infty$, the capacity of noncoherent MIMO channels approaches that of coherent MIMO channels. On the other hand, when $T < \zeta + N_r$, the capacity still increases with the signal-to-noise ratio. In both cases, the increase in capacity is

$$\zeta^* \left(1 - \frac{\zeta^*}{T}\right) \quad \text{bits per channel use,} \tag{8.8}$$

for every 3 dB increase in signal-to-noise ratio, where $\zeta^* \triangleq \min(N_t, N_r, \lfloor T/2 \rfloor)$ and $\lfloor x \rfloor$ denotes the integer part of x.

A few observations can be made from (8.7) and (8.8). First, ζ^* represents the number of degrees of freedom of the noncoherent MIMO channel, whereas ζ represents the number of degrees of freedom of the coherent MIMO channel. Second, the increase in capacity for the noncoherent case approaches that of the coherent case as T grows very large. Lastly, at high signal-to-noise ratios, it would be optimal to use only $N_t = \zeta^*$ since the capacity does not increase any further by using more transmit antennas.

We plot in Figure 8.1 the approximate capacity achieved (in bits per channel use) against ρ in dB for a $N_t = N_r = 8$ MIMO system when $\boldsymbol{H}$ is known and when it is unknown. Two values of T are considered, namely, $T = 16$ and 40. We observe from the figure

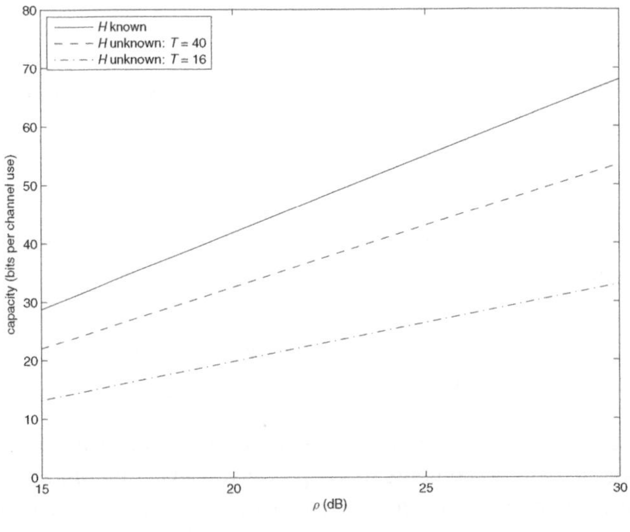

Figure 8.1 Capacity comparisons of an $N_t = N_r = 8$ MIMO channel for the coherent and noncoherent cases with $T = 16$ and 40.

that as T increases, the gap between the capacity of coherent and noncoherent channels decreases, and both capacities approach each other as T goes to ∞ and/or as ρ goes to ∞. We remark that the results reported in the figure are valid for very high signal-to-noise ratios and without any restriction on the signal constellation.

8.1.2 Capacity Achieving Signals

It is shown by Marzetta and Hochwald (1999) that signals that asymptotically achieve the capacity defined by (8.7) should be in the form

$$\boldsymbol{X} = \boldsymbol{\Phi}\boldsymbol{V}, \tag{8.9}$$

where $\boldsymbol{\Phi}$ is an isotropically distributed $T \times N_t$ matrix whose columns are orthonormal, and $\boldsymbol{V}$ is an independent $N_t \times N_t$ real, non-negative, diagonal matrix. An *isotropically distributed* matrix has a probability density that is unchanged when the matrix is left-multiplied by any deterministic unitary matrix. The matrix $\boldsymbol{\Phi}$ also has the property that $\boldsymbol{\Phi}^H\boldsymbol{\Phi} = \boldsymbol{I}_{N_t}$ and its p.d.f. does not change when left-multiplied by any $N_t \times N_t$ unitary matrix. Such signals are normally referred to as *unitary space-time signals*.

Let the diagonal elements of $\boldsymbol{V}$ be v_i, $i = 1, 2, \ldots, N_t$. Consequently, $\boldsymbol{X}$ can be expressed as

$$\boldsymbol{X} = \begin{bmatrix} v_1\boldsymbol{\phi}_1 & v_2\boldsymbol{\phi}_2 & \cdots & v_{N_t}\boldsymbol{\phi}_{N_t} \end{bmatrix}, \tag{8.10}$$

where $\boldsymbol{\phi}_l$ is the lth column of $\boldsymbol{\Phi}$. Furthermore, when $T \gg N_t$ or the signal-to-noise ratio is very high with $T > N_t$, by setting $v_1 = v_2 = \cdots = v_{N_t} = \sqrt{T}$, the signal $\boldsymbol{X}$ still achieves capacity. Note that matrix $\boldsymbol{V}$ determines the magnitude of $\boldsymbol{X}$ whereas $\boldsymbol{\Phi}$ determines its direction.

8.2 Unitary Space-Time Codes

8.2.1 USTC Encoder

A block diagram of the USTC encoder is shown in Figure 8.2. As shown in the figure, the primitive data stream is demultiplexed into parallel TR bit substreams. The unitary space-time modulator accepts TR bits at a time and maps them onto one of the complex

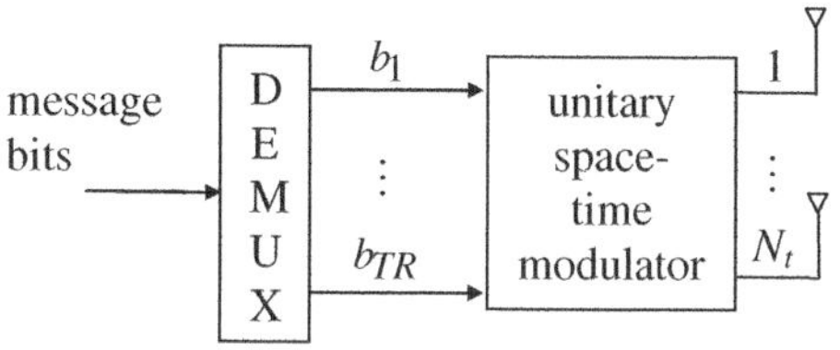

Figure 8.2 USTC encoder.

unitary space-time signals, whose elements are consequently transmitted from the available N_t antennas over T symbol intervals. Note that, to achieve a transmission rate of R bits per channel use, the constellation size should be $L = 2^{TR}$. Later in this chapter we will show how such signals are generated and decoded. Note that when channel coding is used, which will be treated later, the actual transmission rate is reduced to $R_c R$ where R_c is the rate of the channel code. The input to the demultiplexer in this case will be a coded sequence.

8.2.2 ML Detection of USTCs

Assume that the signal constellation size is L and that the signals have the structure

$$X_k = \sqrt{T}\boldsymbol{\Phi}_k, \quad k = 1, 2, \ldots, L \tag{8.11}$$

where $\boldsymbol{\Phi}_k$ are $T \times N_t$ complex matrices satisfying $\boldsymbol{\Phi}_k^H \boldsymbol{\Phi}_k = \boldsymbol{I}_{N_t}$. Note that to obtain (8.11) from (8.9), we set $v_1 = v_2 = \cdots = v_{N_t} = \sqrt{T}$, where this is motivated by the fact that such signals asymptotically achieve the capacity given by (8.7).

When $\boldsymbol{H}$ is unknown at the receiver, given the received signal $\boldsymbol{Y}$, the ML decoder will select $\boldsymbol{\Phi}_k$ that maximizes $P(\boldsymbol{Y}|\boldsymbol{\Phi}_k)$ (Hochwald and Marzetta (2000)). That is,

$$\boldsymbol{\Phi}_{ML} = \underset{\boldsymbol{\Phi}_k \in \{\boldsymbol{\Phi}_1, \ldots, \boldsymbol{\Phi}_L\}}{\arg\max} \; p(\boldsymbol{Y}|\boldsymbol{\Phi}_k) \tag{8.12}$$

$$= \underset{\boldsymbol{\Phi}_k \in \{\boldsymbol{\Phi}_1, \ldots, \boldsymbol{\Phi}_L\}}{\arg\max} \; \frac{\exp\left(-\mathrm{trace}\left\{\boldsymbol{\Lambda}^{-1}\boldsymbol{Y}\boldsymbol{Y}^H\right\}\right)}{\pi^{TN_r}(\det \boldsymbol{\Lambda})^{N_r}} \tag{8.13}$$

$$= \underset{\boldsymbol{\Phi}_k \in \{\boldsymbol{\Phi}_1, \ldots, \boldsymbol{\Phi}_L\}}{\arg\max} \; \frac{\exp\left(-\mathrm{trace}\left\{\left[\boldsymbol{I}_T - \frac{1}{1+1/(\rho T)}\boldsymbol{\Phi}_k\boldsymbol{\Phi}_k^H\right]\boldsymbol{Y}\boldsymbol{Y}^H\right\}\right)}{\pi^{TN_r}(1+\rho T)^{N_t N_r}} \tag{8.14}$$

$$= \underset{\boldsymbol{\Phi}_k \in \{\boldsymbol{\Phi}_1, \ldots, \boldsymbol{\Phi}_L\}}{\arg\max} \; \mathrm{trace}\left\{\boldsymbol{Y}^H\boldsymbol{\Phi}_k\boldsymbol{\Phi}_k^H\boldsymbol{Y}\right\}. \tag{8.15}$$

Expression (8.14) is obtained from (8.13) by using the identities

$$(\boldsymbol{A} + \boldsymbol{BCD})^{-1} = \boldsymbol{A}^{-1} - \boldsymbol{A}^{-1}\boldsymbol{B}\left(\boldsymbol{C}^{-1} + \boldsymbol{DA}^{-1}\boldsymbol{B}\right)^{-1}\boldsymbol{DA}^{-1},$$

$$\mathrm{trace}(\boldsymbol{AB}) = \mathrm{trace}(\boldsymbol{BA}),$$

and

$$\det(\boldsymbol{I} + \boldsymbol{AB}) = \det(\boldsymbol{I} + \boldsymbol{BA}).$$

We now give an intuitive explanation as to why the noncoherent receiver can detect the transmitted signal effectively at high signal-to-noise ratio, as long as it has the structure described above. Recall that $\boldsymbol{\Phi}_k$ is a $T \times N_t$ complex matrix. As such, the columns of $\boldsymbol{\Phi}_k$ span an N_t-dimensional subspace in the T-dimensional complex vector space. The subspace in this vector space is a hyperplane. This suggests that different signals $\boldsymbol{\Phi}_i$ and $\boldsymbol{\Phi}_j$ generating nonidentical subspaces yield two distinct hyperplanes that intersect on some lower dimensional hyperplane. The probability that the received signal lands on one of these intersections is zero. Therefore, as long as the columns of $\boldsymbol{\Phi}_i$ and $\boldsymbol{\Phi}_j$ span different subspaces, the receiver can perfectly distinguish $\boldsymbol{\Phi}_i$ from $\boldsymbol{\Phi}_j$ independently of $\boldsymbol{H}$.

When $\boldsymbol{H}$ is known at the receiver, the received signal can be detected coherently. In this case, the conditional p.d.f. can be expressed as

$$p(\boldsymbol{Y}|\boldsymbol{\Phi}_k, \boldsymbol{H}) = \frac{\exp\left(-\text{trace}\left\{\left(\boldsymbol{Y} - \sqrt{\rho T}\boldsymbol{\Phi}_k\boldsymbol{H}\right)\left(\boldsymbol{Y} - \sqrt{\rho T}\boldsymbol{\Phi}_k\boldsymbol{H}\right)^H\right\}\right)}{\pi^{TN_r}}.$$

Therefore, the ML detector will reduce to the minimum Euclidean distance detector implemented as

$$\begin{aligned}
\boldsymbol{\Phi}_{ML}^{coherent} &= \underset{\boldsymbol{\Phi}_k \in \left\{\boldsymbol{\Phi}_1,\ldots,\boldsymbol{\Phi}_L\right\}}{\arg\max} \quad p(\boldsymbol{Y}|\boldsymbol{\Phi}_k, \boldsymbol{H}) \\
&= \underset{\boldsymbol{\Phi}_k \in \left\{\boldsymbol{\Phi}_1,\ldots,\boldsymbol{\Phi}_L\right\}}{\arg\min} \quad \text{trace}\left\{\left(\boldsymbol{Y} - \sqrt{\rho T}\boldsymbol{\Phi}_k\boldsymbol{H}\right)\left(\boldsymbol{Y} - \sqrt{\rho T}\boldsymbol{\Phi}_k\boldsymbol{H}\right)^H\right\} \\
&= \underset{\boldsymbol{\Phi}_k \in \left\{\boldsymbol{\Phi}_1,\ldots,\boldsymbol{\Phi}_L\right\}}{\arg\max} \quad \text{Re}\left\{\text{trace}\left\{\boldsymbol{\Phi}_k\boldsymbol{H}\boldsymbol{Y}^H\right\}\right\}. \qquad (8.16)
\end{aligned}$$

By comparing (8.15) and (8.16), it is clear that the noncoherent detector is able to perform data detection without the knowledge of the CSI, whereas the coherent detector requires the knowledge of the CSI. Apart from the complexity associated with estimating $\boldsymbol{H}$, both detectors have similar complexities as they both require to consider all possible candidate signals.

8.2.3 Performance Analysis

We now examine the probability of error performance of USTCs in the absence of the CSI. Since all the transmitted signals are assumed to be equally likely, the probability of error, denoted by P_e, can be upper bounded using the union bound argument as

$$\begin{aligned}
P_e &= \frac{1}{L}\sum_{k=1}^{L} P(\text{error}|\boldsymbol{\Phi}_k \text{ transmitted}), \\
&\leq \frac{1}{L}\sum_{k=1}^{L}\sum_{k' \neq k} \Pr\left(\boldsymbol{\Phi}_k \to \boldsymbol{\Phi}_{k'}\right), \qquad (8.17)
\end{aligned}$$

where $\Pr\left(\boldsymbol{\Phi}_k \to \boldsymbol{\Phi}_{k'}\right)$ denotes the probability that the receiver will erroneously pick $\boldsymbol{\Phi}_{k'}$ over $\boldsymbol{\Phi}_k$, which is essentially the pairwise error probability. Obtaining a closed-form expression for $\Pr\left(\boldsymbol{\Phi}_k \to \boldsymbol{\Phi}_{k'}\right)$ for a general constellation of unitary space-time signals is cumbersome. As an alternative, one may use the Chernoff bound to find an upper bound, which can be expressed as (Hughes (2000)),

$$\Pr\left(\boldsymbol{\Phi}_k \to \boldsymbol{\Phi}_{k'}\right) \leq \frac{1}{\det\left(\boldsymbol{I}_{N_t} + \frac{(\rho T)^2}{4(1+\rho T)}\left[\boldsymbol{I}_{N_t} - \boldsymbol{\Phi}_{k'}^H\boldsymbol{\Phi}_k\boldsymbol{\Phi}_k^H\boldsymbol{\Phi}_{k'}\right]\right)^{N_r}}. \qquad (8.18)$$

Clearly, the diversity order is dictated by the rank of matrix $\boldsymbol{I}_{N_t} - \boldsymbol{\Phi}_{k'}^H\boldsymbol{\Phi}_k\boldsymbol{\Phi}_k^H\boldsymbol{\Phi}_{k'}$, whereas the coding gain depends on the nonzero eigenvalues of this matrix. As such, to achieve the maximum diversity, this matrix has to be full rank, i.e., N_t, and the product of its nonzero eigenvalues should be maximum. Although this design criterion resembles that of STTCs, it is somewhat computationally complex, especially since the number of matrices generated is normally very large.

Another approach to upper bounding the pairwise error probability was introduced by Hochwald et al. (2000), which leads to a simpler code design criterion. The result is

$$\Pr(\boldsymbol{\Phi}_k \rightarrow \boldsymbol{\Phi}_{k'}) \leq \frac{1}{2}\prod_{k=1}^{N_t}\left[\frac{1}{1+\frac{(\rho T)^2\left(1-\upsilon_k^2\right)}{4(1+\rho T)}}\right]^{N_r}, \tag{8.19}$$

where $1 \geq \upsilon_1 \geq \upsilon_2 \geq \cdots \geq \upsilon_{N_t} \geq 0$ are the singular values of the $N_t \times N_t$ correlation matrix $\boldsymbol{\Phi}_k^H \boldsymbol{\Phi}_{k'}$.

By examining the expression in (8.19), we can see that the Chernoff bound is lowest, and so is P_e, when the singular values υ_k are all zeros, and it is highest when the υ_k are all ones. The values of υ_k reflect the amount of overlap between the subspaces generated by $\boldsymbol{\Phi}_k$ and $\boldsymbol{\Phi}_{k'}$. Ideally, we would like this overlap to be null, and this can be achieved only when all the N_t columns of $\boldsymbol{\Phi}_k$ are orthogonal to all the N_t columns of $\boldsymbol{\Phi}_{k'}$. More importantly, the maximum spatial diversity that can be achieved is $N_t N_r$, which is similar to the diversity achieved when $\boldsymbol{H}$ is known at the receiver and coherent detection is performed. However, the condition for the maximum diversity to be possible is that all the singular values υ_k be strictly less than one. This is clearly possible only if $T \geq 2N_t$.

8.2.4 Construction of Unitary Space-Time Signals

To achieve R bits per channel use and for a given coherence time T, one needs to generate and store $L = 2^{RT}$ complex matrices each of size $T \times N_t$. In addition, these matrices have to achieve a low probability of error. Obviously this is a tedious task, especially if there is no structure for generating such matrices. In the following, we shall describe a simple *Fourier-based* method for constructing such matrices (Hochwald et al. (2000)). We remark that there is an algebraic approach proposed by Hochwald et al. (2000) to generate such matrices that is equivalent to the Fourier-based approach, but we shall focus on the Fourier-based one for its simplicity. The Fourier-based approach involves generating only one $T \times N_t$ matrix and a $T \times T$ diagonal matrix. The rest of the matrices are generated using these two matrices. We shall begin with the special case when $N_t = 1$ and then generalize the code construction to any N_t.

Signal Construction for $N_t = 1$

When $N_t = 1$, we need to generate L unit vectors in a T-dimensional complex space where it is assumed that $L \gg T$. The Fourier-based construction method is inspired by the theory of *frames* introduced by Duffin and Schaeffer (1952). A collection of L vectors $\boldsymbol{\Phi}_l$ in a T-dimensional space form a *tight frame* if all the eigenvalues of the $T \times T$ matrix $\sum_{k=1}^{L} \boldsymbol{\Phi}_k \boldsymbol{\Phi}_k^H$ are equal, i.e.,

$$\sum_{k=1}^{L} \boldsymbol{\Phi}_k \boldsymbol{\Phi}_k^H = \xi \boldsymbol{I}_T,$$

where ξ is the frame constant.

The construction of tight frames can be achieved by projecting an L-dimensional discrete Fourier transform (DFT) basis onto a T-dimensional space. This projection results in retaining the first T components of the L-dimensional vectors. Following this approach, the resulting L signals are of the form

$$\mathbf{\Phi}_k = \frac{1}{\sqrt{T}} \begin{bmatrix} 1 \\ e^{j\frac{2\pi}{L}(k-1)} \\ e^{j\frac{2\pi}{L}2(k-1)} \\ \vdots \\ e^{j\frac{2\pi}{L}(T-1)(k-1)} \end{bmatrix}, \tag{8.20}$$

for $k = 1, 2, \ldots, L$. The correlation between these signals is given as

$$\begin{aligned} \upsilon_1 &= \left|\mathbf{\Phi}_k^H \mathbf{\Phi}_{k'}\right|, \\ &= \begin{cases} 1, & k = k' \\ \frac{1}{T}\left|\sum_{t=1}^{T} e^{j\frac{2\pi}{L}(t-1)(k'-k)}\right|, & k \neq k', \end{cases} \end{aligned}$$

which can be expressed as

$$\upsilon_1 = \begin{cases} 1, & k = k' \\ \left|\frac{\sin(\pi(k'-k)T/L)}{T\sin(\pi(k'-k)/L)}\right|, & k \neq k'. \end{cases} \tag{8.21}$$

Obviously the performance depends on this correlation where as υ_1 decreases the performance improves and vice versa. Furthermore, the correlation between $\mathbf{\Phi}_k$ and $\mathbf{\Phi}_{k'}$ depends only on $(k' - k) \bmod L$, suggesting that the correlation structure is circulant and thus it suffices to consider only $\left|\mathbf{\Phi}_1^H \mathbf{\Phi}_k\right|$ for $k = 2, 3, \ldots, L$.

We plot in Figure 8.3 the correlation of the signals defined by (8.20) as a function of $\left|k' - k\right|$ for $T = 6$ and $L = 64$. These signals are generated by picking the first $T = 6$ components of each row of the 64×64 DFT matrix. As predicted by (8.21), the correlation follows the sinc function where the maximum correlation is 0.986 which is obtained when $\left|k' - k\right| = 1$. In this example, the transmission rate is $R = 1$ bits per channel use.

Recall from the foregoing discussion that the performance of USTCs is highly dependent upon the correlation among these signals. Thus, this correlation should be minimized as much as possible. Based on the results shown in Figure 8.3, it is clear that some of the signals will be almost indistinguishable from each other due to the high correlation among them, and consequently, the error rate performance is expected to be poor. Therefore, it is not a good idea to pick the first T columns of the $L \times L$ DFT matrix. Instead, one should pick the T columns that yield the smallest correlation possible.

Let the indices of the best columns be $u_1, u_2, \ldots, u_T$. As such, the corresponding signals will have the form

$$\mathbf{\Phi}_k = \frac{1}{\sqrt{T}} \begin{bmatrix} e^{j\frac{2\pi}{L}u_1(k-1)} \\ e^{j\frac{2\pi}{L}u_2(k-1)} \\ \vdots \\ e^{j\frac{2\pi}{L}u_T(k-1)} \end{bmatrix}, \tag{8.22}$$

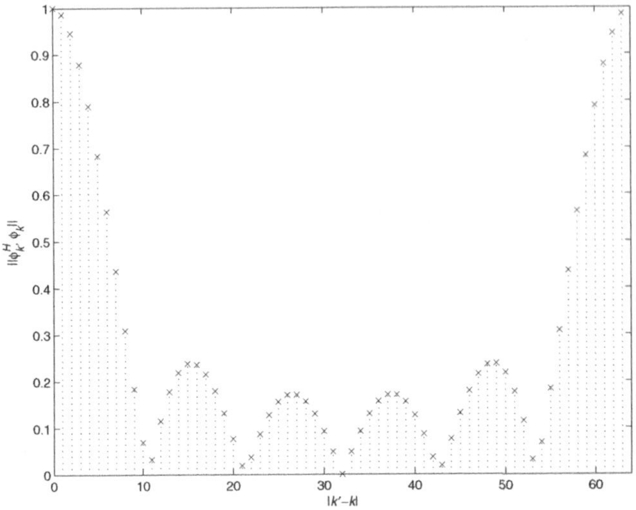

Figure 8.3 Correlation of the signals defined by (8.20) as a function of $|k' - k|$ for $T = 6$ and $L = 64$ resulting from picking the first six columns of the 64×64 DFT matrix.

where without loss of generality $0 \leq u_1, \ldots, u_T \leq L - 1$. These indices are chosen according to the criterion

$$\min_{0 \leq u_1,\ldots,u_T \leq L-1} \max_{k=2,\ldots,L} \frac{1}{T} \sum_{t=1}^{T} \exp\left(j \frac{2\pi}{L} u_t (k-1) \right) = \min_{0 \leq u_1,\ldots,u_T \leq L-1} \delta, \tag{8.23}$$

where $\delta \triangleq \max_{1<k\leq L} \left\| \boldsymbol{\Phi}_1^H \boldsymbol{\Phi}_k \right\|$. To relate δ to the singular values of $\left\| \boldsymbol{\Phi}_1^H \boldsymbol{\Phi}_k \right\|$, we first define μ to be the average of the squares of these singular values, that is,

$$\begin{aligned} \mu &= \frac{1}{N_t} \sum_{i=1}^{N_t} \upsilon_i^2, \\ &= \left\| \boldsymbol{\Phi}_1^H \boldsymbol{\Phi}_k \right\|^2, \end{aligned} \tag{8.24}$$

where $\|\cdot\|^2$ denotes the squared Frobenius norm of a matrix. As such $\delta = \sqrt{\mu}$. For a given fixed average μ, the probability of error performance is minimized when the singular values are equal, i.e., uniformly distributed. Therefore, in generating the indices $u_1, \ldots, u_T$, if two sets of indices have the same δ, the set for which the indices are more uniform should be selected.

Once the column indices $u_1, \ldots, u_T$ are determined, the generation of the rest of the signal matrices is straightforward. To elaborate, let $\boldsymbol{\Omega}$ be a $T \times T$ diagonal matrix defined as

$$\boldsymbol{\Omega} = \begin{bmatrix} e^{j\frac{2\pi}{L}u_1} & 0 & \cdots \\ & \ddots & \\ 0 & \cdots & e^{j\frac{2\pi}{L}u_T} \end{bmatrix}. \tag{8.25}$$

Given that the first signal $\boldsymbol{\Phi}_1$ can be expressed as

$$\boldsymbol{\Phi}_1 = \frac{1}{\sqrt{T}} \begin{bmatrix} 1 \\ 1 \\ \vdots \\ 1 \end{bmatrix}, \tag{8.26}$$

the kth signal can be written in terms of $\boldsymbol{\Phi}_1$ as

$$\boldsymbol{\Phi}_k = \boldsymbol{\Omega}^{k-1}\boldsymbol{\Phi}_1. \tag{8.27}$$

Note that $\boldsymbol{\Omega}^L = \boldsymbol{I}_L$, which guarantees that the generated L matrices are distinct.

When the signal matrices are generated according to the design criterion given by (8.23), as opposed to picking the first T columns of the DFT matrix, the correlation between these signals can be reduced significantly. For example, we plot in Figure 8.4 the correlation between these signals for the same example whose results are plotted in Figure 8.3 but this time the criterion in (8.23) is applied. As shown in the figure, the resulting correlation

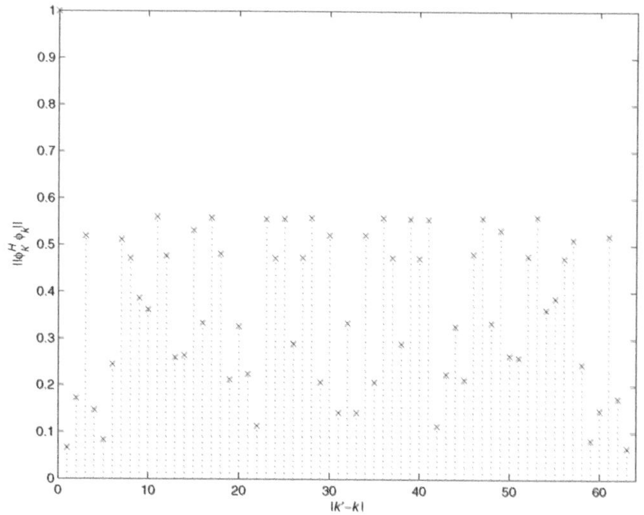

Figure 8.4 Correlation of the signals defined by (8.20) as a function of $|k' - k|$ for $T = 6$ and $L = 64$ generated according to (8.23) (with $\boldsymbol{\Omega}$ generated by random selection).

ranges between about 0.05 and 0.56, which is significantly less than the correlation range shown in Figure 8.3.

In obtaining the set of column indices that result in the lowest correlation, one should ideally perform an exhaustive search among the $\binom{L}{T}$ combinations and pick the best one. However, this approach is computationally complex for large values of L. For example, for $L = 64$ and $T = 6$, the number of combinations that should be searched is $\binom{64}{6} =$ 74,974,368, which is very high. An alternative approach, although suboptimal, is to limit the number of combinations to be considered. In particular, instead of considering all the L columns, one would consider only a small subset of L. Let the size of this subset be $L' < L$. Consequently, the number of resulting combinations to consider would be $\binom{L'}{T}$. The process of picking the subset L' can be repeated randomly several times so as to increase the probability that a good set of columns is found. We refer to this process as *random selection*.

In the above example, we let $L' = 15$, and the experiment is repeated 50 times. Again, in each experiment, the $L' = 15$ columns are selected at random. Obviously the number of combinations to be considered in each experiment is $\binom{15}{6} = 250{,}250$. Therefore, the total number of combinations considered in the 50 experiments is $125{,}125{,}00$, which constitutes only 16.69% of the $\binom{64}{6}$ combinations. Of course the greater the number of experiments the more likely that the best set of columns will be found. In this example, the best combination of column indices among the 125,125,00 combinations was found to be $[1, 18, 23, 39, 46, 57]$. The resulting correlation results are plotted in Figure 8.4 where it is clear that a significant reduction in the correlation among signals is obtained.

Signal Construction for $N_t > 1$

When $N_t > 1$, the constellation generation is somewhat similar to the $N_t = 1$ case except that the initial matrix $\mathbf{\Phi}_1$ is now a $T \times N_t$ complex matrix with $\mathbf{\Phi}_1^H \mathbf{\Phi}_1 = \mathbf{I}_{N_t}$. The same is true for the other matrices where $\mathbf{\Phi}_k^H \mathbf{\Phi}_k = I_{N_t}$. Once $\mathbf{\Phi}_1$ and $\mathbf{\Omega}$ are generated, the generation of the rest of the matrices follows (8.27). Matrix $\mathbf{\Phi}_1$ is generated by picking N_t distinct columns from the $T \times T$ DFT matrix, whereas matrix $\mathbf{\Omega}$ is generated by random selection as explained above.

To show the importance of random selection, we plot in Figure 8.5 the correlation among the signals generated for the $N_t = 3$ case with $L = 64$ and $T = 6$. In this case, $\mathbf{\Phi}_1$ is generated by picking the first three (distinct) columns of a 6×6 DFT matrix, whereas $\mathbf{\Omega}$ is formed by picking the first six columns of the same DFT matrix, i.e., $[u_1, u_2, \ldots, u_6] = [1, 2, \ldots, 6]$. It is clear from the figure that the correlation among signals is quite high. On the other hand, when the columns are selected by random selection, the resulting correlation is reduced, as shown in Figure 8.6. In this example, the indices of the best columns are found to be $[23, 34, 39, 51, 54, 64]$. We finally note that the correlation among constellation signals increases as the number of transmit antennas increases (see for example Figures 8.4 and 8.6). The reason is that, for fixed L and T, as N_t increases, the number of subspaces spanned by the constellation signals increases, and thus there will be more overlap between these subspaces. This consequently translates to higher correlation among these signals.

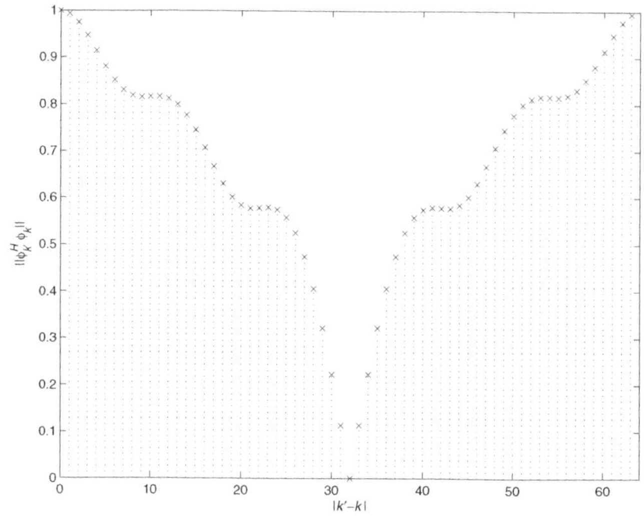

Figure 8.5 Correlation of the signals defined by (8.20) as a function of $|k' - k|$ for the $N_t = 3$ case with $T = 6$ and $L = 64$. These signals resulted from picking the first six columns of the 64×64 DFT matrix.

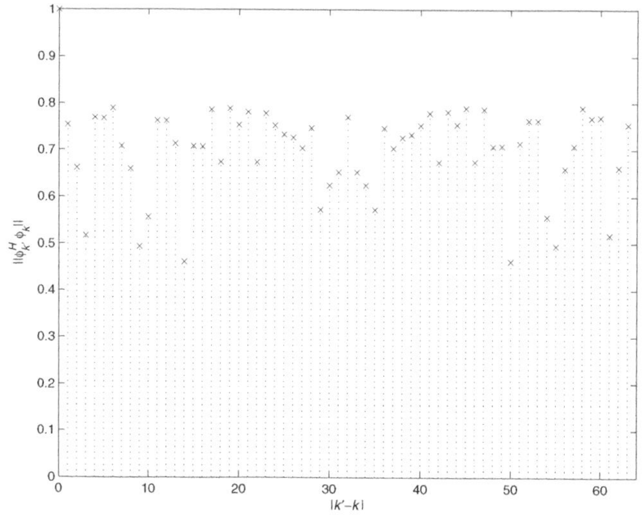

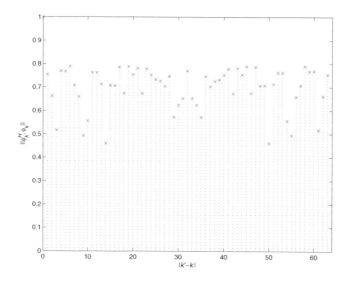

Figure 8.6 Correlation of the signals defined by (8.20) as a function of $|k' - k|$ for the $N_t = 3$ case with $T = 6$ and $L = 64$. These signals are generated according to (8.23) (with $\mathbf{\Omega}$ generated by random selection).

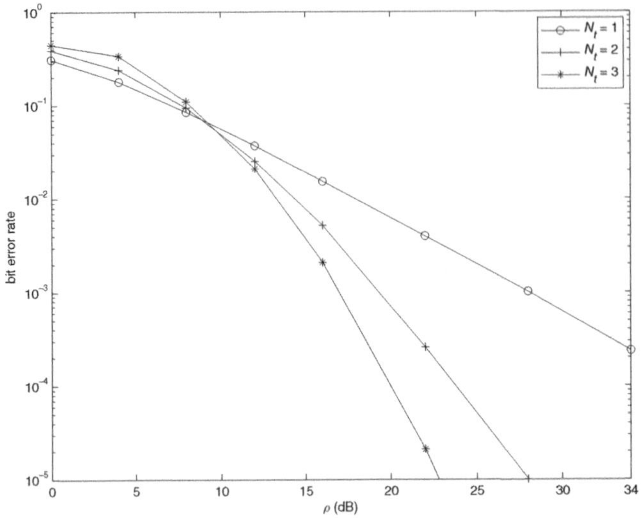

Figure 8.7 Bit error rate performance of a USTC with parameters $L = 64$ and $T = 6$ for $N_t = 1, 2, 3$ and $N_r = 1$.

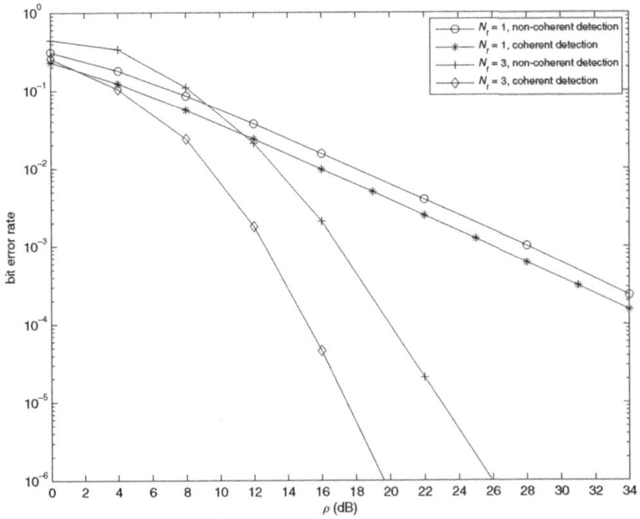

Figure 8.8 Bit error rate performance comparison between the coherent and noncoherent receivers for a USTC with $L = 64$ and $T = 6$ for $N_r = 1$.

8.2.5 Examples

In this section, we examine the bit error rate performance of USTCs via several examples. We plot in Figure 8.7 the bit error rate performance for the USTC mentioned above with parameters $L = 64$ and $T = 6$. We consider three cases of transmit antennas, namely, $N_t = 1, 2, 3$. Binary, i.e., natural, mapping is used for the constellation signal matrices, which is not necessarily optimal. Other mappings such as Gray mapping may be used as well. We observe from the figure the improvement in performance as N_t increases, as expected. In Figure 8.8, we plot the bit error rate performance of the scenarios when the receiver performs coherent detection using (8.16) and noncoherent detection using (8.15). The results show that coherent detection performs better than noncoherent detection by about 2.0 dB for the $N_t = 1$, whereas the gap widens to around 5.0 dB at a bit error rate of 10^{-5}.

8.3 Differential Space-Time Codes

When channel estimation cannot be accomplished at the receiver, an alternative to unitary space-time signals, which are decoded using noncoherent techniques, is the differential space-time coding schemes, which are decoded using differentially coherent receiver approaches. To motivate the use of differential space-time codes, we first discuss how differential encoding and differentially coherent detection can be used in single antenna systems. We then generalize the discussion to MIMO systems.

8.3.1 Differential Space-Time Coding for Single Antenna Systems

Consider a PSK constellation of size $M = 2^b$. The PSK modulator groups every b bits at its input and maps them onto one of the constellation symbols. It is assumed here that all signals have unit energy; this assumption is not necessary but it is used for notational convenience. The outputs of the PSK modulator are then used to generate the differential PSK (DPSK) modulated signal, as shown in Figure 8.9.

The DPSK signal at time k, denoted by $x(k)$, is generated according to

$$x(k) = x(k-1)s(k). \tag{8.28}$$

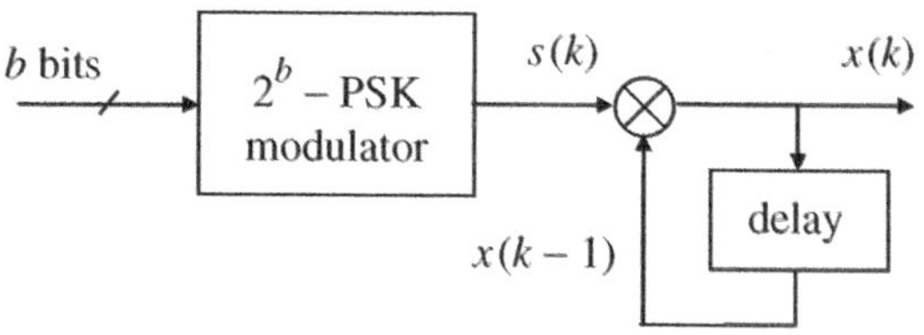

Figure 8.9 DPSK modulator.

Note that the transmitted DPSK signal is a function of the previous transmitted DPSK signal and the present PSK signal. The transmission process begins with an arbitrary, predetermined signal $s(0)$ that is known to the receiver.

In a flat fading environment, the received signal at time k can be expressed as

$$y(k) = \sqrt{\rho} h x(k) + n(k), \tag{8.29}$$

where h is the fading coefficient between the transmit and receive antennas which is modeled as a complex Gaussian random variable with zero mean and variance $1/2$ per dimension; $n(k)$ is the noise sample at time k which is modeled as a complex Gaussian random variable with zero mean and variance $1/2$ per dimension. To recover $s(k)$, the receiver computes $y(k)y^*(k-1)$ which can be expressed as

$$\begin{aligned} y(k)y^*(k-1) &= \left(\sqrt{\rho} h x(k) + n(k)\right)\left(\sqrt{\rho} h^* x^*(k-1) + n^*(k-1)\right) && (8.30)\\ &= |h|^2\left(\rho x(k)x^*(k-1)\right) + \sqrt{\rho} h x(k) n^*(k-1) \\ &\quad + \sqrt{\rho} h^* x^*(k-1) n(k) + n(k) n^*(k-1) && (8.31)\\ &\approx \rho |h|^2 |x(k-1)|^2 s(k) + \sqrt{\rho} h x(k) n^*(k-1) \\ &\quad + \sqrt{\rho} h^* x^*(k-1) n(k) && (8.32)\\ &= \rho |h|^2 s(k) + \sqrt{\rho} h x(k) n^*(k-1) + \sqrt{\rho} h^* x^*(k-1) n(k). && (8.33) \end{aligned}$$

In (8.30) it is assumed that h remains constant over two consecutive symbol intervals, hence the terminology differential coherent detection as opposed to noncoherent detection. Expression (8.32) is obtained from (8.31) by ignoring the noise term $n(k)n^*(k-1)$ which is a valid assumption at high signal-to-noise ratios. The last two terms in (8.33) can be regarded as a single noise term with zero mean and variance $2|h|^2\rho$.

Since h is unknown at the receiver, and given that the optimal decision regions for any PSK constellation depend on the phase of the signal points and not on their magnitude, the differential receiver recovers $s(k)$ according to the decision rule

$$\hat{s}(k) = \arg\min_{s(k)} \left|\angle y(k)y^*(k-1) - \angle s(k)\right|, \tag{8.34}$$

where $\angle a$ denotes the phase of a.

In contrast, in the coherent detection case, i.e., when h is known at the receiver, to recover $s(k)$, the receiver first computes $h^* y(k)$, which can be expressed as

$$h^* y(k) = \sqrt{\rho} |h|^2 x(k) + h^* n(k). \tag{8.35}$$

Note that the noise term in (8.35) has variance $|h|^2$. Also, the received signal, after being multiplied by h^*, is still a function of $x(k)$ where $s(k)$ does not appear explicitly in the expression. This is not a problem because the receiver in this case would recover $x(k)$ instead, and since $s(k-1)$ is already known to the receiver, $s(k)$ can be easily recovered.

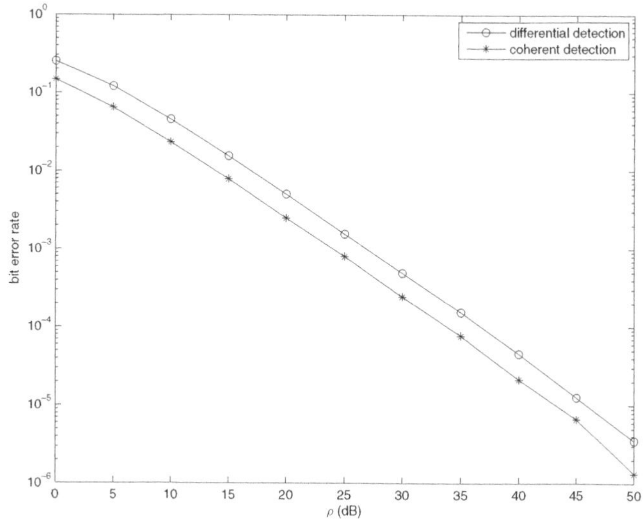

Figure 8.10 Bit error rate performance comparison between coherent and differential detection for the differential BPSK scheme with $N_t = N_r = 1$.

Therefore, the optimal receiver would pick $x(k)$ that minimizes

$$\hat{x}(k) = \arg\min_{x(k)} \left| h^* y(k) - \sqrt{\rho}\, |h|^2 x(k) \right|^2 . \tag{8.36}$$

The penalty for performing differential detection as compared with coherent detection can be assessed by evaluating the overall signal-to-noise ratio in both cases. As per (8.33), the received signal power is $|h|^4 \rho^2$, whereas the noise power is $2\,|h|^2 \rho$. Thus, the received signal-to-noise ratio is $|h|^2 \rho/2$. On the other hand, with coherent detection, the received signal power is $\rho\,|h|^4$ and the noise power is $|h|^2$. Thus the corresponding signal-to-noise ratio is $|h|^2 \rho$. Upon comparing the two cases, it is clear that differential detection degrades the performance by about 3 dB.

In Figure 8.10, we plot the bit error rate performance of the differential BPSK scheme using coherent and differential detection. The simulated system uses a single antenna at the transmitter and a single antenna at the receiver. The channel is assumed to remain fixed over two consecutive symbol intervals. No error propagation is assumed in the simulations. As shown in the figure, the coherent performance is superior to the differential performance by 3 dB as predicted above.

We remark that the degradation resulting from differential detection for DPSK over AWGN channels is slightly different than that for Rayleigh fading channels. For example, as shown in Proakis (2001), for binary DPSK over AWGN, the degradation is around 1 dB and it is around 2.5 dB for two-phase DPSK, all at a bit error rate of 10^{-5}. The degradation approaches 3 dB for $M > 4$ (i.e., for higher order modulation schemes).

8.3.2 Differential Space-Time Coding for MIMO Systems

In this section, we consider DSTCs designed for MIMO systems. We shall consider the DSTCs designed by Hughes (2000) because they are general in the sense that they apply to any number of transmit and receive antennas and to any constellation. Similar codes are proposed by Hochwald and Sweldens (2000). We remark that the DSTCs proposed by Tarokh and Jafarkhani (2000) are designed for two transmit antennas, and they are extended by Jafarkhani (2005) to four transmit antennas. These codes are based on orthogonal STBCs and DPSK modulation. Consequently, they achieve full-rate and full diversity for $N_t \leq 2$ for complex constellations, or $N_t \leq 8$ for real constellations, which are the same constraints imposed by the orthogonal STBC design guidelines.

Signal Structure of Differential Space-Time Codes

Let $\mathcal{C}$ denote the signal constellation employed, where there is no restriction on $\mathcal{C}$. Define $\mathcal{S}$ to be any group of $T \times T$ unitary matrices, i.e., these matrices satisfy the property $\boldsymbol{S}\boldsymbol{S}^H = \boldsymbol{S}^H\boldsymbol{S} = \boldsymbol{I}_T$ for all $\boldsymbol{S} \in \mathcal{S}$. Since $\mathcal{S}$ is a group, it must satisfy the properties of a group, that is, it is closed under matrix multiplication, each element has a multiplicative inverse and it contains the multiplicative identity, which is in this case $\boldsymbol{I}_T$. Let $\boldsymbol{D}$ be a $T \times N_t$ matrix such that $\boldsymbol{S}\boldsymbol{D} \in \mathcal{C}^{T \times N_t}$ for all $\boldsymbol{S} \in \mathcal{S}$.

Define $\mathcal{M}$ to be the set of matrices resulting by multiplying $\boldsymbol{D}$ by $\mathcal{S}$, that is,

$$\mathcal{M} \triangleq \{\boldsymbol{S}\boldsymbol{D} : \boldsymbol{S} \in \mathcal{S}\}.$$

The set $\mathcal{M}$ essentially represents a multichannel group code of length T over the constellation $\mathcal{C}$. To make the group code $\mathcal{M}$ unitary, $\boldsymbol{D}$ must be unitary, i.e., $\boldsymbol{D}^H\boldsymbol{D} = T\boldsymbol{I}_{N_t}$. To construct $\mathcal{M}$, one may pick $\boldsymbol{D}$ to be in $\mathcal{C}$ and $\mathcal{S}$ to be any group of $T \times T$ permutation matrices. When the size of $\mathcal{S}$ is $L = |\mathcal{S}|$, then the corresponding transmission rate is $R = \frac{1}{T}\log_2 L$ bits per channel use. Alternatively, to achieve rate R bits per channel use, we must have $L = 2^{RT}$.

The resulting unitary group code can be differentially encoded and transmitted similar to the DPSK case. As such, the differential encoding is done based on the underlying fading channel. There are two cases to consider: *non-overlap* and *overlap*. In the non-overlap case, successive coherence times are assumed to be statistically independent. Consequently, a reference (known) matrix must be transmitted at the beginning of each new coherence time, followed by an information matrix, whereas in the overlap case, the coherence time changes slowly such that it extends over several matrix transmission periods. In this case, the transmission begins with a reference matrix, followed by several information matrix transmissions.

To make this more concrete, assume, without loss of generality, that K information matrices are to be transmitted. Let $\boldsymbol{D}$ denote the reference signal. Accordingly, the differential encoding in the non-overlap case proceeds as

$$\boldsymbol{X}_k = \begin{cases} \boldsymbol{D}, & k = 1, 3, 5, \ldots, 2K-1 \\ \boldsymbol{S}_k\boldsymbol{X}_{k-1}, & k = 2, 4, 6, \ldots, 2K, \end{cases} \tag{8.37}$$

whereas in the overlap case, it proceeds as

$$\boldsymbol{X}_k = \begin{cases} \boldsymbol{D}, & k = 0 \\ \boldsymbol{S}_k\boldsymbol{X}_{k-1}, & k = 1, 2, 3, \ldots, K. \end{cases} \tag{8.38}$$

Note that the spectral efficiency in the overlap case is almost double that of the non-overlap case.

Example

In this example, we consider a DSTC designed for $N_t = T = 2$. Define $\mathcal{S}$ as

$$\mathcal{S} = \{S_0, S_1, \ldots, S_7\},$$

where

$$S_0 = \begin{bmatrix} 1 & 0 \\ 0 & 1 \end{bmatrix}, \quad S_2 = \begin{bmatrix} j & 0 \\ 0 & -j \end{bmatrix},$$

$$S_4 = \begin{bmatrix} 0 & -j \\ j & 0 \end{bmatrix}, \quad S_6 = \begin{bmatrix} 0 & 1 \\ -1 & 0 \end{bmatrix},$$

and $S_1 = -S_0$, $S_3 = -S_2$, $S_5 = -S_4$, $S_7 = -S_6$. Also, define

$$D = \begin{bmatrix} 1 & -1 \\ 1 & 1 \end{bmatrix}.$$

This is a group code $\mathcal{M}$ over the QPSK constellation $\mathcal{C} = \{\pm 1, \pm j\}$.

Differential Detection of DSTCs

The decoding of the received signal obviously depends on whether the differential encoding scheme is non-overlap or overlap. In this section, we shall assume the non-overlap scenario. Furthermore, we assume that $N_t = T$, following the notation used by Hughes (2000).

Let Z denote the matrix comprising all $K + 1$ received blocks, i.e.,

$$Z = \begin{bmatrix} Y_0 & Y_1 & \cdots & Y_K \end{bmatrix}.$$

Note that Y_k can be expressed as

$$Y_k = \sqrt{\rho} X_k H + N_k, \quad k = 0, 1, 2, \ldots, K.$$

To decode S_k, according to the differential encoding process described by (8.37), the differential receiver should be based on the last two blocks of Y, namely Y_{k-1} and Y_k. In the overlap case, the differential receiver should be based on the entire received sequence. However, this is not possible to implement because of the associated high computational complexity, even for moderate values of N_t and K. As an alternative, one may use the last two blocks of Z to decode S_k, similar to the non-overlap case.

Define

$$\mathbf{Y}_k \triangleq \begin{bmatrix} Y_{k-1} \\ Y_k \end{bmatrix}. \tag{8.39}$$

The code matrices that affect $\mathbf{Y}_k$ are

$$\bar{C}_{S_k} = \begin{bmatrix} X_{k-1} \\ S_k X_{k-1} \end{bmatrix}. \tag{8.40}$$

Note that $N_t = T$ and $\boldsymbol{D}\boldsymbol{D}^H = \boldsymbol{D}^H\boldsymbol{D} = N_t\boldsymbol{I}_T = T\boldsymbol{I}_{N_t}$. Using this result, and the fact that $\boldsymbol{X}_0 = \boldsymbol{D}$ and $\boldsymbol{S}_k^H\boldsymbol{S}_k = \boldsymbol{S}_k\boldsymbol{S}_k^H = \boldsymbol{I}_T$, we have

$$\begin{aligned}\boldsymbol{X}_k\boldsymbol{X}_k^H &= (\boldsymbol{S}_k\boldsymbol{X}_{k-1})(\boldsymbol{S}_k\boldsymbol{X}_{k-1})^H,\\ &= \boldsymbol{S}_k\boldsymbol{X}_{k-1}\boldsymbol{X}_{k-1}^H\boldsymbol{S}_k^H,\\ &\vdots\\ &= N_t\boldsymbol{I}_{N_t},\\ &= \boldsymbol{X}_k^H\boldsymbol{X}_k,\end{aligned}$$

for all $\boldsymbol{X}_k \in \mathcal{M}$. Consequently, all matrices $\bar{\boldsymbol{C}}_{\boldsymbol{S}}\bar{\boldsymbol{C}}_{\boldsymbol{S}}^H$ can be expressed as

$$\begin{aligned}\bar{\boldsymbol{C}}_{\boldsymbol{S}}^H\bar{\boldsymbol{C}}_{\boldsymbol{S}} &= \begin{bmatrix} \boldsymbol{X}_{k-1}^H & \boldsymbol{X}_{k-1}^H\boldsymbol{S}_k^H \end{bmatrix}\begin{bmatrix} \boldsymbol{X}_{k-1} \\ \boldsymbol{S}_k\boldsymbol{X}_{k-1} \end{bmatrix},\\ &= 2N_t\boldsymbol{I}_{N_t},\end{aligned}$$

for all $\boldsymbol{X}_{k-1} \in \mathcal{M}$, and thus the set comprising these matrices can be regarded as a unitary block code of length $2N_t$.

When $\boldsymbol{X}_{k-1}$ is known to the receiver, the optimal decoder for this block code would reduce to

$$\hat{\boldsymbol{S}}_k = \arg\max_{\boldsymbol{S}_k\in\mathcal{S}} \text{trace}\left\{\boldsymbol{Y}_k^H\bar{\boldsymbol{C}}_{\boldsymbol{S}_k}\bar{\boldsymbol{C}}_{\boldsymbol{S}_k}^H\boldsymbol{Y}_k\right\}, \tag{8.41}$$

which is analogous to the decoder defined by (8.15). Since

$$\bar{\boldsymbol{C}}_{\boldsymbol{S}_k}\bar{\boldsymbol{C}}_{\boldsymbol{S}_k}^H = \begin{bmatrix} N_t\boldsymbol{I}_{N_t} & N_t\boldsymbol{S}_k^H \\ N_t\boldsymbol{S}_k & N_t\boldsymbol{I}_{N_t} \end{bmatrix},$$

we can re-write (8.41) as

$$\hat{\boldsymbol{S}}_k = \arg\max_{\boldsymbol{S}_k\in\mathcal{S}} \text{Re}\left\{\text{trace}\left\{\boldsymbol{Y}_k^H\boldsymbol{S}_k\boldsymbol{Y}_{k-1}\right\}\right\}, \tag{8.42}$$

$$= \arg\max_{\boldsymbol{S}_k\in\mathcal{S}} \text{Re}\left\{\text{trace}\left\{\boldsymbol{S}_k\boldsymbol{Y}_k^H\boldsymbol{Y}_{k-1}\right\}\right\}, \tag{8.43}$$

where the last step follows from trace($\boldsymbol{A}\boldsymbol{B}$) = trace($\boldsymbol{B}\boldsymbol{A}$).

When $\boldsymbol{H}$ is available at the receiver, it can be exploited in the detection process where coherent detection is performed instead. In this case, the ML decoding rule would be

$$\hat{\boldsymbol{S}}_k^{coherent} = \arg\max_{\boldsymbol{S}_k\in\mathcal{S}} p(\boldsymbol{Y}_k|\bar{\boldsymbol{C}}_{\boldsymbol{S}_k}, \boldsymbol{H}), \tag{8.44}$$

$$= \arg\min_{\boldsymbol{S}_k\in\mathcal{S}} \text{trace}\left\{\left(\boldsymbol{Y}_k - \sqrt{\rho}\bar{\boldsymbol{C}}_{\boldsymbol{S}_k}\boldsymbol{H}\right)\left(\boldsymbol{Y}_k - \sqrt{\rho}\bar{\boldsymbol{C}}_{\boldsymbol{S}_k}\boldsymbol{H}\right)^H\right\}, \tag{8.45}$$

$$= \arg\max_{\boldsymbol{S}_k\in\mathcal{S}} \text{Re}\left\{\text{trace}\left\{\bar{\boldsymbol{C}}_{\boldsymbol{S}_k}\boldsymbol{H}\boldsymbol{Y}_k^H\right\}\right\}, \tag{8.46}$$

$$= \arg\max_{\boldsymbol{S}_k\in\mathcal{S}} \text{Re}\left\{\text{trace}\left\{\boldsymbol{S}_k\boldsymbol{X}_{k-1}\boldsymbol{H}\boldsymbol{Y}_k^H\right\}\right\}, \tag{8.47}$$

where (8.47) is obtained from (8.46) since the term $X_{k-1}Y_{k-1}^H$ is not a function of S_k, and X_{k-1} is assumed to be known at the receiver.

Performance Analysis and Design Criteria

The performance of DSTCs is similar to that of USTCs simply because both are based on unitary matrices and both employ the same receiver structure. To elaborate, consider the DSTC case in which the detection of S_k given by (8.41) is based on $\bar{C}_{S_k}$, which is a unitary block code of length $2N_t$. On the other hand, in the USTC case, the detection of S_k given by (8.15) is based on Φ_k, which is also a unitary block code of length N_t. Thus, the upper bound on $\Pr(\Phi_k \rightarrow \Phi_{k'})$ given by (8.18) extends to the DSTC in a straightforward manner. The only difference here is that $\Pr\left(\bar{C}_{S_k} \rightarrow \bar{C}_{S_{k'}}\right)$ for DSTCs is defined as the probability that receiver will erroneously decide on $\bar{C}_{S_{k'}}$ given that $\bar{C}_{S_k}$ was transmitted. Accordingly, we have

$$\Pr\left(\bar{C}_{S_k} \rightarrow \bar{C}_{S_{k'}}\right) \leq \frac{1}{\det\left(I_{N_t} + \frac{N_t^2\rho^2}{(1+N_t\rho)}\left[I_{N_t} - \frac{1}{4T^2}\bar{C}_{S_{k'}}^H\bar{C}_{S_k}\bar{C}_{S_k}^H\bar{C}_{S_{k'}}\right]\right)^{N_r}}. \tag{8.48}$$

Using the result

$$I_{N_t} - \frac{1}{4T^2}\bar{C}_{S_{k'}}^H\bar{C}_{S_k}\bar{C}_{S_k}^H\bar{C}_{S_{k'}} = \frac{1}{4T}\left(C_{S_k} - C_{S_{k'}}\right)^H\left(C_{S_k} - C_{S_{k'}}\right)$$

where $C_{S_k} \triangleq S_k X_{k-1}$ and $C_{S_{k'}}$ is similarly defined, (8.48) can be expressed as

$$\Pr\left(\bar{C}_{S_k} \rightarrow \bar{C}_{S_{k'}}\right) \leq \frac{1}{\det\left(I_{N_t} + \frac{N_t^2\rho^2}{4T(1+N_t\rho)}\left(C_{S_k} - C_{S_{k'}}\right)^H\left(C_{S_k} - C_{S_{k'}}\right)\right)^{N_r}}. \tag{8.49}$$

To optimize the performance, the upper bound in (8.49) suggests that the code design criterion should be such that the rank of $C_{S_k} - C_{S_{k'}}$ is maximized, where the maximum in this case is N_t, and the product of its nonzero eigenvalues should be maximized so as to maximize the coding gain. This design criterion appears to be somewhat analogous to that of STTCs, but it is more involved because it depends on T. Optimal codes for two transmit antennas can be found in the paper by Hughes (2000), where it is assumed that $N_t = T = 2$.

Examples

In this section, we examine the performance of the DSTC mentioned above which was designed for $N_t = T = 2$. We consider both the overlap and non-overlap cases. For the overlap case, we assume that the coherence time of the channel remains fixed over 12 symbol intervals, which implies that six matrices see the same fade. Out of these six matrices, the first matrix is known to the receiver whereas the rest are information matrices. The encoding is done according to (8.38). The bit error performance of this code is plotted in Figure 8.11 for coherent and differential detection. Coherent detection is performed according to (8.47) and differential detection is performed according to (8.43). As shown

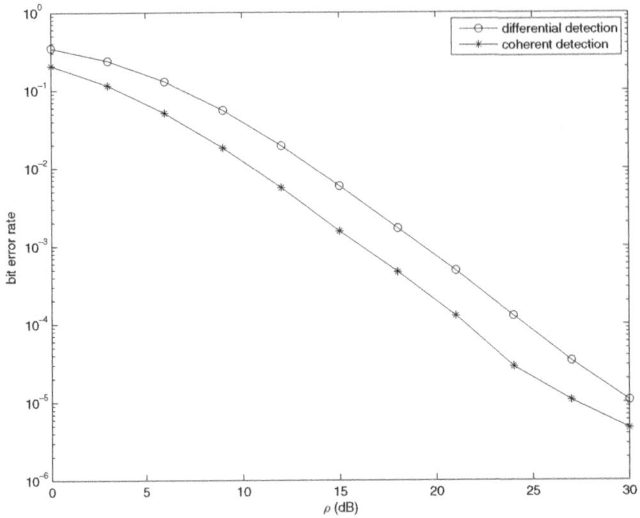

Figure 8.11 Bit error rate performance for the DSTC example for the overlap case using coherent and noncoherent detection.

in the figure, the coherent detector outperforms the differential detector by about 2 dB at a bit error rate of 10^{-5}.

In Figure 8.12, we repeat the same experiment but now for the non-overlap case. The channel coherence time in this case is fixed over four consecutive symbol intervals, spanning two consecutive matrices. The first matrix is a reference matrix, whereas the second one is an information matrix. We also consider here coherent and differential detection. As shown in the figure, the performance of the coherent detector is better than that of the differential detector by about 2.5 dB at a bit error rate of 10^{-5}. By comparing the results in Figures 8.11 and 8.12, we can see that the results for the former are a bit worse than those for the latter. This difference is attributed to the error propagation that the overlap case suffers, whereas this error propagation is not a problem for the non-overlap case. Actually, in the absence of error propagation, both schemes should yield identical performances, despite having different bandwidth efficiencies.

8.4 Turbo-Coded Unitary Space-Time Codes

Although unitary signal constellations are capacity achieving, using them alone is not sufficient to obtain a very good performance. This is similar to other well-known cases, for example, with BPSK modulation over an AWGN channel. Although it is easy to show that the supported information rates by BPSK modulation are very close to the channel capacity (for low rates of transmission), the uncoded BPSK performance is significantly far from those limits, and channel coding is necessary to obtain an acceptable performance. With

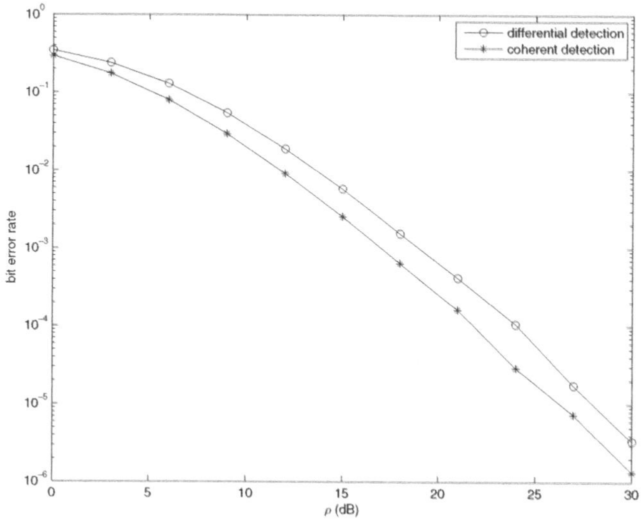

Figure 8.12 Bit error rate performance for the DSTC example for the non-overlap case using coherent and noncoherent detection.

this observation, channel coding ideas which introduce redundancy between consecutive unitary signals transmitted should be developed. In this section, we consider the serial concatenation of a turbo code and a USTC (Bahceci and Duman (2002)). We refer to this scheme as TC-USTC. In the next section, we consider the serial concatenation scheme in which trellis-coded modulation with unitary signal constellations is employed.

8.4.1 Encoder Structure

A block diagram of the TC-USTC encoder is shown in Figure 8.13. As shown in the figure, the primitive data is first encoded by a turbo encoder. The turbo-coded sequence is interleaved, where interleaving is employed to decorrelate the adjacent bits and distribute the errors due to deep fades. The interleaved sequenced is demultiplexed into TR parallel bit streams. These parallel streams are fed into the unitary space-time modulator, which maps every TR bits onto one of the $L = 2^{TR}$ complex matrices, each of size $T \times N_t$. These matrices are generated according to the design criterion described in Section 8.2.4. Note that the transmission rate in this case is $R_c R$ bits per channel use where R_c denotes the code rate of the turbo code. To achieve higher spectral efficiencies, while keeping L fixed, puncturing may be used.

8.4.2 Noncoherent Iterative Decoder

Let $\boldsymbol{b}$ denote the kth vector of TR bits that are fed into the unitary space-time modulator, and let $\boldsymbol{X}$ denote the corresponding unitary space-time matrix. The received kth signal,

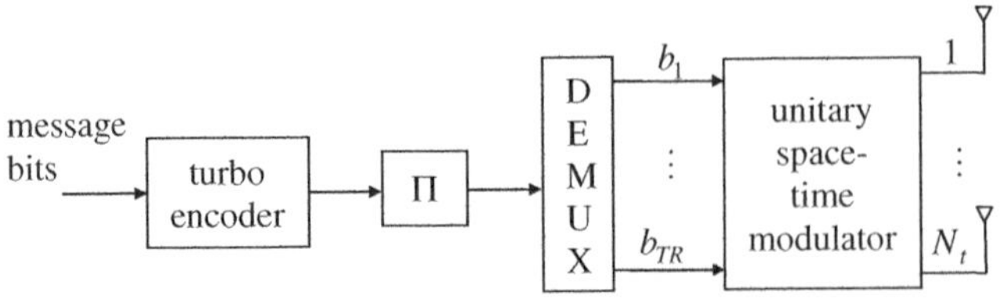

Figure 8.13 TC-USTC encoder.

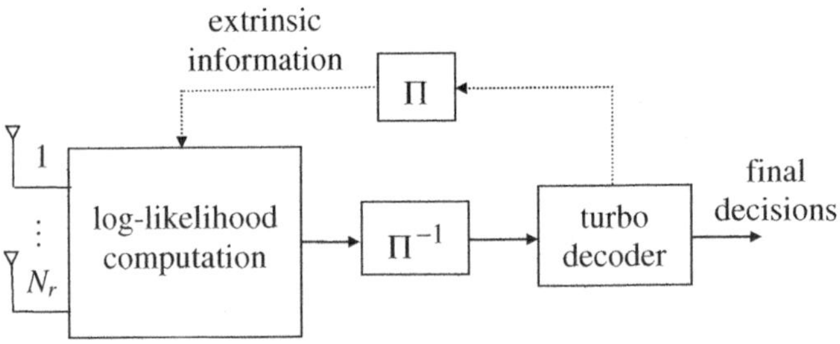

Figure 8.14 Noncoherent iterative TC-USTC decoder.

denoted by $\boldsymbol{Y}$, is expressed as (see (8.2))

$$\boldsymbol{Y} = \sqrt{\rho}\boldsymbol{X}\boldsymbol{H} + \boldsymbol{N}.$$

The TC-USTC decoder, shown in Figure 8.14, consists of two main modules: the channel detector and the turbo decoder. The channel detector uses the received signal $\boldsymbol{Y}$ to compute log-likelihood ratios for the bits comprising $\boldsymbol{b}$. These likelihoods are deinterleaved and passed to the turbo decoder for further processing. The turbo decoder consists of two SISO decoders that work iteratively between themselves, where each SISO decoder is matched to one of the component encoders of the turbo code.

The log-likelihood for the kth bit in $\boldsymbol{b}$ is given as

$$L(b_k) = \log \frac{P(b_k = 1|\boldsymbol{Y})}{P(b_k = 0|\boldsymbol{Y})}, \tag{8.50}$$

$$= \log \frac{P(b_k = 1, \boldsymbol{Y})}{P(b_k = 0, \boldsymbol{Y})}. \tag{8.51}$$

Since there is a one-to-one correspondence between $\boldsymbol{b}$ and $\boldsymbol{X}$, we can write (8.51) as

$$L(b_k) = \log \frac{\sum_{\boldsymbol{X}:\boldsymbol{X}=f(\boldsymbol{b}),b_k=1} p(\boldsymbol{Y}, \boldsymbol{X})}{\sum_{\boldsymbol{X}:\boldsymbol{X}=f(\boldsymbol{b}),b_k=0} p(\boldsymbol{Y}, \boldsymbol{X})}, \tag{8.52}$$

where $f(\cdot)$ is the mapping from $\boldsymbol{b}$ to $\boldsymbol{X}$. Assuming that the transmitted signals are equally likely, the joint p.d.f. in (8.52) can be expressed as

$$L(b_k) = \log \frac{\sum_{\boldsymbol{X}:\boldsymbol{X}=f(\boldsymbol{b}),b_k=1} p(\boldsymbol{Y}|\boldsymbol{X})}{\sum_{\boldsymbol{X}:\boldsymbol{X}=f(\boldsymbol{b}),b_k=0} p(\boldsymbol{Y}|\boldsymbol{X})}, \tag{8.53}$$

where $p(\boldsymbol{Y}|\boldsymbol{X})$ is defined by (8.4). By using the definition of $p(\boldsymbol{Y}|\boldsymbol{X})$, the expression in (8.53), using simple algebra, can be further simplified to

$$L(b_k) = \log \frac{\sum_{\Phi_k=\boldsymbol{X}/\sqrt{T}:\boldsymbol{X}=f(\boldsymbol{b}),b_k=1} \exp\left(\text{trace}\left\{\frac{1}{1+1/(\rho T)}\boldsymbol{Y}^H\boldsymbol{\Phi}_k\boldsymbol{\Phi}_k^H\boldsymbol{Y}\right\}\right)}{\sum_{\Phi_k=\boldsymbol{X}/\sqrt{T}:\boldsymbol{X}=f(\boldsymbol{b}),b_k=0} \exp\left(\text{trace}\left\{\frac{1}{1+1/(\rho T)}\boldsymbol{Y}^H\boldsymbol{\Phi}_k\boldsymbol{\Phi}_k^H\boldsymbol{Y}\right\}\right)}. \tag{8.54}$$

Once the likelihoods in (8.54) are computed, they are passed, after being deinterleaved, to the turbo decoder. They are used by the turbo decoder as if they are the likelihoods of an observation sequence from a BPSK modulated signal over an AWGN channel. The turbo decoder runs for several iterations and at the end produces its own extrinsic information about the received bits. The final decisions can then be obtained by applying simple threshold detection at the output of the turbo decoder.

Depending on the complexity and performance requirements, it is possible to have the channel detector and turbo decoder work together iteratively. In this case, the extrinsic information produced by the turbo decoder is fedback to the channel detector to refine its log-likelihoods for the next iteration. Specifically, the likelihoods in (8.53) are updated after the first iteration to utilize the extrinsic information produced by the turbo decoder as

$$L(b_k) = \log \frac{\sum_{\boldsymbol{X}:\boldsymbol{X}=f(\boldsymbol{b}),b_k=1} p(\boldsymbol{Y}|\boldsymbol{X})P(\boldsymbol{X})}{\sum_{\boldsymbol{X}:\boldsymbol{X}=f(\boldsymbol{b}),b_k=0} p(\boldsymbol{Y}|\boldsymbol{X})P(\boldsymbol{X})}. \tag{8.55}$$

Note that the turbo decoder normally computes extrinsic information about the input to its encoder, i.e., the uncoded sequence. But here the decoder must compute extrinsic information about the output of its encoder, which includes both information and parity bits, i.e., the coded sequence. Since the code is systematic, it is easy to accomplish this, as explained in Chapter 7.

Again, since there is a one-to-one correspondence between $\boldsymbol{X}$ and $\boldsymbol{b}$, we can write $P(\boldsymbol{X})$ in (8.55) as

$$\begin{aligned} P(\boldsymbol{X}) &= P(b_1, b_2, \ldots, b_{k-1}, b_{k+1}, \ldots, b_{TR}), \\ &= \prod_{i=1,i\neq k}^{TR} P(b_i), \end{aligned}$$

where it is implied here that $b_k = 1$ for the numerator and $b_k = 0$ for the denominator in (8.55). The probability of each bit being 1 or 0 can be expressed in terms of its likelihood value as

$$P(b_i = 1) = \frac{e^{L(b_i)}}{1 + e^{L(b_i)}} \text{ and } P(b_i = 0) = \frac{1}{1 + e^{L(b_i)}}.$$

To summarize, with iterative demodulation/decoding, in the beginning, the channel detector computes the sequence of log-likelihoods according to (8.53) where the transmitted bits are assumed to be equally likely. The computed likelihoods are passed to the turbo decoder, which produces its own extrinsic information about the transmitted sequence. This information is fedback to the channel detector to compute the new likelihoods according to (8.55). The updated likelihoods are then passed to the turbo decoder, and so on. This process repeats for a certain number of iterations before final decisions are made. This iterative decoding increases the computational complexity, but it results in improved performance.

8.4.3 Example

In this example, we consider the TC-USTC scheme with $N_t = 2$ and $N_r = 1$. The turbo encoder uses two identical recursive systematic convolutional encoders, each employing the generator polynomials $(g_1, g_2) = (21, 37)_{octal}$, where g_1 is the feedforward polynomial and g_2 is the feedback polynomial. All simulations were done with pseudo-randomly-generated interleavers of size 140, 3500, and 14,000. The rate of the turbo encoder is $7/8$, which is achieved by keeping the $(14n + 1)$st parity bits of the first component encoder and the $(14n + 9)$th parity bits of the second component encoder $(n = 0, 1, 2, \ldots)$. We assume $T = 8$. As such, we generate $L = 2^8 = 256$ complex matrices, each of size 8×2. The first

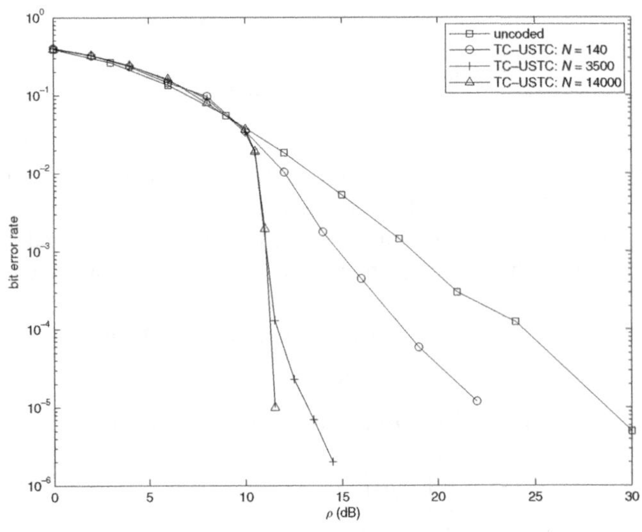

Figure 8.15 Performance of the TC-USTC scheme with various interleaver sizes.

matrix, Φ_1, is expressed as

$$\Phi_1 = \frac{1}{\sqrt{8}} \begin{bmatrix} 1 & 1 \\ 1 & e^{j\frac{2\pi}{8}} \\ \vdots & \vdots \\ 1 & e^{j\frac{14\pi}{8}} \end{bmatrix}.$$

The rest of the matrices are generated according to (8.27), where $\mathbf{\Omega}$ is defined by (8.25) with $[u_1, \ldots, u_8] = [1, 7, 60, 79, 187, 125, 198, 154]$.

The bit error performance for the above scheme for four different cases is shown in Figure 8.15. The uncoded case corresponds to unitary space-time coding only (without turbo coding), and the other three cases correspond to the serial concatenation of the turbo code and USTC for different interleaver sizes. The number of iterations, where applicable, is set to eight. The transmission rate (or bandwidth efficiency) for the uncoded case is $R = 1$, whereas it is $7/8$ for the turbo-coded system. As we can see from the figure, there is a significant improvement in performance by employing the turbo code, where this improvement increases as the interleaver size increases. For example, at a bit error rate of 10^{-5}, the improvement in signal-to-noise ratio is about 15 dB for interleaver size 3500 as compared with the uncoded system. This improvement comes at the expense of expanding the bandwidth by a factor of $8/7$, which is not significant.

8.5 Trellis-Coded Unitary Space-Time Codes

In this section, we consider the serial concatenation of an outer trellis-coded modulation code and an inner unitary space-time code proposed by Bahceci and Duman (2004). The basic idea is very similar to the usual trellis coded modulation (Ungerboeck (1982)). The basic difference is that the modulation scheme used is not PSK or QAM, but instead unitary signals, described earlier in the chapter, are employed. A block diagram of the system is shown in Figure 8.16.

For a rate of R bits per channel use transmission, there are 2^{RT} branches emanating from each state of the trellis (where T is the duration of each unitary signal). The signal constellation is expanded (e.g., to 2^{2RT}), and depending on the specific transition, one of these signals is selected at each step. The generation of the "larger" unitary signal constellation is done using similar principles as before (e.g., using the Fourier-based construction). A similar method to Ungerboeck's set partitioning can be employed with the δ-metric described earlier (Bahceci and Duman (2004)).

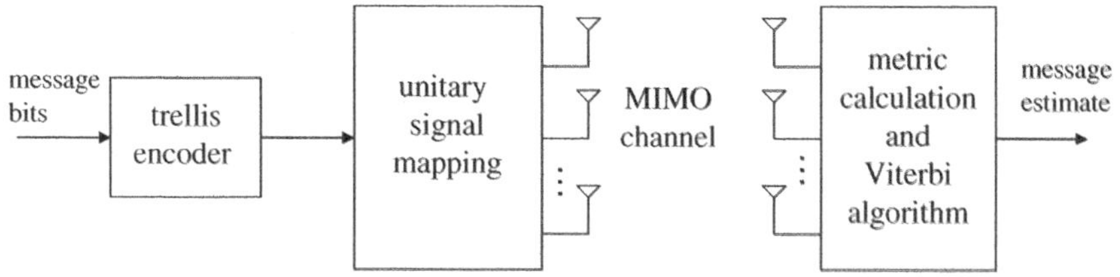

Figure 8.16 Block diagram of the trellis-coded unitary signaling scheme.

Unlike the uncoded use of unitary signals, the receiver should operate on sequences of received signals, and perform a sequence detection. It is easy to show that the joint p.d.f. of the received sequence $\boldsymbol{Y}_1, \boldsymbol{Y}_2, \ldots, \boldsymbol{Y}_N$ given the sequence of transmitted unitary signals $\boldsymbol{\Phi}_1, \boldsymbol{\Phi}_2, \ldots, \boldsymbol{\Phi}_N$ is given by

$$p(\boldsymbol{Y}_1, \ldots, \boldsymbol{Y}_N|\boldsymbol{\Phi}_1, \ldots, \boldsymbol{\Phi}_N) = \prod_{n=1}^{N} p(\boldsymbol{Y}_n|\boldsymbol{\Phi}_n), \tag{8.56}$$

$$= \sum_{n=1}^{N} \frac{\exp\left\{-\text{trace}\left[\boldsymbol{I}_T - \frac{1}{1+\frac{1}{\rho T}}\boldsymbol{\Phi}_n\boldsymbol{\Phi}_n^H\right]\boldsymbol{Y}_n\boldsymbol{Y}_n^H\right\}}{\pi^{TN}(1+\rho T)^{N_t N_r}}, \tag{8.57}$$

where we have assumed that fading is independent from one block to the next, i.e., consecutive transmitted unitary signals see different fades. Taking the logarithm of this expression, and ignoring the identical terms (for different sequences of unitary signals), we find that the ML decoder should maximize

$$\sum_{n=1}^{N} \text{trace}\left\{\boldsymbol{Y}_n^H\boldsymbol{\Phi}_n\boldsymbol{\Phi}_n^H\boldsymbol{Y}_n\right\}, \tag{8.58}$$

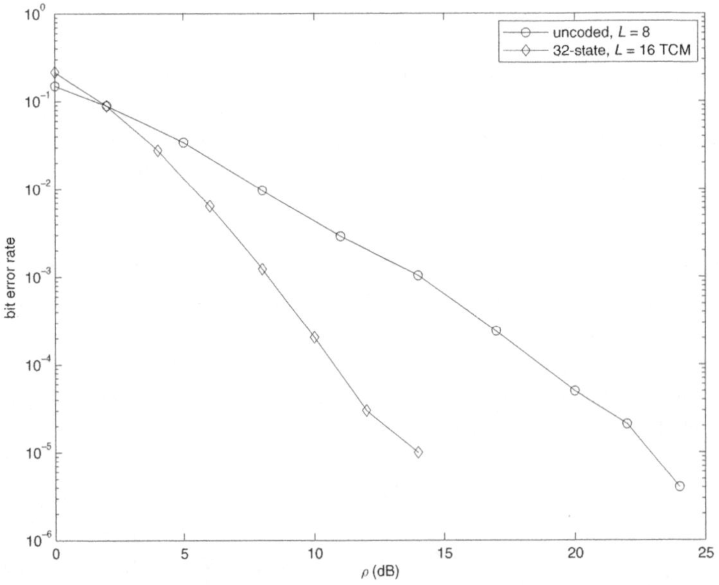

Figure 8.17 Bit error rate performance comparison between TCM coded and uncoded unitary space-time systems with $N_t = 2$ and $N_r = 1$. The signal constellation size for the coded system is $L = 16$ and it is $L = 8$ for the uncoded system, thus achieving a spectral efficiency of 3/8 bits per channel use.

over all possible sequences of unitary signals $\mathbf{\Phi}_1, \mathbf{\Phi}_2, \ldots, \mathbf{\Phi}_N$. To summarize, the ML decoding is achieved by maximizing the sum of trace metrics over the entire sequence, as opposed to the maximization of a single trace metric. This can be done efficiently using the Viterbi algorithm owing to the fact that the metric is additive.

In trellis-coded unitary signaling, since the signal constellation is expanded, the unitary signals are closer to each other, hence the performance without trellis coding would deteriorate. However, the channel coding introduced increases the "equivalent distances" between transmitted sequences, and when properly designed could give large gains over the direct use of unitary constellations as reported by Bahceci and Duman (2004).

As an example, we consider the performance of a trellis-coded unitary space-time system with $N_t = 2$ and $N_r = 1$ with a 32-state trellis. We compare the bit error rate performance of this scheme with that of an uncoded unitary space-time system with the same number of antennas. To make the comparisons fair, we design both systems in such a way that they achieve the same spectral efficiency. For the coded system, the unitary space-time signal constellation size is $L = 16$, whereas it is $L = 8$ for the uncoded system. Thus, the spectral efficiency is $3/8$ bits per channel use for both schemes. The bit error rate performance for both schemes is plotted in Figure 8.17. As shown in the figure, the coded system provides a coding gain of about 8 dB at a bit error rate of 10^{-5} over the uncoded case.

8.6 Turbo-Coded Differential Space-Time Codes

Concatenated coding and iterative decoding can also be applied to differential space-time code. In this section, we consider the serial concatenation of an outer channel code and an inner DSTC, which we refer to as the turbo-coded differential space-time code (TC-DSTC) scheme (Schlegel and Grant (2003)). Although the outer code is not limited to turbo codes, we refer to this scheme as such because it is decoded iteratively in a way similar to the decoding of turbo codes.

8.6.1 Encoder Structure

The TC-DSTC encoder is shown in Figure 8.18. As shown in the figure, the primitive data sequence is encoded by the outer channel encoder. The coded sequence is interleaved where interleaving is done on a block-by-block basis. For example, if the code rate of the outer code is $R_c = b/n$, then every consecutive n coded bits are grouped and interleaved as one block. As such there will be 2^n distinct n-bit patterns. The interleaved blocks are

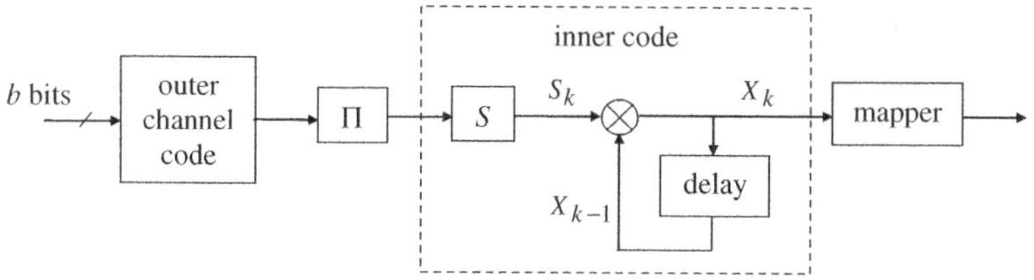

Figure 8.18 TC-DSTC encoder.

then mapped to their corresponding matrices from the set $\mathcal{S}$. Note that since there is a one-to-one correspondence between the coded blocks and the DSTC matrices, we must have $|\mathcal{S}| = 2^n$. For example, when $T = 2$ and $R_c = 2/3$, then $|\mathcal{S}| = 8$. The size of each of these matrices is $T \times N_t$ and they are generated according to the design criterion introduced in Section 8.3.2. The transmission rate in this example is 1 bit per channel use. Then $\boldsymbol{S}_k$ is differentially encoded according to (8.37) or (8.38), depending on the underlying channel, and the resulting $\boldsymbol{X}_k$ is transmitted from the available N_t antennas over T consecutive symbol intervals.

8.6.2 Iterative Detectors

This code can be detected differentially or near-differentially. In the former case, the structure of the detector is shown in Figure 8.19. In this case, the front-end receiver uses the last two blocks, $\boldsymbol{Y}_k$, to provide estimates of the transmitted signal, $\boldsymbol{S}_k$, according to (8.43). These estimates are deinterleaved and passed to the outer decoder. The outer decoder produces its own extrinsic information, which is consequently used to make the final decisions. Similar to the decoding of the TC-USTC scheme, to improve the performance further, it is possible to have the outer decoder and inner detector exchange extrinsic information back and forth for a number of times in an iterative fashion before final decisions are made.

If channel estimates are available at the receiver, detection can be done near-differentially (Schlegel and Grant (2003)), as shown in Figure 8.20. Such channel estimates can be made available by exploiting the differential property of the inner code where it provides *a priori* estimates of the CSI, which can be used in an iterative setting. The iterative decoding proceeds as follows. In the first iteration, the soft differential encoder uses the last two received matrices $\boldsymbol{Y}_k$ to produce extrinsic information on the transmitted signal, according to (8.43). This extrinsic information is then fed to the inner decoder. During the first iteration, the output and input of the inner decoder are identical because this decoder cannot improve the symbol probabilities since it has no knowledge of the CSI. The output of the inner decoder is deinterleaved and passed to the outer decoder. The same information is passed to the channel estimator. The outer decoder produces its own extrinsic information and feeds it back to the inner decoder. This completes the first iteration.

The inner decoder uses the channel estimates from the channel estimator, the extrinsic information from the outer decoder and the received signal to produce refined extrinsic information. Note that the output of the soft differential encoder (dashed line) is used only

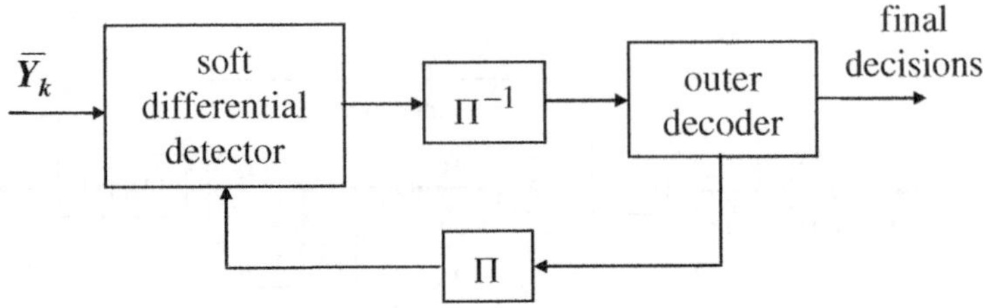

Figure 8.19 Differential TC-DSTC detector.

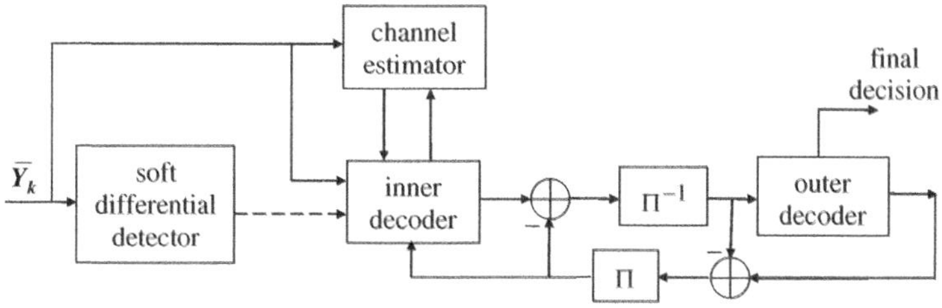

Figure 8.20 Near-differential TC-DSTC detector.

once, because it contains no additional information from the previous iteration. The refined extrinsic information is then passed to the outer decoder, which in turn produces its own refined extrinsic information. This completes the second iteration. This process continues for a number of times before final decisions are made. To obtain relatively accurate channel estimates, it is required in this scheme to send pilot symbols periodically, especially if the channel changes very fast, and hence the near-coherent performance.

8.7 Chapter Summary and Further Reading

We have seen in this chapter various aspects of noncoherent MIMO channels. In particular, we have seen that the capacity of noncoherent MIMO channels approach that of coherent channels when the signal-to-noise ratio is relatively high and the coherence time of the channel is long. We have discussed the structure of signals that would asymptotically achieve such capacity, namely USTCs and DSTCs. We have considered the design, encoding and decoding of these coding schemes. Concatenated coding schemes involving USTCs and DSTCs have also been discussed, where it was shown that substantial performance improvements are possible.

The subject of unitary and differential space-time coding has received considerable attention from the research community in this area, and thus many research papers have appeared in the past few years. We point out a few references that one may consider reading to learn more about this subject.

An excellent reference on differential space-time codes based on orthogonal STBCs and DPSK is Jafarkhani (2005). Liang and Xia (2002) design unitary signal constellations for differential space-time modulation optimized for two transmit antennas. Unitary space-time codes designed for frequency selective fading channels with blind equalization are proposed by Aktas and Mitra (2003). Differential space-time modulation for frequency selective fading channels are introduced by Li (2005), Ma et al. (2005) and Diggavi et al. (2002). Liu et al. (2001b) propose a double differential space-time block coding scheme for time-selective fading channels. Differential space-time codes over spatially correlated channels are introduced by Nguyen (2007) and Cai and Giannakis (2006).

Signal constellation design for trellis coded unitary space-time modulation systems is introduced by Zhao et al. (2004) and Wu et al. (2006). Chen et al. (2006) consider the design of unitary signal constellations for differential space-time modulation systems decoded via parallel sphere decoders. A serial concatenation of a convolutional code with a differential space-time code with bit interleaving is proposed by Lampe and Schober (2002). Two receivers for differential space-time modulation systems that are robust against fast fading are introduced by Schober and Lampe (2002). Cheung and Schober (2006) consider differential space-time modulations that are suitable for systems with a large number of transmit antennas. In this case, the transmit antennas are divided in subgroups and a differential space-time code is applied to each group. The receiver combines group interference cancellation and decision feedback equalization.

Grassmann space-time codes for noncoherent MIMO systems are proposed by Zheng and Tse (2002) and Kammoun and Belfiore (2003). Tao and Cheng (2003) consider trellis-coded schemes based on differential unitary space-time modulation. McCloud et al. (2002) consider signal design and convolutional coding for noncoherent space-time coded MIMO systems.

Problems

8.1 Derive Equation (8.4).

8.2 Derive Equation (8.15).

8.3 Using (8.19), argue that the maximum diversity of a unitary space-time coded MIMO system, i.e., $N_t N_r$, is possible to achieve if $T \geq 2N_t$, where T is the coherence time of the channel.

8.4 Reproduce Figures 8.3 and 8.4.

8.5 Show that the bit error probability for a single antenna system using DPSK signaling over a Rayleigh fading channel is given by

$$P_b = \frac{1}{2(1+\rho)},$$

where ρ is the average signal-to-noise ratio per information bit at the receiver.

8.6 Derive Equation (8.47).

8.7 Reproduce Figure 8.17.

9

Space-Time Coding for Frequency Selective Fading Channels

In the previous chapters, we have concentrated mainly on flat fading channel models for wireless communications. However, if the symbol durations are decreased in order to increase the transmission rates, particularly beyond the multipath spread of the channel, flat fading channel models will no longer be appropriate. In such a case, the channel will be frequency selective and will exhibit ISI. Hence, it is of importance to consider the use of MIMO communication systems over frequency selective fading channels as well.

Our objective in this chapter is several fold. We review some of the information-theoretical aspects of single-input single-output and MIMO communication systems. Then we consider several space-time coding approaches for frequency selective fading. These include space-time block codes, space-time trellis codes, as well as concatenated coding approaches. After considering the single carrier approach, we switch our attention to a different way of handling the ISI, namely, the use of multi-carrier communications (specifically, orthogonal frequency division multiplexing (OFDM)) for MIMO systems.

The chapter is organized as follows. We first consider MIMO frequency selective channel models, study their capacity, and compute constrained information rates. We then look at the problem from a different perspective and derive full diversity space-time code design principles. We consider several coding approaches over such channels, and give examples. Finally, we briefly describe the idea of multi-carrier modulation and OFDM for MIMO channels.

9.1 MIMO Frequency Selective Channels

Let us now start with a generalization of the tapped delay line model for frequency selective channels (briefly described in Chapter 2) to the case of MIMO systems. Consider a system

Coding for MIMO Communication Systems Tolga M. Duman and Ali Ghrayeb

with N_t transmit and N_r receive antennas, and assume that the channel has L ISI taps. The received signal at the jth receive antenna at time k is given by

$$y_j(k) = \sqrt{\rho}\sum_{l=0}^{L-1}\sum_{i=1}^{N_t} h_{i,j}^{(l)}(k)x_i(k-l) + n_j(k), \tag{9.1}$$

where $x_i(k)$ denotes the transmitted signal from antenna i at time k, $h_{i,j}^{(l)}(k)$ is the channel coefficient for the lth path from transmit antenna i to receive antenna j at time k, and $n_j(k)$ is the additive white Gaussian noise (both spatially and temporally) with a variance of $1/2$ per dimension. The total signal energy for each use of the channel is normalized to unity. The constant ρ can be used as the effective signal-to-noise ratio at the receiver after proper normalization of the channel gains.

We need to differentiate several cases. For baseband MIMO ISI channels, the channel coefficients are real numbers. For deterministic bandpass channels (which could be useful for fixed wireless communications), the channel coefficients are in general complex, but they still can be well modeled as constants. For mobile communications, the channel tap coefficients are random variables. For instance, if the wireless channel is very slowly varying, e.g., quasi-static fading, the tap coefficients remain constant for each frame of data. However, for ergodic channels, they vary with time. As an example, for block fading channels, they change independently from one block of data to the next.

For Rayleigh fading channels, the channel tap coefficients are modeled as zero mean complex Gaussian random variables. Different channel taps are usually assumed to be independent. The average channel gains for different paths are determined from the power delay profile of the wireless channel. For instance, for the uniform power delay profile, all the channel gains (for different paths) have equal average channel power. As another example, for the exponential power delay profile, the channel tap powers decay exponentially.

9.2 Capacity and Information Rates of MIMO Frequency Selective Fading Channels

We now consider a brief information-theoretic exposition of MIMO FS channels. We first consider the channel capacity for the case of deterministic and quasi-static fading channels. Then, we review the achievable information rates under less restrictive assumptions.

9.2.1 Information Rates with Gaussian Inputs

For deterministic MIMO ISI channels, El Gamal et al. (2003) provide a method of evaluating the mutual information obtained when the inputs are independent and identically distributed circularly symmetrical complex Gaussian random variables across space and time. Their method extends the technique derived by Hirt and Massey (1988) for single-input single-output systems, and uses DFT techniques. The resulting information rate is given by

$$I_{Gauss} = \int_0^{\infty} \log\left(\det\left(\boldsymbol{I}_{N_r} + \frac{\rho}{N_t}\tilde{\boldsymbol{H}}_f^T\tilde{\boldsymbol{H}}_f^H\right)\right) df, \tag{9.2}$$

where

$$\tilde{H}_f = \sum_{l=0}^{L-1} H_l \exp(-j2\pi fl), \tag{9.3}$$

with

$$H_l = \begin{bmatrix} h_{1,1}^{(l)} & h_{1,2}^{(l)} & \cdots & h_{1,N_r}^{(l)} \\ h_{2,1}^{(l)} & h_{2,2}^{(l)} & \cdots & h_{2,N_r}^{(l)} \\ \vdots & \vdots & \ddots & \\ h_{N_t,1}^{(l)} & h_{N_t,2}^{(l)} & \cdots & h_{N_t,N_r}^{(l)} \end{bmatrix}, \tag{9.4}$$

where the time dependence of the fading coefficients is dropped since they are constants.

It is easy to extend this result for quasi-static fading channels as well (assuming that the receiver knows the channel coefficients). In this case, the expression in (9.2), for a given channel matrix, should be interpreted as the achievable instantaneous information rate with independent Gaussian inputs, which is clearly random. Therefore, for a given transmission rate, one can compute the relevant outage probability, or for a given outage level, one can calculate the maximum achievable information rate. We will provide examples a little later in this section.

We point out that the result above can be interpreted as a lower bound on the actual Shannon type capacity (for the deterministic case), and the outage capacity (for the quasi-static fading scenario). This is not a complete capacity characterization, however it is clearly a useful quantity that can be used for system design. Furthermore, we emphasize that the above expression is not useful for ergodic capacity calculations.

9.2.2 Achievable Information Rates with Practical Constellations

Clearly, using Gaussian inputs for a practical system is out of question. Therefore, it is of interest to compute the achievable information rates using more practical input constraints. Let us now assume that a specific signal constellation is employed (e.g., BPSK, QAM, etc.), and compute the corresponding achievable information rates. Our exposition is based on the work of Zhang et al. (2004). By using this technique, it will also be possible to consider ergodic fading channels, which was omitted from the discussion in the previous subsection.

Consider a general MIMO FS channel model, and assume that the channel gains are known at the receiver (they can be deterministic or random). Then, the achievable information rate can be written as

$$I = \lim_{N\to\infty} \frac{1}{N} I(X(1), X(2), \ldots, X(N); Y(1), Y(2), \ldots, Y(N)), \tag{9.5}$$

where $X(1), X(2), \ldots, X(N)$ is the sequence of channel inputs, and $Y(1), Y(2), \ldots, Y(N)$ is the set of channel outputs. We note that we have omitted necessary conditioning on the channel gains for simplicity of notation (for the random channel case).

We now present a way of computing this limit using a simulation-based method to estimate the information rates of MIMO FS channels. Deterministic and random channel cases will be treated in parallel, and distinctions will be made when necessary.

Independent Uniformly Distributed Inputs from a Given Constellation

Let us compute the information rates under the constraint that the inputs are i.u.d. and are picked from a finite constellation such as BPSK. The only uncertainty that remains in the sequence of output vectors (given the channel coefficients) is due to the noise terms. Therefore, we can write the mutual information expression above as

$$I(\boldsymbol{X}(1), \ldots, \boldsymbol{X}(N); \boldsymbol{Y}(1), \ldots, \boldsymbol{Y}(N)) = H(\boldsymbol{Y}(1), \ldots, \boldsymbol{Y}(N)) - H(\boldsymbol{N}(1), \ldots, \boldsymbol{N}(N)), \tag{9.6}$$

where $\boldsymbol{N}(1), \boldsymbol{N}(2), \ldots, \boldsymbol{X}(N)$ are the noise vectors. Thus, we can easily write

$$I = \lim_{N \to \infty} \left(\frac{1}{N} H(\boldsymbol{Y}(1), \ldots, \boldsymbol{Y}(N)) - \frac{1}{N} H(\boldsymbol{N}(1), \ldots, \boldsymbol{N}(N)) \right). \tag{9.7}$$

Clearly, the second term (entropy of the noise) can be easily computed as we simply have a sequence of independent complex Gaussian noise vectors. However, the main problem is the computation of the first term since it is more involved than its counterpart considered in Chapter 3 in the context of MIMO flat fading channels. This is because, for the case of flat fading, given the channel coefficients, if the input symbols are independent from one time instant to the next, then the output symbols are independent as well, and the entropy can be estimated by considering the entropy of a single output vector at a time. However, due to the presence of ISI, we do not have such a simplification here and we need to resort to another technique.

By definition, we have

$$\frac{1}{N} H(\boldsymbol{Y}(1), \ldots, \boldsymbol{Y}(N)) = -\frac{1}{N} \mathrm{E}\left[\log\left(p(\boldsymbol{Y}(1), \ldots, \boldsymbol{Y}(N))\right)\right], \tag{9.8}$$

that is, we simply need to estimate the expected value of the logarithm of the channel output sequence probability. In order to do this calculation, we can resort to Monte Carlo techniques, that is, we can generate a large number of sequences of channel outputs using the mathematical model for the MIMO FS channel, compute the joint probability of each sequence, take the logarithm, and average the results over all realizations of the experiment. With a large number of trials, one would obtain an accurate estimate of the average entropy of the output sequence, and thus the achievable information rates.

This computation looks very cumbersome as a large number of realizations of a long sequence are needed to perform the averaging. However, note that with i.u.d. inputs, the output is stationary and ergodic when the channel is ergodic fading (or deterministic). Therefore, the Shannon–McMillan–Breiman theorem applies (see Cover and Thomas (2006)), and we can simply perform averaging over a single long run of the sequence of channel outputs. Therefore, for deterministic or ergodic fading cases, the necessary calculation is simplified considerably.

The problem that remains is the calculation of the logarithm of the joint probability of a given output sequence. This can be done by applying the BCJR algorithm (Bahl et al. (1974)) as described next.

Calculation of $P(y(1), y(2), \cdots, y(N))$

Assume that the size of the signal constellation is M. Since there are N_t transmit antennas, and there are L ISI taps for each pair of antennas, one can describe the MIMO ISI channel using an $n_s = M^{N_t(L-1)}$ state trellis diagram. For a given sequence of channel inputs, the noiseless channel outputs form a path through this trellis. Let us denote the state of the trellis at time k by S_k.

Our objective is to illustrate that the well-known BCJR algorithm used in the context of MAP decoding of convolutional codes can be adopted to compute the joint probability of a given sequence of channel outputs (see Arnold and Loeliger (2001); Arnold et al. (2006); Pfister et al. (2001)). To this end, define

$$\alpha_k(m) = P(\mathbf{y}(1), \ldots, \mathbf{y}(k), S_k = m), \tag{9.9}$$

and

$$\gamma_k(m', m) = p(\mathbf{y}(k) \mid S_k = m, S_{k-1} = m') \cdot P(S_k = m | S_{k-1} = m'), \tag{9.10}$$

where $0 \leq m', m \leq n_s - 1$.

We can then use the forward recursion of the BCJR algorithm to write

$$\begin{aligned}
\alpha_k(m) &= \sum_{m'=0}^{n_s-1} P(\mathbf{y}(1), \ldots, \mathbf{y}(k), S_k = m, S_{k-1} = m'), \\
&= \sum_{m'=0}^{n_s-1} P(\mathbf{y}(k)|S_k = m, S_{k-1} = m')P(\mathbf{y}(1), \ldots, \mathbf{y}(k-1), S_k = m, S_{k-1} = m'), \\
&= \sum_{m'=0}^{n_s-1} P(\mathbf{y}(k)|S_k = m, S_{k-1} = m')P(S_k = m|S_{k-1} = m', \mathbf{y}(1), \ldots, \mathbf{y}(k-1)) \\
&\quad \cdot \alpha_{k-1}(m'), \\
&= \sum_{m'=0}^{n_s-1} \gamma_k(m', m)\alpha_{k-1}(m').
\end{aligned} \tag{9.11}$$

We note that, for i.u.d. inputs, we have

$$p(S_k = m \mid S_{k-1} = m') = \frac{1}{M^{N_t}}, \tag{9.12}$$

whenever there is a valid transition between the two states. If there is no valid transition, this conditional probability is simply zero.

Furthermore, for a given branch of the trellis, the elements of

$$\mathbf{y}(k) = [y_1(k)\ y_2(k)\ \ldots\ y_{N_r}(k)]^T$$

are complex Gaussian. They are independent with means determined by the channel coefficients and the inputs corresponding to the transition, and variances specified by the noise statistics. Therefore, the γ_k terms in Equation (9.10) can be computed easily.

The only other issue that remains is the initialization of the algorithm. If initially encoding starts from the zero state, we can easily set $\alpha_0(0) = 1$ and $\alpha_0(m) = 0$ if $m \neq 0$. To summarize, with this initialization and the explanations above, a recursive algorithm is used step by step to calculate $\alpha_n(m)$ for all m. The desired joint probability is then given by the sum of the metrics over all trellis states, that is,

$$p(\mathbf{y}(1), \mathbf{y}(2), \ldots, \mathbf{y}(N)) = \sum_{m=0}^{n_s-1} \alpha_N(m). \tag{9.13}$$

We note that, for stochastic channels, for simplicity of notation, we have omitted the necessary conditioning on the channel coefficients. However, we assume that the receiver has access to them, and they are treated as constants at every stage of the trellis processing.

To summarize: in order to estimate the information rates of deterministic or random (but ergodic) MIMO FS channels, we simply generate a large number of realizations, calculate the information rate corresponding to each realization, and average the results. In each realization, we use a long sequence of i.u.d. channel inputs, find the corresponding channel outputs, and use the forward recursion of the BCJR algorithm as described above to compute the joint probability of the channel outputs generated. It is also clear that for the case of deterministic channels or ergodic fading channels, we can simply resort to the Shannon–McMillan–Breiman theorem, and use a single long simulation.

Non-Ergodic Case and Outage Information Rates

As in the case of channel capacity, if the fading channel is non-ergodic, then the Shannon type information rate is simply zero. Instead, we need to talk about outage information rates. The computation of this quantity for a fixed outage level, or the outage probability for a fixed rate (with specific i.u.d. channel inputs) is done in a similar manner. For instance, for a quasi-static MIMO FS channel, we simply generate realizations of the channel inputs and outputs, but the channel remains constant for each such realization. The calculation of the joint probability of the specific output sequence follows the same steps as above.

One fundamental difference in this case is that the channel is not ergodic, hence the Shannon–McMillan–Breiman theorem does not hold. This clearly means that we cannot rely on a single long sequence of channel outputs to estimate the outage information rates as we did in the deterministic case. We have to resort to a large number of simulations (where each run corresponds to a specific channel realization) to estimate them given a desired transmission rate.

Markov Inputs from a Finite Constellation

In the previous subsection, we presented a way of computing the achievable information rates of MIMO FS channels when the inputs were i.u.d. and drawn from a finite constellation. This assumption can be relaxed, and a similar technique can be used for Markov inputs as well. In this case, the only difference is in the description of the input–output relationship using a trellis. One needs to take into account the Markov source statistics in addition to the channel structure in setting up the suitable trellis diagram. If the Markov source has a memory of V, then the number of states of the resulting trellis is $M^{N_t \cdot \max(V, L-1)}$. The algorithm is applied in a similar way, with the modification that, in the computation of the γ terms, the conditional probability $p(S_k = m \mid S_{k-1} = m')$ depends on the distribution of the Markov inputs.

Maximization of the Mutual Information over Markov Inputs

We conclude this section by noting that one can also perform "waterfilling" with constrained inputs by maximizing the information rates over the M-ary Markov inputs with a certain memory. This will result in an increased information rate compared with the i.u.d. inputs because of the underlying frequency selectivity of the channel. Details of this maximization process can be found in Zhang et al. (2004).

9.2.3 Examples

We now present several examples. In Figure 9.1, we consider an $N_t = N_r = 2$ MIMO system for both Rayleigh flat fading and frequency selective fading with $L = 2$ and $L = 3$ taps. We assume a uniform average power delay profile, and that the channel is ergodic. We show the information rates achievable with BPSK modulation for the three different channels, as well as the channel capacity (obtained with Gaussian inputs) for the case of flat fading. We observe that, as expected, for large signal-to-noise ratios, the information rates are limited to two bits/channel use, whereas the channel capacity is much larger. Also, we see that there is not much difference between the information rates and the channel capacity for low signal-to-noise ratios. Furthermore, for frequency selective fading channels, the information rates increase (compared with the flat fading case) due to the additional multipath diversity. However, the increase becomes marginal with increasing number of taps.

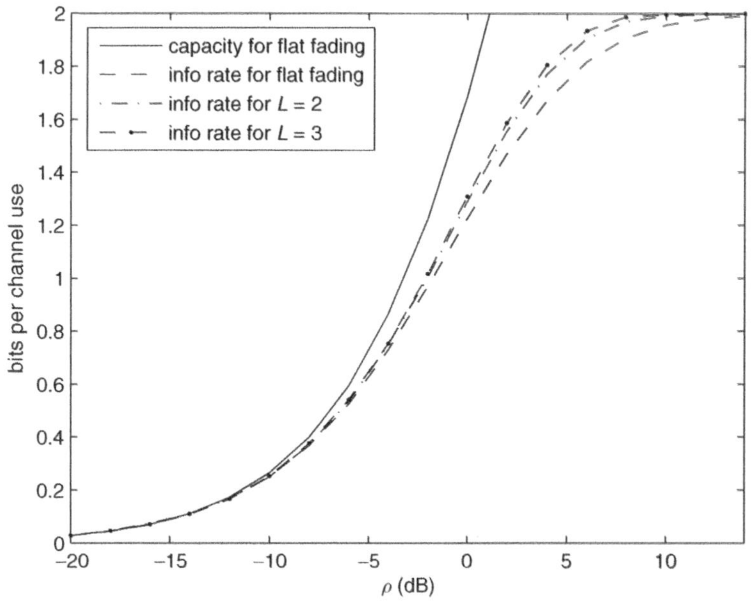

Figure 9.1 Information rates with BPSK over ergodic MIMO FS channels with two transmit and two receive antennas.

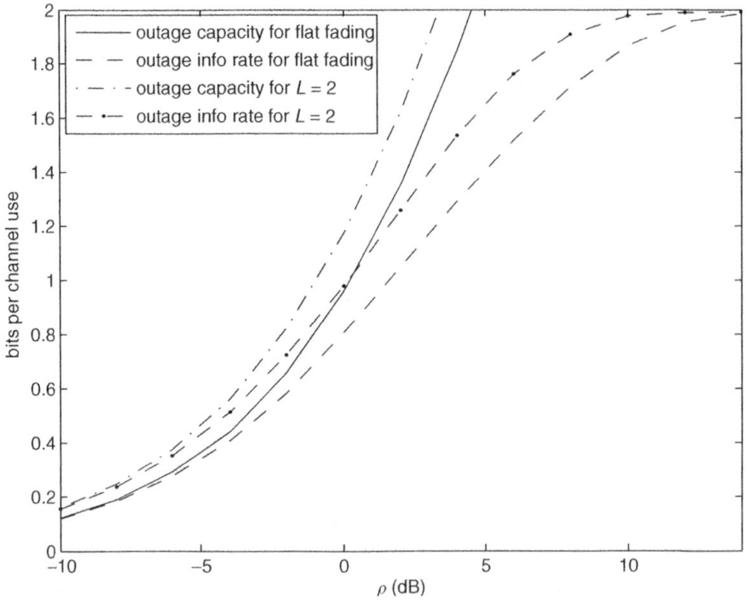

Figure 9.2 Gaussian input outage capacity and outage information rates with BPSK modulation for quasi-static MIMO FS channels with two transmit and two receive antennas (10% outage level).

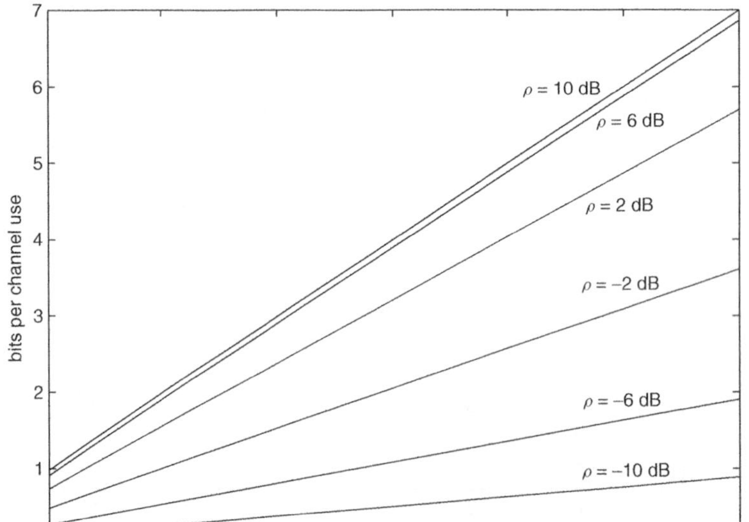

Figure 9.3 Ergodic BPSK information rates for MIMO FS channels (with two equal average power taps) as a function of $N_t = N_r = n$.

In Figure 9.2, we again consider an $N_t = N_r = 2$ system over Rayleigh fading (with equal average power taps for the case of frequency selective channel), but we assume that the fading is non-ergodic. Hence the outage information rates for BPSK and outage capacity are plotted. An outage probability of 10% is assumed. In this case, along with the information rates with BPSK inputs, the information rates with independent Gaussian inputs are shown. The observations are similar to the case of ergodic fading.

As a final example, we consider an ergodic MIMO FS channel with $N_t = N_r = n$ and two equal average power taps. We show the information rates that can be obtained with i.u.d. BPSK inputs, as a function of the number of antennas n for several signal-to-noise ratios in Figure 9.3. The result is very interesting in the sense that what was already confirmed for flat fading channels seems to hold for frequency selective channels as well. That is, the information rates increase linearly with the number of antennas, even at very low signal-to-noise ratios. Hence, there is a tremendous increase with the use of MIMO transmission without requiring additional bandwidth or power for frequency selective MIMO channels as well.

9.3 Space-Time Coding for MIMO FS Channels

In this section, we change our focus to practical channel coding approaches over MIMO frequency selective fading channels. As in most of our coding coverage, we consider quasi-static fading channels. That is, the fading coefficients remain constant for the entire codeword. We first consider an interpretation of MIMO FS channels, and then derive the code design criteria accordingly. Finally, we review several specific coding approaches.

9.3.1 Interpretation of MIMO FS Channels Using Virtual Antennas

Assume that the space-time codeword length is N, and additional $L-1$ zeros are appended at the end of each codeword to clear the channel. Then, the input–output relationship of the MIMO FS channel can be written as

$$\boldsymbol{Y} = \sqrt{\rho}\boldsymbol{X}_{eq}\boldsymbol{H}_{eq} + \boldsymbol{N}, \tag{9.14}$$

where the equivalent space-time codeword transmitted is an $(N+L-1) \times N_t \cdot L$ matrix given by

$$\boldsymbol{X}_{eq} = \begin{bmatrix} x_1(1) & 0 & \cdots & 0 & \cdots & x_{N_t}(1) & 0 & \cdots & 0 \\ x_1(2) & x_1(1) & \cdots & 0 & \cdots & x_{N_t}(2) & x_{N_t}(1) & \cdots & 0 \\ & \vdots & & & \vdots & & & \vdots & \\ x_1(N) & x_1(N-1) & \cdots & 0 & \cdots & x_{N_t}(N) & x_{N_t}(N-1) & \cdots & 0 \\ 0 & x_1(N) & \cdots & 0 & \cdots & 0 & x_{N_t}(N) & \cdots & 0 \\ & \vdots & & & \cdots & & & \vdots & \\ 0 & 0 & \cdots & x_1(1) & \cdots & 0 & 0 & \cdots & x_{N_t}(1) \end{bmatrix},$$

the equivalent channel coefficient matrix is

$$
\boldsymbol{H}_{eq} = \begin{bmatrix}
h_{1,1}^{(0)} & h_{1,2}^{(0)} & \cdots & h_{1,N_r}^{(0)} \\
h_{1,1}^{(1)} & h_{1,2}^{(1)} & \cdots & h_{1,N_r}^{(1)} \\
\vdots & \vdots & \ddots & \vdots \\
h_{1,1}^{(L-1)} & h_{1,2}^{(L-1)} & \cdots & h_{1,N_r}^{(L-1)} \\
h_{2,1}^{(0)} & h_{2,2}^{(0)} & \cdots & h_{2,N_r}^{(0)} \\
h_{2,1}^{(1)} & h_{2,2}^{(1)} & \cdots & h_{2,N_r}^{(1)} \\
\vdots & \vdots & \ddots & \vdots \\
h_{2,1}^{(L-1)} & h_{2,2}^{(L-1)} & \cdots & h_{2,N_r}^{(L-1)} \\
\vdots & \vdots & \vdots & \vdots \\
h_{N_t,1}^{(0)} & h_{N_t,2}^{(0)} & \cdots & h_{N_t,N_r}^{(0)} \\
h_{N_t,1}^{(1)} & h_{N_t,2}^{(1)} & \cdots & h_{N_t,N_r}^{(1)} \\
\vdots & \vdots & \ddots & \vdots \\
h_{N_t,1}^{(L-1)} & h_{N_t,2}^{(L-1)} & \cdots & h_{N_t,N_r}^{(L-1)}
\end{bmatrix}, \tag{9.15}
$$

and the matrices $\boldsymbol{Y}$ and $\boldsymbol{N}$ are $(N + L - 1) \times N_r$ matrices of the received signals and noise terms whose (k, j)th element denotes the corresponding quantity for the kth time instant and jth antenna.

Writing the input–output relationship for a MIMO FS fading channel has an interesting interpretation. Going back to the MIMO flat fading channel model studied throughout the book, we see that, mathematically, we have the same input–output expression. The only difference is the special structure of the input matrix (that determines what is being transmitted from which antenna at what time), and the fact that the new channel matrix consists of $N_t N_r L$ entries. For independent Rayleigh fading, the channel coefficients are independent for different antenna pairs and for different channel taps.

By examining the equivalent space-time code matrix, $\boldsymbol{X}_{eq}$, we see that it is as if we have $L - 1$ additional "virtual" antennas corresponding to each actual transmit antenna that emit delayed versions of transmitted sequence. This is a useful observation in designing codes for MIMO FS channels. We illustrate this interpretation and the resulting transmission scheme in Figure 9.4.

It is easy to observe the maximum diversity order achievable for a MIMO FS fading channel. With the quasi-static Rayleigh fading channel model, there are $N_t N_r L$ independent fading coefficients, therefore by a proper design of the space-time code, it is possible to obtain a diversity of order $N_t N_r L$. Clearly, the presence of multipath, in fact, provides us with a better performance compared with the flat fading scenario. Going back to Chapter 2 and also from the information rate results of the previous section, we know that this is expected as the presence of multipath propagation provides another form of diversity.

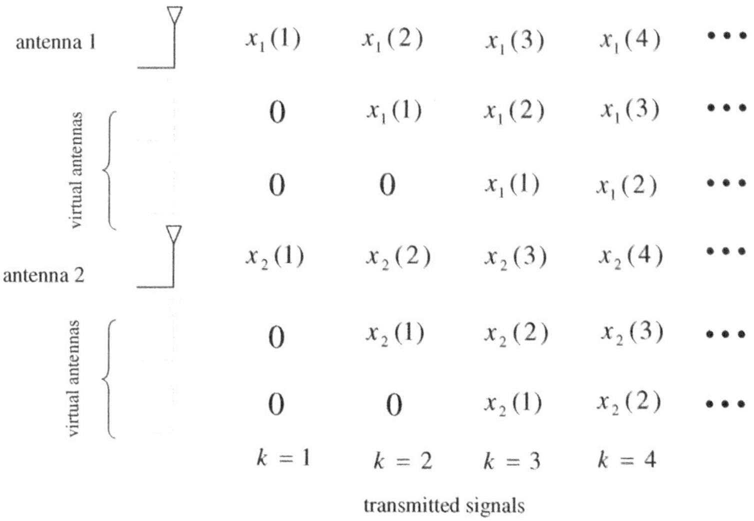

Figure 9.4 Virtual antenna interpretation of transmission over MIMO FS channels (with $N_t = 2$ and $L = 3$).

Although the available diversity order may be high, care must be taken in code design to exploit it. We now describe the rank criterion for achieving full diversity (over quasi-static) MIMO FS channels by extending the corresponding criterion derived in Chapter 5 in the context of MIMO flat fading channels.

Criterion for Achieving Full Diversity

A space-time code achieves full diversity over a MIMO FS channel if for any pair of codewords (of the form $\boldsymbol{X}_{eq}$ whose form is given above), the corresponding codeword difference matrices are of full column rank.

We conclude this subsection by noting that it is easy to write down this principle, but it may be tedious to test whether a given code will satisfy it or not.

9.3.2 A Simple Full Diversity Code for MIMO FS Channels

Before we describe a simple code that achieves full diversity for MIMO FS channels, let us consider the idea of delay diversity for the flat fading scenario. For an N_t transmit antenna system, a delay diversity code is obtained by simply transmitting delayed versions of the sequence on the first transmit antenna from the other antennas (Gore et al. (2001)). That is, the ith transmit antenna emits a delayed version of the sequence (by $i - 1$ symbols) on the first transmit antenna. By this method, it is easy to see that for flat fading, space-time codewords are of the form

$$\begin{bmatrix} x(1) & 0 & 0 & \cdots & 0 \\ x(2) & x(1) & 0 & \cdots & 0 \\ x(3) & x(2) & x(1) & \cdots & 0 \\ & \vdots & & \vdots & \\ x(N_t) & x(N_t-1) & x(N_t-2) & \cdots & x(1) \\ x(N_t+1) & x(N_t) & x(N_t-1) & \cdots & x(2) \\ & \vdots & & \vdots & \\ x(N) & x(N-1) & x(N-2) & \cdots & x(N-N_t+1) \end{bmatrix}. \tag{9.16}$$

Therefore, the codeword difference matrices will clearly be full column rank if a distinct pair of codewords are considered. Hence, the delay diversity code will achieve a full rank of $N_t N_r$ over flat fading.

To extend this idea to the case of MIMO FS channels, we need a slight modification. It is easy to see that delaying by a single symbol for each transmit antenna will not work, because some columns of the equivalent space-time codeword matrix (with the use of virtual antenna interpretation given in the previous section), will be identical, hence the code will not be of full rank. However, it is straightforward to see that, if we delay the transmitted signal of the first antenna by the number of ISI taps L instead of by a single symbol, we obtain the equivalent space-time codeword (for the MIMO FS channel) given as

$$\begin{bmatrix} x(1) & 0 & 0 & \cdots & 0 \\ x(2) & x(1) & 0 & \cdots & 0 \\ x(3) & x(2) & x(1) & \cdots & 0 \\ & \vdots & & \vdots & \\ x(N_tL) & x(N_tL-1) & x(N_tL-2) & \cdots & x(1) \\ x(N_tL+1) & x(N_tL) & x(N_tL-1) & \cdots & x(2) \\ & \vdots & & \vdots & \\ x(N) & x(N-1) & x(N-2) & \cdots & x(N-N_tL+1) \end{bmatrix}. \tag{9.17}$$

We can then easily verify that the idea of delaying by L satisfying the full rank criterion given in the previous section, hence it provides a diversity of order $N_t N_r L$.

We do not go into the details here, but this delay diversity code can be described using a trellis diagram, hence decoding can be performed using the Viterbi algorithm. This point will be illustrated further in the next section when we describe a more general class of codes for MIMO FS channels.

9.3.3 Space-Time Trellis Codes for MIMO FS Channels

We would first like to point out that it is possible to use space-time block codes over MIMO FS channels. However, much of the benefits of space-time block coding, particularly in the decoding process, are lost. This is because, with the presence of ISI, it not trivial to perform simple linear processing to decouple the decisions of different symbols in ML decoding.

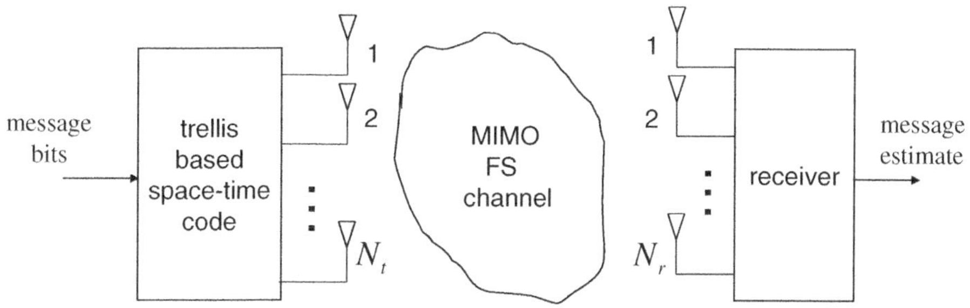

Figure 9.5 Block diagram of space-time trellis coding over a MIMO FS channel.

Therefore, we do not describe them here, instead, we directly talk about the extension of space-time trellis codes to MIMO FS channels.

We have already described space-time trellis codes for MIMO flat fading channels in Chapter 5. These are simply extensions of trellis-coded modulation schemes to the multiple antenna context, and can be designed using the diversity and rank criteria, and can be decoded efficiently using the Viterbi algorithm as described in detail in Chapter 5.

Let us now focus on the use of space-time trellis codes over MIMO FS Channels. In Figure 9.5, we show the system block diagram. The only difference in principle is the presence of the ISI channel as opposed to the flat fading channel considered earlier.

Both the space-time code and the channel can be described by their corresponding trellis diagrams. Hence, it is possible to describe their "concatenation" by an equivalent trellis. Let us illustrate this point by a simple example.

Example

Consider a simple two-antenna four-state space-time code (of rate 1 bits per channel use) with BPSK described by two convolutional codes with connection polynomials $1 + D^2$ (D refers to the delay element) for the first antenna, and $1 + D + D^2$ for the second antenna, i.e., $(5, 7)_{octal}$ convolutional code. Clearly, this describes a four-state space-time code. Let us assume that each sub-channel has two taps. For instance, if there is a single receive antenna, the received signal is given by (ignoring noise)

$$y(n) = h_1^{(1)} x_1(n) + h_1^{(2)} x_1(n-1) + h_2^{(1)} x_2(n) + h_2^{(2)} x_2(n-1), \tag{9.18}$$

where x_1 and x_2 denote the transmitted signals from the two antennas. Clearly, if the sequence of bits being encoded by the space-time code are represented by $s(n)$, we have $x_1(n) = s(n) \oplus s(n-2)$ and $x_2(n) = s(n) \oplus s(n-1) \oplus s(n-2)$. Combining these with the expression in Equation (9.18), and assuming that the channel state information is available at the receiver, it is clear that one can describe the "noiseless" received signal using an eight-state trellis diagram where the states are formed by the three previous input bits ($s(n-1), s(n-2)$ and $s(n-3)$). The resulting trellis diagram is shown in Figure 9.6 where we have used a vector of length four to illustrate the branch labels. These denote the signals $x_1(n), x_1(n-1), x_2(n), x_2(n-1)$. As an example, corresponding to the label

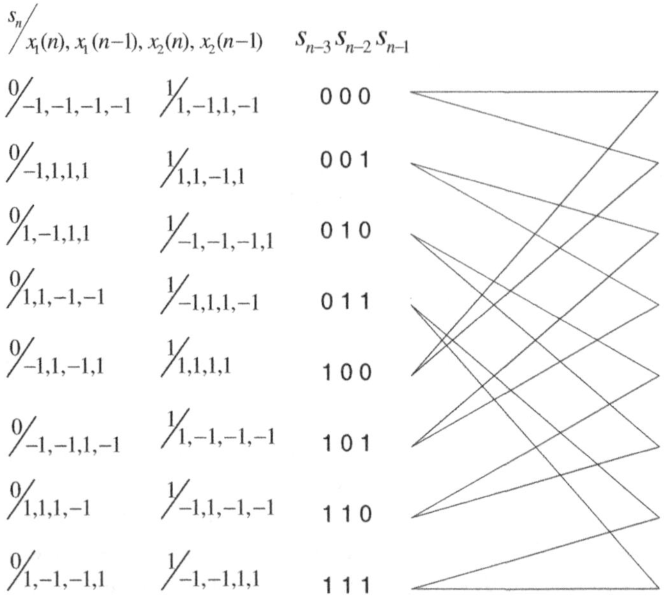

Figure 9.6 Combined code and channel trellis for $(5, 7)_{octal}$ convolutional code over a two-input two-tap FS channel.

$(-1, +1, -1, -1)$, the received signal in the absence of noise is $-h_1^{(1)} + h_1^{(2)} - h_2^{(1)} - h_2^{(2)}$. The same representation can be used for more than one receive antenna as well in a straightforward manner. We further note that this is by no means a unique way of representing the space-time code together with the ISI channel; it only represents one possible approach.

The joint trellis representation of the space-time trellis code and the frequency selective channel is useful in the decoding of the code. Since the channel state information is assumed to be known at the receiver, and the noise terms are independent both spatially and temporally, minimization of the squared Euclidean distance between the noiseless sequences through the joint trellis and the received signal sequence will minimize the sequence error probability, hence it is optimal. Furthermore, this can be implemented using the well-known Viterbi algorithm.

Examples of good space-time trellis codes for MIMO FS channels with BPSK and QPSK modulation schemes are provided by El Gamal et al. (2003). We do not go into a detailed description of these codes here.

As a simple space-time trellis code example, consider the convolutional codes $(5, 7)_{octal}$ and $(64, 74)_{octal}$ with BPSK modulation for two transmit antennas. Clearly, the first one is a four-state code, and the second one is an eight-state code. When combined with a channel of length L, the number of states becomes 2^{L+1} and 2^{L+2}, respectively. In Figure 9.7, we show the frame error rates (for a frame length of 100) over a quasi-static Rayleigh fading channel with number of ISI taps $L = 1, 2$ and 3. For the flat fading case, i.e., with $L = 1$, both codes achieve a diversity of order two, while the second code provides a larger coding gain. For the case of $L = 2$, i.e., two-tap uniform power delay profile FS channel, they

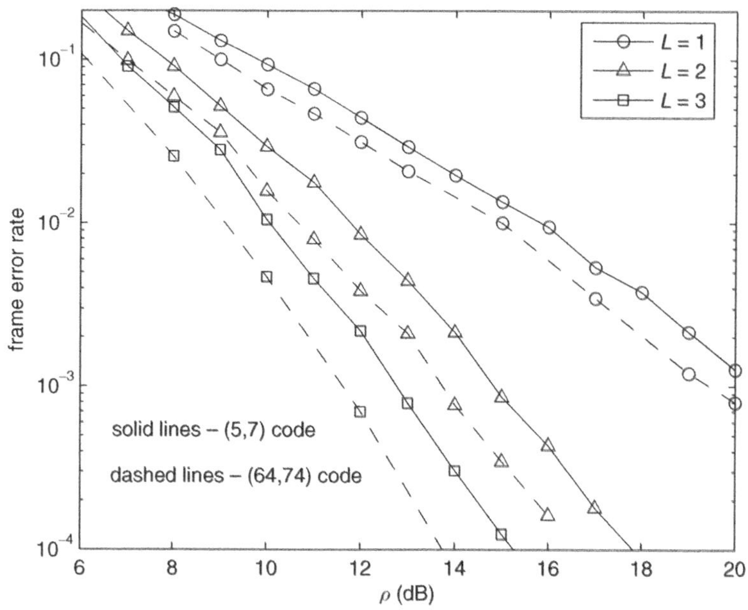

Figure 9.7 Bit error rates of two convolutional codes with BPSK over quasi-static MIMO FS channels (with two transmit and one receive antennas).

achieve a diversity of order four, and for $L = 3$ (again for a uniform power delay profile), they achieve a diversity of order five and six, respectively. The $(5, 7)_{octal}$ code is unable to achieve the full diversity for $L = 3$, because its memory length is not sufficient.

We consider the same two codes for the case of two receive antennas in Figure 9.8. Our observations are similar, except that the diversity orders are now twice the ones obtained for $N_r = 1$. We note that, to see them clearly, we should show the error rates for larger signal-to-noise ratios.

9.3.4 Concatenated Coding for MIMO FS Channels

In Chapter 7, we have talked about concatenated coding and iterative decoding techniques for MIMO channels for frequency flat fading. Similar ideas can be extended to the case of frequency selective fading as well. One basic difference is that the frequency selective fading channel can itself be used as an "inner code" in code concatenation.

The basic block diagram of a concatenated coding approach for MIMO FS channels is shown in Figure 9.9. An outer code which can be a space-time trellis code or a convolutional code is concatenated with an M-ary modulator through an interleaver with a certain length, and then the modulated symbols are transmitted through the ISI channel. This is basically the same approach we have used over MIMO flat fading channels.

For demodulation/decoding, the MIMO channel is equalized using a soft-output equalizer, for instance, implemented via the MAP algorithm. The soft-output equalizer produces likelihoods of the M-ary symbols transmitted. The symbol level likelihood information is

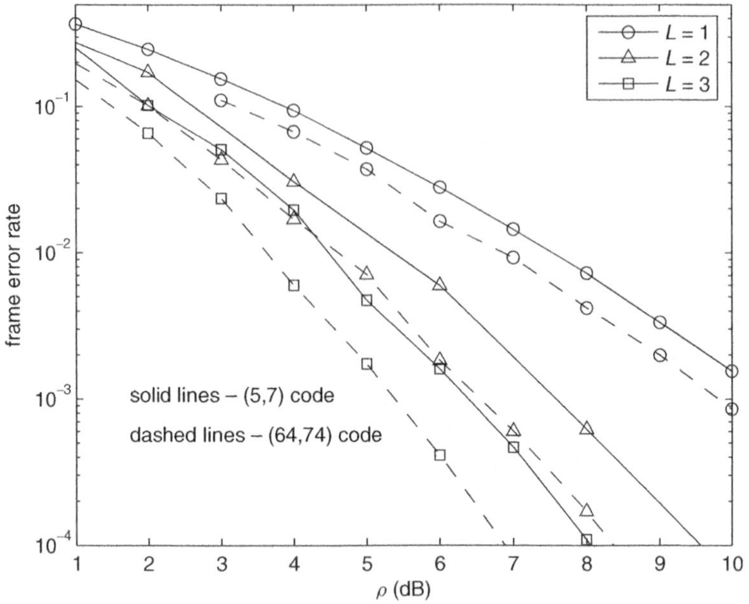

Figure 9.8 Bit error rates of two convolutional codes with BPSK over quasi-static MIMO FS channels (with two transmit antennas and two receive antennas).

then converted to bit level likelihoods, and these are used in the decoding of the information bits using the code constraints. The only difference in this scheme, compared with the frequency flat fading case, is the presence of the soft-output equalizer. The demodulation/decoding is shown in Figure 9.9 as well.

Since magnetic recording channels exhibit intersymbol interference, concatenated coding schemes are thoroughly investigated in this context (see Ghrayeb and Ryan (2001); Kaynak et al. (2004)). The majority of these investigations apply for single-input single-output systems. What we briefly describe here is a simple extension of these to MIMO FS channels.

The outer code may take many different forms. For instance, it can be a convolutional code, a parallel concatenated convolutional code (Berrou et al. (1993)), a serial concatenated convolutional code (Benedetto et al. (1998)), a space-time trellis code, or even an LDPC code (Gallager (1962); MacKay (1999)). For the case of a convolutional code or a space-time trellis code, we simply have its concatenation with an ISI channel which may be referred to as a serial concatenation scheme. We note that in such a scheme using a precoder after interleaving may improve the performance by effectively making the channel constraints "recursive" (as in serially concatenated convolutional code in the standard turbo-coding context). The outer decoder, in this case, can be implemented using the Viterbi algorithm, or a MAP type algorithm. For the case of a turbo code, as the decoder of the

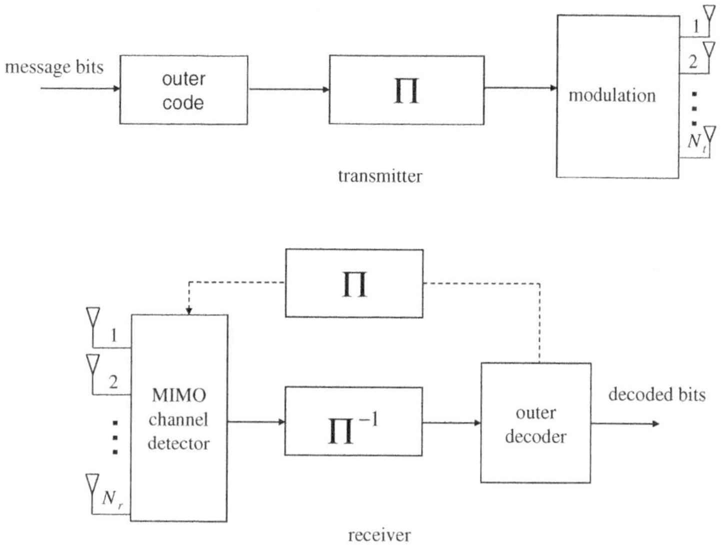

Figure 9.9 Block diagram of the concatenated coding approach.

outer code, we would need an iterative algorithm (implemented using component MAP type decoders). For the case of an outer LDPC code, decoding can be performed using the usual message-passing algorithm.

It is also possible to use the idea of iterative "turbo" equalization at the receiver to improve the error rate performance of the system. Basically, the outer channel decoder can be implemented using a soft-output algorithm, and the extrinsic (new) information that it produces can be passed back to the MIMO channel equalizer. This information can be used to update the symbol likelihoods, hence the bit likelihoods, to be used in the next iteration of the channel decoder. This process is illustrated using the dashed lines in Figure 9.9.

As a simple illustration, let us consider the use of a $(5, 7)_{octal}$ convolutional code over a two transmit antenna system. In Figure 9.10, we show the bit error rate obtained with BPSK modulation over a three-tap (quasi-static) frequency selective fading channel with a frame length of 1000 symbols. The power delay profile of the channel is assumed to be [0.16 0.68 0.16]. For code concatenation, we employ a pseudo-random interleaver with a length of 1000. The underlying code is of rate 1/2 and the transmission rate is 1 bit per channel use (with two transmit antennas). We show the error rate results for both $N_r = 2$ and $N_r = 3$, and with a single and three decoder iterations.

Another example for basically the same set-up, but with QPSK modulation is given in Figure 9.11. The transmission rate is 2 bits per channel use. Our observations are similar to the previous case.

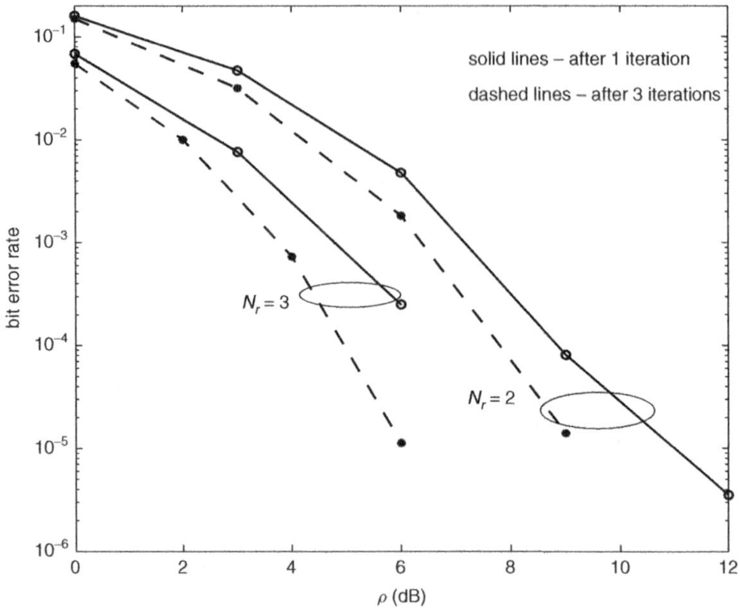

Figure 9.10 Bit error rates of $(5, 7)_{octal}$ convolutional code with BPSK concatenated with a three-tap quasi-static MIMO FS channel (with two transmit antennas).

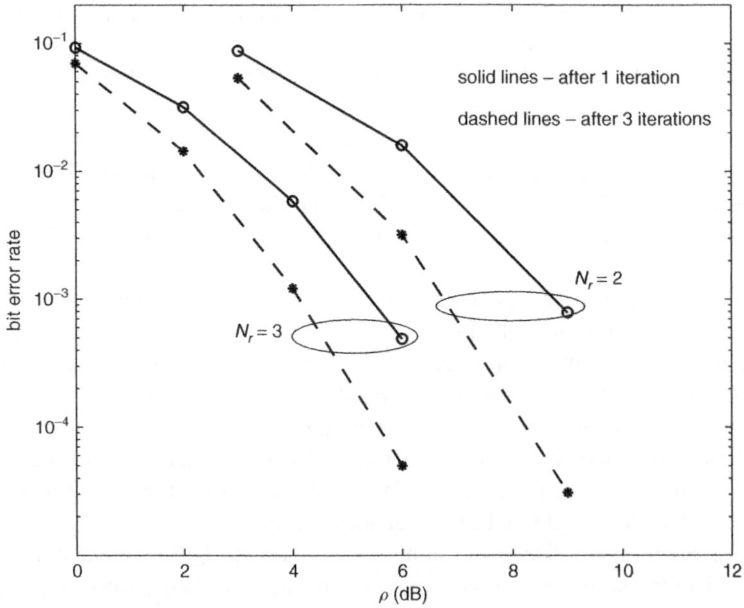

Figure 9.11 Bit error rates of $(5, 7)_{octal}$ convolutional code with QPSK modulation concatenated with a three-tap quasi-static MIMO FS channel (with two transmit antennas).

9.3.5 Spatial Multiplexing for MIMO FS Channels

If we are interested in increasing the transmission rates (with possibly a penalty in the diversity order provided by the system), we can employ spatial multiplexing schemes (described in Chapter 6) for MIMO FS channels as well. There is basically no difference in the encoding approaches. Whatever can be used for the case of flat fading will also be easily extended to the case of frequency selective fading channels. On the other hand, there are differences in the receiver design. In the case of flat fading, the receiver algorithms basically spatially equalize the received signal using different algorithms. For instance, zero forcing or MMSE receivers are employed. In the case of MIMO FS channels, the receiver works in the same way, however, in addition to the spatial equalization, it performs temporal equalization.

In the following section, we briefly describe several approaches for signal detection over MIMO FS channels.

9.4 Channel Detection for MIMO FS Channels

In this section we concentrate on MIMO FS channel equalization, and describe several alternatives with varying levels of complexity. In addition to channel equalizers for uncoded MIMO systems, we also consider soft-input soft-output type MIMO equalizers to be used in conjunction with channel coding. The fundamental problem is almost the same as the standard equalization problem thoroughly studied in the wireless communication literature, except that in this case, we need to do equalization both temporally and spatially due to the simultaneous transmission of different data streams from different antennas.

MIMO FS channels can be described by a trellis with $M^{N_t(L-1)}$ states where M is the constellation size. If uncoded data streams are being transmitted simultaneously from different antennas, there are N_t symbols for each period. Hence the number of branches emanating from each state is M^{N_t}. Once the trellis diagram is obtained using standard methods, the optimal detection problem is easy. If our objective is to minimize the sequence error probability, we can simply use the Viterbi algorithm. If we would like to minimize the symbol error probability, we can resort to the MAP algorithm (which is the same one used in the turbo-decoding process as a component decoder for each constituent convolutional code).

One problem with the full complexity algorithms, e.g., the Viterbi or MAP algorithm, is the associated complexity. The number of states in the trellis grows exponentially with the product of the number of transmit antennas and the length of the ISI channel. This quickly becomes prohibitively complex even with relatively short ISI levels. For instance, with BPSK modulation, if the number of antennas is four, with an ISI length of $L = 5$, the number of states needed for optimal equalization is $2^{16} = 65{,}536$ which is too large. Therefore, we need other simplified approaches for channel detection.

We note that the need for reduced complexity receivers for MIMO FS channels should not come as a surprise. Even in the case of frequency flat fading, when the number of antennas is increased, we need to resort to a BLAST type of transmission scheme (i.e., spatial multiplexing), and apply reduced complexity detection algorithms. The problem with frequency selectivity becomes even more complicated.

As reduced complexity MIMO FS channel equalizers, we consider linear equalizers and decision feedback equalizers.

9.4.1 Linear Equalization for MIMO FS Channels

Consider an uncoded VBLAST transmission over a MIMO FS channel, and assume that the channel coefficients for all the antenna pairs and for all the channel taps are known. A simple way of estimating the transmitted bits based on the observations at the receiver is to employ a spatially and temporally linear equalizer.

The estimate of the kth symbol transmitted from the ith antenna is given by

$$\hat{x}_i(k) = \sum_{j=1}^{N_r} \sum_{n=-N_1}^{N_2} a_{j,n} y_j(k-n), \tag{9.19}$$

where $a_{j,n}$ are the set of filter coefficients and $N_1 + N_2 + 1$ is the length of the filter used. Given the fading channel coefficients, the equalizer coefficients can be found using different constraints, such as zero forcing, or MMSE. The solutions will be very similar to the case of single-transmit single-receive antenna systems.

In practice, if the channel coefficients are not available, adaptive solutions can also be derived. The ideas are again similar to the case of single-input single-output systems studied in standard digital communications textbooks (Proakis (2001)).

9.4.2 Decision Feedback Equalization for MIMO FS Channels

As an alternative to linear equalization, decision feedback equalization (see Proakis (2001)) can also be extended to the case of MIMO communications. The approach in this case is to use two sets of filters (feedforward and feedback), where the feedback filter operates on already detected symbols with the objective of canceling out their contribution from the observation vector. Due to the non-linearity in the decision process (on the prior symbols), the overall equalizer is no longer linear. A block diagram of the MIMO decision feedback equalizer (DFE) is shown in Figure 9.12.

As in the case of a linear equalizer, different criteria with different restrictions on the feedforward and feedback filters can be employed, such as the use of finite-length filters with MMSE. Also, adaptive solutions can be derived using a training-based approach if the channel coefficients are not known at the receiver. Furthermore, we emphasize that the block diagram shown in Figure 9.12 is only one approach; many other approaches are possible (see Lozano and Papadias (2002)).

9.4.3 Soft-Input Soft-Output Channel Detection

Consider coded transmission over MIMO FS channels. In particular, for the case of concatenated coding (with an interleaver) over MIMO FS channels, in order to be able to run the outer decoder we often need soft outputs about the coded bits. This is directly provided by the MAP type equalization algorithms. However, for other equalization approaches they may not be readily available.

Both linear and decision feedback equalizers can be made to accept soft information about the input symbols, and produce soft outputs. Therefore, instead of the MAP type

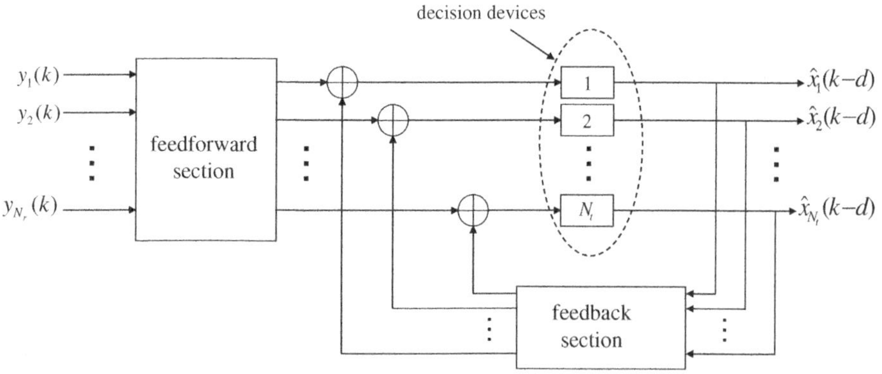

Figure 9.12 Block diagram of a MIMO-DFE.

optimal (soft-output) channel equalizer, these suboptimal solutions can be used. After the likelihood information about the symbols transmitted are obtained, we can compute the likelihoods of the coded bits (by simply using the approaches described in Chapter 7), then these can be used in decoding the information bits using the outer code constraints. Therefore, reduced complexity linear and DFE based MIMO equalizers can be incorporated in equalization/decoding of coded systems (with or without turbo equalization).

9.4.4 Other Reduced Complexity Approaches

Many other reduced complexity MIMO FS equalization methods exist. For example, instead of the full complexity MAP type algorithm, we can employ the soft-output Viterbi algorithm. Or, we can further simplify the receiver structure by using other trellis-based schemes such as the M-algorithm and T-algorithm and their soft-output versions. In the M-algorithm, the idea is to extend paths through only a limited (but fixed) number of states, as opposed to all the states. In the T-algorithm, the approach is similar, but only the paths with metrics below a threshold are extended.

Another approach is to employ sequential decoding-based equalizers, e.g., stack equalizer. This algorithm can also be made to accept soft inputs and produce soft outputs, and provide a viable alternative to full complexity MIMO equalization (Gucluoglu and Duman (2005)). In addition, a Kalman filtering based soft-input soft-output equalization algorithm can also be used as it is shown that it provides a good complexity performance trade-off (Roy and Duman (2004, 2007)).

Finally, we can also employ the BP algorithm (normally used in the decoding of LDPC codes) for MIMO FS channels. In this case, the approach is to represent the MIMO FS channel constraints using a factor graph, and use a message-passing algorithm for channel equalization (Kaynak et al. (2005, 2006)). By nature, this is a soft-input soft-output algorithm, and it gives excellent performance over quasi-static or ergodic (fading) channels. For fixed channels, the performance may be degraded as there may be many loops with short cycles in the factor graph representation, and there is no randomization to help break them.

9.5 MIMO OFDM Systems

Until now, we have focused on single carrier transmission over frequency selective fading channels, which requires sophisticated equalization techniques to overcome the intersymbol interference problem. Another way to deal with such channels and to avoid relatively complex signal processing algorithms is to employ multi-carrier transmissions used in the same frequency band. For example, assume that the overall available bandwidth is W. The idea is to split the overall band available into N frequency sub-bands so that each sub-band (of bandwidth W/N) experiences flat fading as illustrated in Figure 9.13. The symbol duration on each multi-carrier component is $\sim N/W$ which can be made much larger than the multipath spread of the channel, hence the subchannels do not experience ISI (or, the effects of ISI are reduced to a desired level). The overall transmission rate is still $\sim N \cdot W/N = W$ symbols per second, i.e., the same as the original single-carrier system.

A simple and efficient way of achieving multi-carrier modulation is to employ orthogonal sub-carriers (that are separated by $1/T_s$ Hz where T_s is the symbol duration), and use the DFT pair (more precisely, fast Fourier transform (FFT) and its inverse) for implementation. The idea is to realize that with $1/T_s$ frequency separation, samples of the overall signal to be transmitted over N sub-carriers (at a rate $1/T_s$) can be obtained using the N-point inverse DFT (IDFT) or inverse FFT (IFFT) of the data sequence of the N sub-carriers. At the receiver the DFT (or FFT) is applied, and the standard receiver algorithms are used. The resulting scheme is called orthogonal frequency division multiplexing (Boelcskei (2006); Stuber et al. (2004); Su et al. (2005)).

The idea of multi-carrier modulation, or OFDM, can be applied to MIMO FS channels as well, as shown in Figure 9.14. For each transmit antenna element, the inverse DFT of N (uncoded or coded) M-ary symbols, denoted by

$$X_i(0), X_i(1), X_i(2), \ldots, X_i(N-1)$$

is first computed. Then a cyclic prefix is appended to each of these sequences, and the resulting signal is transmitted (after digital-to-analog conversion). The objective in adding the

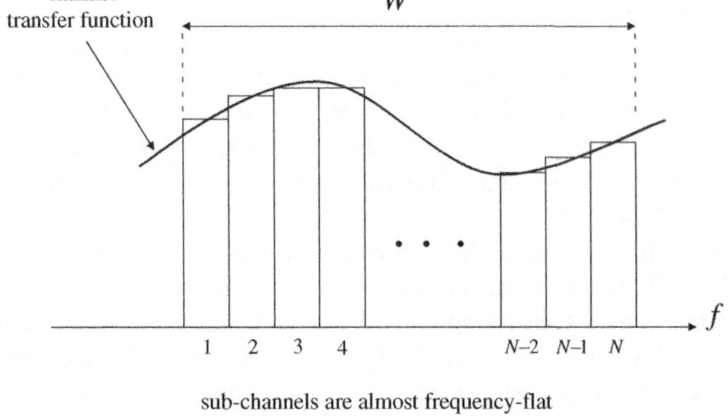

Figure 9.13 Illustration of multi-carrier communications in the frequency domain.

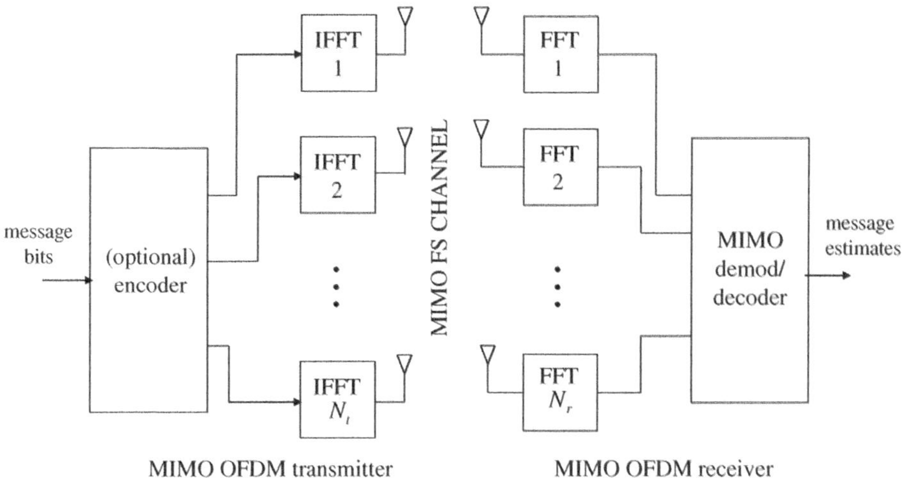

Figure 9.14 Block diagram of a MIMO-OFDM system.

cyclic prefix is to remove possible interference between two consecutive OFDM symbols, and to make sure that the equivalent channel (after DFT at the receiver) has a simple form which allows for easier processing. For each of the received signals (at the N_r antenna elements), the DFT of the aggregate received signal (superposition of all the OFDM words from each of the transmitted antennas) is calculated, and the cyclic prefix is removed. The resulting set of signals is then used for demodulation/decoding.

Let us now review the resulting channel model, possible coding approaches, as well as challenges in MIMO-OFDM.

9.5.1 MIMO-OFDM Channel Model

Consider a MIMO-OFDM system with N subcarriers used over a MIMO FS channel with L ISI taps. Assume that the fading coefficients are spatially uncorrelated, and that they remain constant over one OFDM symbol. The transmitted signal over N_t transmit antennas can be represented by an $N \times N_t$ matrix, $\boldsymbol{X}_{OFDM}$ whose (n, i)th element is the symbol transmitted at subcarrier n on transmit antenna i, $X_i(n)$. We assume that there is a power constraint over the MIMO-OFDM word such that

$$\sum_{n=0}^{N-1} \sum_{i=1}^{N_t} E[|X_i(n)|^2] = N.$$

The transmitter simply takes the IFFT of the columns of the codeword matrix $\boldsymbol{X}_{OFDM}$, appends cyclic prefix, and transmits the resulting sequence. The role of the cyclic prefix is to convert the circular convolution to linear convolution as described below. Also, it can be used for synchronization purposed. Obviously, there is some overhead involved with the inclusion the cyclic prefix, however, the length of the ISI is usually much smaller than the

OFDM word, thus the overhead required is small. At the jth receive antenna, after applying the FFT and removing of the cyclic prefix, the resulting signal for the nth subcarrier is given by

$$y_j(n) = \sqrt{\rho} \sum_{i=1}^{N_t} X_i(n) H_{i,j}(n) + n_j(n), \tag{9.20}$$

where the equivalent channel coefficient from the ith to the jth antenna for the nth subcarrier is given by

$$H_{i,j}(n) = \sum_{l=0}^{L-1} h_{i,j}^{(l)} e^{-j2\pi nl/T}. \tag{9.21}$$

Denoting the channel coefficients for the (i, j)th antenna pair by

$$\boldsymbol{H}_{i,j} = [H_{i,j}(0) \quad H_{i,j}(1) \quad \cdots \quad H_{i,j}(N-1)], \tag{9.22}$$

we can see that

$$\boldsymbol{H}_{i,j} = \boldsymbol{W}\boldsymbol{A}_{i,j}, \tag{9.23}$$

where $\boldsymbol{W}$ is the DFT (or FFT) matrix, and

$$\boldsymbol{A}_{i,j} = [h_{i,j}^{(0)} \quad h_{i,j}^{(1)} \quad \cdots \quad h_{i,j}^{(L-1)}],$$

i.e., the vector of L independent fading coefficients (for the L different paths of the ISI channel). Since we have a total of N equivalent channel coefficients derived from L independent random variables (with N typically much larger than L), it is clear that in the equivalent channel model above the fading coefficients for distinct subcarriers are different but dependent. The maximum rank of the channel gain matrix (in the frequency domain) is the number of ISI taps L, hence it is a low-rank matrix in general. This is the basic difference with most of the models adopted in the previous parts of the book. For instance, for quasi-static fading channels, the coefficients were all identical (over a time frame), and for fully interleaved fading, they were independent of each other. Only in Chapter 10, we make reference to time varying channels (which have similar models with the MIMO-OFDM links). Obviously, the channel models look identical mathematically, however the reason of the different, but dependent, fading coefficients is the intersymbol interference here.

9.5.2 Space-Frequency Coding

It is clear from the above description that the role of the (possible) channel code before the OFDM modulation is to perform coding both spatially and across different sub-carriers, i.e., space-frequency coding. This is to be contrasted with the ideas in space-time coding studied in detail in this book.

Based on the channel model derived in the previous section, and by utilizing the multipath delay profiles of subchannels to determine the correlation structures, it is possible to derive code design principles (based on pairs of codewords) similar to the rank and determinant criteria for space-time code design. We do not go into the details of this approach here. We note that the maximum available diversity order is the same as the single-frequency approach, namely, $N_t N_r L$.

Many alternatives are possible in selecting the specific space-frequency coding scheme. For instance, we can simply use uncoded sequences, and obtain "spatial multiplexing" combined with OFDM. Or the Alamouti scheme can be employed as a space-frequency code. We can also employ bit interleaved coded modulation (BICM) which may make it easier to obtain diversity. Since the fading coefficients of nearby sub-carriers are highly correlated, interleaving across subcarriers may be essential to achieve the maximum possible diversity. Finally, we note that space-frequency codes can also be designed using a systematic approach to get all or part of the total available diversity. We provide a list of references on space-frequency coding at the end of the chapter.

9.5.3 Challenges in MIMO-OFDM

So far, we have focused on the benefits of MIMO-OFDM, namely, in helping us deal with the frequency selectivity of wideband wireless communication channels. Clearly, avoiding expensive equalization algorithms, and achieving this at a reasonable complexity that can be affordable using modern digital signal processing techniques is very desirable. However, there are also challenges with the use of OFDM, specifically, with MIMO-OFDM.

Here we will not be comprehensive about the difficulties involved, but we will highlight some of the issues. MIMO-OFDM symbols are typically long (for the same data rates, N times more than the single-carrier counterparts), therefore, in a practical system, even if fading is slow, there may be some channel variations over one OFDM word. This may destroy the orthogonality of the subcarriers at the receiver, and may cause interchannel interference, degrading the system performance. Also, possible frequency offset will deteriorate the performance if it is not properly addressed. Carrier frequency offset estimation for MIMO-OFDM is considered by Zeng and Ghrayeb (2007).

Another potential problem is the possibility of large peak to average power ratios (PAPRs) in the transmitted signal as a result of the addition of many sinusoidal signals with different frequencies. With large PAPRs, the non-linear amplifiers needed for inexpensive implementation may introduce significant distortion into the transmitted signal. Approaches helping to reduce the PAPR are clearly necessary. Remedies to this problem are reported in Reddy and Duman (2003); Tan et al. (2005); Venkataraman et al. (2006).

9.6 Chapter Summary and Further Reading

In this chapter, we have studied MIMO communications over frequency selective fading channels. We have described a way of computing achievable information rates of MIMO FS channels with constrained signaling. We then discussed several coding approaches for single-carrier communications, together with possible equalization algorithms with varying complexity levels. Finally we have concentrated on MIMO transmission with multi-carrier modulation (with specific attention to MIMO-OFDM) as an alternative to the single-carrier approach. The latter provides a way of alleviating the need for equalizers with a large number of taps.

There is an extensive literature on space-time coding over MIMO FS channels. Schober et al. (2004); Zhou and Giannakis (2001, 2003) consider several space-time coding techniques over frequency selective fading channels. Equalizers for MIMO FS channels are designed in Al-Dhahir (2001); Al-Dhahir and Sayed (2000); Al-Dhahir et al. (2001); Bauch and Al-Dhahir (2002); Hedayat et al. (2005a); Lozano and Papadias (2002); So and Cheng (2006); Voulgarelis et al. (2003); Zhu and Murch (2003). Soft-output channel equalizers can be found in Tuchler et al. (2002) (for single-input single-output systems, but can be easily extended for the MIMO case), Gucluoglu and Duman (2005); Kaynak et al. (2005); Liu and Tian (2004); Roy and Duman (2004, 2007). Underwater acoustic communications with MIMO technology has been studied in McDonald et al. (2006); Roy et al. (2004, 2006, 2007).

Agrawal et al. (1998); Blum et al. (2001); Boelcskei (2006); Yang (2005) give a nice introduction to MIMO-OFDM schemes. Space-frequency coding for MIMO-OFDM schemes are described in Boelcskei et al. (2003); Gong and Letaief (2003); Molisch et al. (2002); Shao and Roy (2005). Akay and Ayanoglu (2006) employ bit interleaved coded modulation together with OFDM to achieve full diversity in MIMO FS channels. LDPC coded MIMO-OFDM schemes are developed in Lu et al. (2004). Space-time frequency coding schemes are further studied in Fozunbal et al. (2005); Liu et al. (2002b); Su et al. (2005); Zhang et al. (2007). Other schemes for MIMO-OFDM are given in Lee and Williams (2000); Song et al. (2004), and comparisons of space-frequency and space-time codes are made in Bauch (2003). Rende and Wong (2005); Tujkovic et al. (2001) consider bit interleaved coded modulation and turbo based space-frequency coding schemes. Furthermore, several recent papers consider MIMO-OFDM combined with spatial multiplexing (Boelcskei et al. (2002); Boubaker et al. (2001); Kadous (2003); Piechocki et al. (2001)), and Sohn (2003) combines turbo BLAST with space-frequency coding.

Problems

9.1 Using Matlab, implement the appropriate BCJR algorithm, and perform Monte Carlo simulations to reproduce Figure 9.1.

9.2 Using Matlab, implement the appropriate BCJR algorithm, and perform Monte Carlo simulations to reproduce Figure 9.2.

9.3 Consider an $N_t = 3$ MIMO FS channel with $L = 4$ ISI taps. Illustrate the transmitted signals from all the antennas (assuming a VBLAST type transmission – i.e., independent layers from each of the antennas) with the "virtual antenna" interpretation.

9.4 Describe the delay diversity code for an $L = 3$ tap, $N_t = 3$ MIMO FS channel. Demonstrate that the diversity order available (for quasi-static fading) with $N_r = 1$ is 9 using the appropriate rank criterion.

9.5 Consider an $N_t = 2$, $N_r = 1$ MIMO FS channel, and assume that the Alamouti scheme is used. Assuming quasi-static fading, find the diversity order available with ML decoding

a) if the number of taps is $L = 2$

b) if the number of taps is $L = 3$.

9.6 Assume that the space-time trellis code described in Figure 5.1 is used over an $N_t = 2$ MIMO FS channel. Show an appropriate combined code and channel trellis (that can be used at the receiver for ML decoding)

a) if $L = 2$,

b) if $L = 3$.

9.7 Consider VBLAST transmission over MIMO FS channels (with quasi-static fading). Assuming that optimal decoding can be accomplished, what are the diversity orders for

a) $N_t = 3$, $N_r = 2$, $L = 3$,

b) $N_t = 2$, $N_r = 3$, $L = 2$?

9.8 Design a MIMO-OFDM system that achieves an overall rate of 3 Mbps over a bandwidth of 200 kHz. Assume that $N_t = 2$, multipath spread $T_m = 1$ ms and Doppler spread $B_D = 10$ Hz. Specify the OFDM symbol duration, the number of subcarriers, the length of cyclic prefix, and the modulation scheme used.

9.9 How would your design in the previous problem change if the number of antennas is increased to $N_t = 4$?

10

Practical Issues in MIMO Communications

In this chapter, we explore several practical issues encountered in MIMO communication system design. In the earlier chapters of the book, we have assumed relatively idealistic models for wireless MIMO channels. We now intend to introduce some of the problems encountered in practice, and discuss the resulting system performance in the presence of these problems.

We consider two major issues in this chapter. The first one is the estimation of the channel state information and the effects of the estimation error on the system performance. The second subject detailed is the possible sub-channel correlation between the transmit and receive antenna pairs and the performance analysis with the presence of correlation. Finally, we point out some other practical issues briefly, but leave the detailed study to other chapters or omit the details altogether. These include a discussion on how the implementation complexity can be reduced by resorting to antenna selection (at the receiver and/or at the transmitter), the use of space-time coding versus spatial multiplexing algorithms, simplification of receiver algorithms, and signaling for frequency selective MIMO channels.

10.1 Channel State Information Estimation

In earlier chapters, we have assumed the availability of perfect CSI at the receiver for most of the MIMO signaling schemes described. The only exception is the study in Chapter 8, where it is assumed that neither the transmitter nor the receiver has access to the CSI, and unitary and differential coding schemes are described. Therefore, in a practical MIMO system, the estimation of the CSI at the receiver is a critical issue. In this section, we briefly review CSI estimation, and identify the effects of estimation errors on the overall system performance.

Coding for MIMO Communication Systems Tolga M. Duman and Ali Ghrayeb

10.1.1 CSI Estimation Using Pilot Tones

Let us consider the quasi-static MIMO frequency flat fading channel model, and present a simple CSI estimation technique based on the use of pilot tones. The model is the same as before. Namely, the received signal for the jth antenna at time k is given by

$$y_j(k) = \sqrt{\rho}\sum_{i=1}^{N_t} h_{i,j}x_i(k) + n_j(k), \tag{10.1}$$

or

$$\boldsymbol{Y} = \sqrt{\rho}\boldsymbol{X}\boldsymbol{H} + \boldsymbol{N}, \tag{10.2}$$

where all the terms are defined in the same way as in Chapter 5.

The problem to be addressed is the following. The channel coefficients are unknown for each frame of transmission, and we would like to use a set of pilot tones (known to the receiver *a priori*) to estimate these fading coefficients. Assume that pilot sequences of length N_p are used, i.e., we are given an $N_p \times N_t$ training matrix $\boldsymbol{X}_p$, and the observations corresponding to them, i.e., an $N_p \times N_r$ matrix $\boldsymbol{Y}_p$. That is, we have

$$\boldsymbol{Y}_p = \sqrt{\rho}\boldsymbol{X}_p\boldsymbol{H} + \boldsymbol{N}, \tag{10.3}$$

and we would like to find the best estimates of the unknown channel coefficients $\boldsymbol{H}$. Different criteria can be adopted to find the optimal solutions, resulting in different channel estimation approaches. These include ML estimation, least squares (LS) estimation, MAP estimation and linear minimum mean squared error (LMMSE) estimation (Kay (1993)).

Maximum Likelihood Channel Estimation

The ML channel estimator is obtained as follows

$$\hat{\boldsymbol{H}}_{ML} = \arg\max_{\boldsymbol{H}} p(\boldsymbol{Y}_p|\boldsymbol{H}). \tag{10.4}$$

Since the noise is assumed to be independent Gaussian (temporally and spatially), conditioned on the channel matrix, the matrix $\boldsymbol{Y}_p$ is complex Gaussian, that is, we can simplify the above expression to (assuming that $\boldsymbol{X}_p^H\boldsymbol{X}_p$ is full rank so that we have a non-degenerate distribution for the channel output $\boldsymbol{Y}_p$)

$$\hat{\boldsymbol{H}}_{ML} = \arg\min_{\boldsymbol{H}} \|\boldsymbol{Y}_p - \sqrt{\rho}\boldsymbol{X}_p\boldsymbol{H}\|^2, \tag{10.5}$$

where $\|\cdot\|^2$ is the Frobenius norm.

To minimize the above expression, we can take the derivative with respect to $\boldsymbol{H}$ and set it to zero to arrive at the solution

$$\hat{\boldsymbol{H}}_{ML} = \frac{1}{\sqrt{\rho}}(\boldsymbol{X}_p^H\boldsymbol{X}_p)^{-1}\boldsymbol{X}_p^H\boldsymbol{Y}_p. \tag{10.6}$$

Least Squares Estimation

In LS estimation, we find the estimate $\boldsymbol{H}$ that minimizes the squared error between the true observations and the fictitious ones generated by the estimated channel matrix. That is, the LS estimator is given by

$$\hat{\boldsymbol{H}}_{LS} = \arg\min_{\hat{\boldsymbol{H}}} \|\hat{\boldsymbol{Y}}_p - \boldsymbol{Y}_p\|^2, \tag{10.7}$$

$$= \arg\min_{\hat{\boldsymbol{H}}} \|\sqrt{\rho}\boldsymbol{X}_p\hat{\boldsymbol{H}} - \boldsymbol{Y}_p\|^2. \tag{10.8}$$

This can be a natural criterion if we treat the channel coefficients as deterministic, but unknown. To find the optimal $\hat{\boldsymbol{H}}$, we differentiate the above squared error with respect to $\hat{\boldsymbol{H}}$, and set it to zero, to obtain

$$\hat{\boldsymbol{H}}_{LS} = \frac{1}{\sqrt{\rho}}(\boldsymbol{X}_p^H\boldsymbol{X}_p)^{-1}\boldsymbol{X}_p^H\boldsymbol{Y}_p, \tag{10.9}$$

which is the same solution obtained using the ML criterion owing to the fact that the noise is Gaussian. Clearly, we have assumed that the training sequence is selected to be full column rank, i.e., $\boldsymbol{X}_p^H\boldsymbol{X}_p$ is full rank (this inherently assumes that the training length is equal or larger than the number of transmit antennas $N_p \geq N_t$ which is natural).

It is interesting to note that this estimator is unbiased. If we compute its expectation over the white Gaussian noise statistics, we obtain

$$E[\hat{\boldsymbol{H}}] = E\left[\frac{1}{\sqrt{\rho}}(\boldsymbol{X}_p^H\boldsymbol{X}_p)^{-1}\boldsymbol{X}_p^H(\sqrt{\rho}\boldsymbol{X}_p\boldsymbol{H} + N)\right], \tag{10.10}$$

$$= \boldsymbol{H}. \tag{10.11}$$

Furthermore, the resulting mean squared error is minimum over all unbiased estimators (which can be proved by calculating the Cramér–Rao lower bound on the MSE). Therefore, the LS solution is the minimum variance unbiased estimator.

Maximum A Posteriori Estimation

Let us now describe the MAP type channel estimator for MIMO systems in the presence of complex white Gaussian noise. The optimal solution is the one that maximizes the a-posteriori probability, i.e.,

$$\hat{\boldsymbol{H}}_{MAP} = \arg\max_{\boldsymbol{H}} p(\boldsymbol{H}|\boldsymbol{Y}_p), \tag{10.12}$$

$$= \arg\max_{\boldsymbol{H}} p(\boldsymbol{Y}_p|\boldsymbol{H})p(\boldsymbol{H}), \tag{10.13}$$

where the second equality is obtained using Bayes' rule and noting that $p(\boldsymbol{Y}_p)$ is the same for all candidate coefficient matrices. Writing the probability density functions explicitly, ignoring the constants, and taking the natural logarithm of the expression, we arrive at

$$\hat{\boldsymbol{H}}_{MAP} = \arg\min_{\boldsymbol{H}} \left(\|\boldsymbol{Y}_p - \sqrt{\rho}\boldsymbol{X}_p\boldsymbol{H}\|^2 + \|\boldsymbol{H}\|^2\right), \tag{10.14}$$

where we have used the fact that the noise terms and the channel coefficients are zero mean complex white Gaussian with unit variance. Taking the derivative of the expression with respect to $\boldsymbol{H}$ and setting it to zero, we obtain the MAP estimator as

$$\hat{\boldsymbol{H}}_{MAP} = \sqrt{\rho}(\boldsymbol{I}_{N_t} + \rho \boldsymbol{X}_p^H \boldsymbol{X}_p)^{-1} \boldsymbol{X}_p^H \boldsymbol{Y}_p. \tag{10.15}$$

Linear Minimum Mean Squared Error Estimation

We now describe the linear estimator that minimizes the expected value of the estimation error. In other words we would like to find the matrix $\boldsymbol{G}$ such that the mean squared error between the true channel coefficient and the linear estimate $\hat{\boldsymbol{H}} = \boldsymbol{G}\boldsymbol{Y}_p$ is minimized. That is, we would like to minimize

$$\text{MSE} = E\left[\|\boldsymbol{H} - \hat{\boldsymbol{H}}\|^2\right] \tag{10.16}$$

$$= E\left[\|\boldsymbol{H} - \boldsymbol{G}\boldsymbol{Y}_p\|^2\right], \tag{10.17}$$

with respect to $\boldsymbol{G}$, where the expectation is over the noise terms and the channel coefficients.

We can write

$$\text{MSE} = E\left[\|\boldsymbol{H} - \boldsymbol{G}(\sqrt{\rho}\boldsymbol{X}_p\boldsymbol{H} + \boldsymbol{N})\|^2\right], \tag{10.18}$$

which can be simplified to

$$\text{MSE} = \text{trace}\left\{E\left[\left((\boldsymbol{I}_{N_t} - \sqrt{\rho}\boldsymbol{G}\boldsymbol{X}_p)\boldsymbol{H} - \boldsymbol{G}\boldsymbol{N}\right)\left((\boldsymbol{I}_{N_t} - \sqrt{\rho}\boldsymbol{G}\boldsymbol{X}_p)\boldsymbol{H} - \boldsymbol{G}\boldsymbol{N}\right)^H\right]\right\},$$

$$= \text{trace}\left\{(\boldsymbol{I}_{N_t} - \sqrt{\rho}\boldsymbol{G}\boldsymbol{X}_p)E(\boldsymbol{H}\boldsymbol{H}^H)(\boldsymbol{I}_{N_t} - \sqrt{\rho}\boldsymbol{G}\boldsymbol{X}_p)^H + \boldsymbol{G}E(\boldsymbol{N}\boldsymbol{N}^H)\boldsymbol{G}^H\right\}.$$

Using the fact that $E[\boldsymbol{H}\boldsymbol{H}^H] = N_r\boldsymbol{I}_{N_t}$ and $E[\boldsymbol{N}\boldsymbol{N}^H] = N_r\boldsymbol{I}_{N_p}$, we obtain

$$\text{MSE} = N_r\left(\|\boldsymbol{I}_{N_t} - \sqrt{\rho}\boldsymbol{G}\boldsymbol{X}_p\|^2 + \|\boldsymbol{G}\|^2\right). \tag{10.19}$$

Taking the derivative of the MSE with respect to the matrix $\boldsymbol{G}$, and setting it to zero, the optimal choice of the matrix is obtained as $\boldsymbol{G}_{opt} = \sqrt{\rho}(\boldsymbol{I}_{N_t} + \rho\boldsymbol{X}_p^H\boldsymbol{X}_p)^{-1}\boldsymbol{X}_p^H$. Therefore, the LMMSE estimator is given by

$$\hat{\boldsymbol{H}}_{LMMSE} = \sqrt{\rho}(\boldsymbol{I}_{N_t} + \rho\boldsymbol{X}_p^H\boldsymbol{X}_p)^{-1}\boldsymbol{X}_p^H\boldsymbol{Y}_p, \tag{10.20}$$

which incidentally is the same as the MAP estimator since the noise is assumed to be Gaussian.

Training Optimization

We have just seen several possible estimators that can be used. Our assumptions were that the training sequence $\boldsymbol{X}_p$ is selected to be of a certain length N_p and a given total power. In a practical system, all these parameters are to be optimized to get the best performance. That is, we need to optimize the training length, the fraction of power to be allocated for training as opposed to actual data transmission in the rest of the frame, and the specific training sequence. Also, if the fading channel varies with time (even when variations are very slow) the exact placement of the training symbols will also be very important. We do not go into the details of the optimization problems involved. We only note that Hassibi and Hochwald (2003) show that the optimal training length (if the powers of the training and data symbols are allowed to vary) is the same as the number of transmit antennas, but it may be larger if identical power levels are used.

A natural selection for the pilot sequences is orthogonal sequences from a constant energy constellation across all the transmit antennas. As an example, consider BPSK for a two transmit antenna system, with a training length of $N_p = 4$. In this case, one can employ

$$\boldsymbol{X}_p = \frac{1}{\sqrt{2}} \begin{bmatrix} 1 & 1 \\ 1 & -1 \\ 1 & 1 \\ 1 & -1 \end{bmatrix}.$$

With orthogonal sequences from a constant energy constellation from each transmit antenna (assuming that the signal power is normalized to $\frac{1}{N_t}$ per transmit antenna), we have $\boldsymbol{X}_p^H \boldsymbol{X}_p = \frac{N_p}{N_t} \boldsymbol{I}_{N_t}$. Since it is an unbiased estimator, let us consider the LS solution (which is also the same as the ML estimator). We have

$$\hat{\boldsymbol{H}}_{LS} = \boldsymbol{H} + \frac{N_t}{N_p\sqrt{\rho}} \boldsymbol{X}_p^H \boldsymbol{N}. \tag{10.21}$$

Since the noise terms are white Gaussian with unit variance, it is easy to argue that for each channel coefficient, the error is complex Gaussian distributed and independent of each other, and its variance is $\frac{N_t}{2\rho N_p}$ per complex dimension. We note that for the LS estimation, the above training matrix $\boldsymbol{X}_p$ with constant energy orthogonal rows is optimal (see Biguesh and Gershman (2004)).

We observe that the estimation error becomes larger with an increased number of transmit antennas (this is because we have split the total power among different transmit antennas), and it reduces with the signal-to-noise ratio, ρ, as expected, and also it is inversely proportional to the training length. We further emphasize that this estimator is optimal in the sense that it is unbiased and it has the minimum variance possible over all unbiased estimators (achieves the Cramér–Rao lower bound).

10.1.2 What to Do with CSI?

In practical MIMO communications, after obtaining the channel estimates, $\hat{\boldsymbol{H}}$, with training symbols, we simply use them as if they are the perfect channel coefficients for the actual data transmission, and carry out the decoding process as usual. For instance, in the decoding of space-time block codes, we form the channel matrix based on the estimates, and decouple the decisions of symbols transmitted. Similarly, in decoding space-time trellis codes, we use them to form the path metrics for use in the Viterbi algorithm, or in a VBLAST scheme, we use these estimates for nulling and cancellation, etc. Such schemes using the estimated CSI in place of the perfect CSI are referred as "mismatched receivers," and they are clearly sub-optimal. We will give several examples of their performance using space-time block and trellis codes later in this section.

One way to improve the system performance with CSI estimation is to employ joint data detection and CSI estimation. An example is to use the well-known Expectation-Maximization (EM) algorithm in an iterative manner which is successfully demonstrated both in delay diversity systems by Cozzo and Hughes (2003) and for turbo-coded MIMO communications by Ronen et al. (2007). Other receiver structures are also conceivable to improve the system performance. However, it turns out that the performance of the mismatched receiver is not that far from the ideal case even with a small number of training symbols. This is illustrated by several examples in the following subsection.

10.1.3 Space-Time Coding Examples with Estimated CSI

Let us give several examples of space-time coded systems with CSI estimation to evaluate possible performance loss in a practical system. We use orthogonal sequences (with identical power levels in training and actual data transmission) for channel estimation, and employ the LS method.

We show the symbol error rates for the Alamouti scheme (for two transmit antennas) with QPSK signals for $N_r = 1$ and $N_r = 2$ in Figure 10.1. We observe that there is a degradation in the symbol error rate with channel estimation, however it is not very severe. When only two training symbols are used, the loss is about 2.5 dB compared with the perfect CSI case, while the loss reduces to about 1.5 dB when four training symbols are employed. There is no loss in the diversity order achieved which is two for $N_r = 1$, and four when $N_r = 2$.

We present another example in Figure 10.2 where the full-rate four transmit antenna orthogonal space-time block code is employed (described by $\boldsymbol{X}_2$ in Chapter 4). Our observations are similar, that is, imperfect knowledge of the channel coefficients results in some degradation in the bit error rates, however the diversity order achieved does not change.

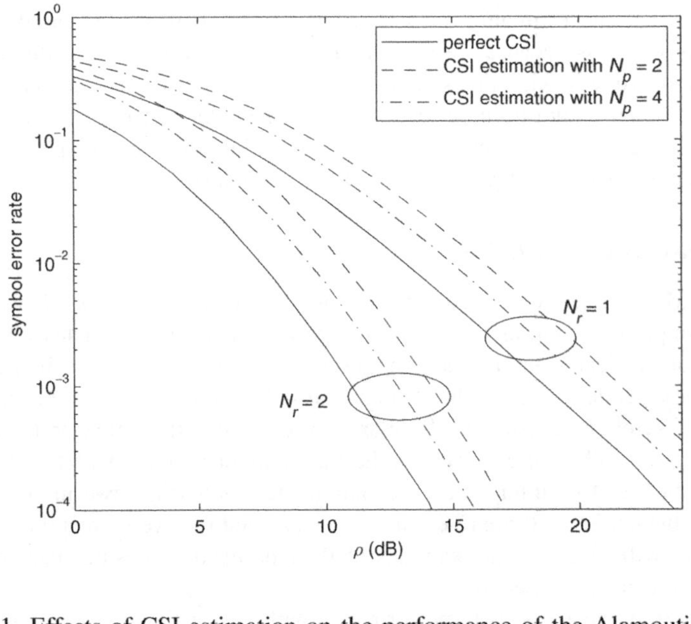

Figure 10.1 Effects of CSI estimation on the performance of the Alamouti scheme using QPSK.

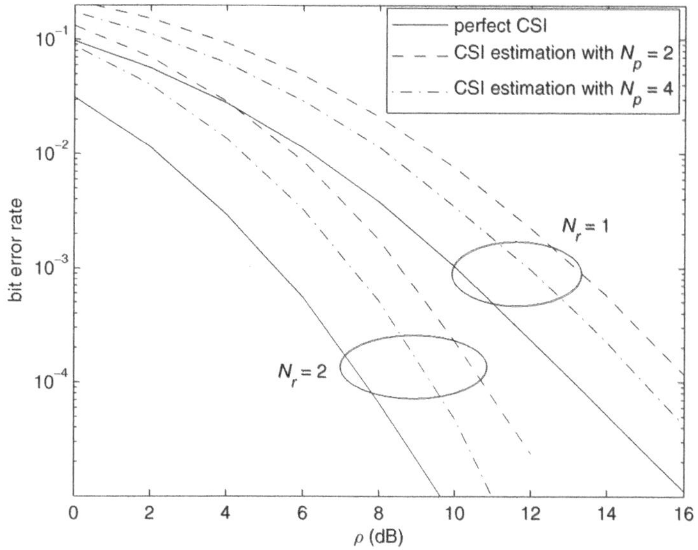

Figure 10.2 Effects of CSI estimation on the performance of an orthogonal space-time block code with BPSK ($N_t = 4$).

We consider the rate 2 bits per channel use four-state and eight-state space-time trellis codes with QPSK (described in Chapter 5). We assume quasi-static Rayleigh fading with a frame length of 130, where $N_p = 8$ symbols are used for training. This amounts to about 6% overhead, hence it not very significant. In Figures 10.3 and 10.4, we demonstrate the frame error rates obtained with and without channel estimation. We observe that even though the overhead added using the pilot tones is very small (about 6 %), the performance degradation is minor.

The examples presented above illustrate that in a practical setting where the channel coefficients need to be estimated using pilot tones, the performance of MIMO systems degrades slightly and much of the benefits are retained. Furthermore, the estimation can be performed with little overhead. This is an important observation which shows that MIMO communications remain useful in a practical setting.

10.2 Spatial Channel Correlation for MIMO Systems

In our study of MIMO systems in earlier chapters, we have exclusively assumed that the sub-channels between different antenna pairs see independent fading. However, this may be difficult to achieve in a practical system. To illustrate this point, let us consider receive diversity only, and assume that scattering is well approximated as uniform, and there is no line of sight. In this case, the signals received by different antennas become (almost) uncorrelated if the antenna separation is larger than half the wavelength of the carrier signal

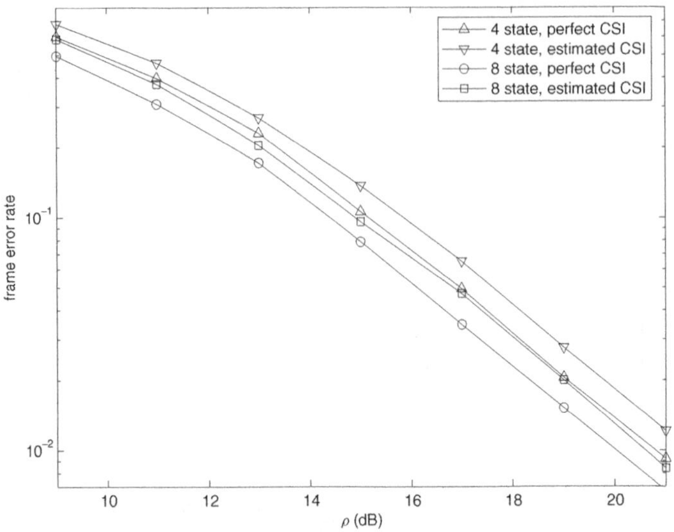

Figure 10.3 Effects of CSI estimation on the performance of an STTC ($N_t = 2$ and $N_r = 1$).

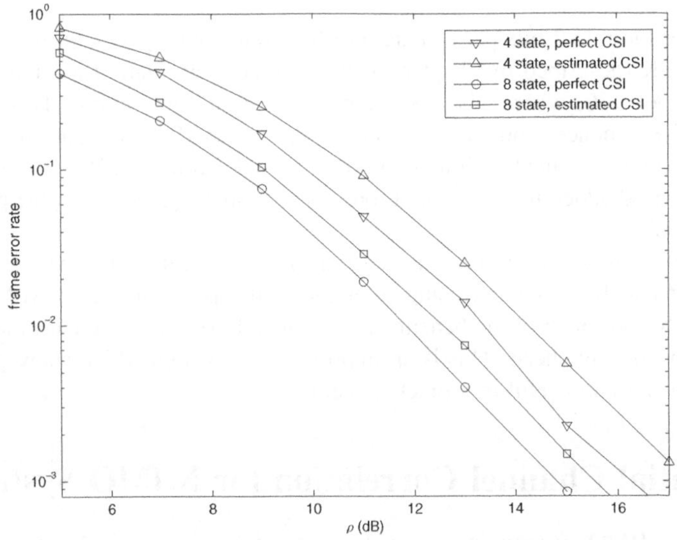

Figure 10.4 Effects of CSI estimation on the performance of an STTC ($N_t = N_r = 2$).

used in transmission. Depending on the carrier frequency used, and the requirements at the receiver, such a separation may be impractical. For instance, if the carrier frequency is 900 MHz, the antenna separation should be about 16 – 17 cm which may be too large for a typical cell phone.

If scattering is not uniform, it will be even more difficult to achieve uncorrelated sub-channels at the receiver, i.e., a larger separation will be needed. Furthermore, at the transmit side the situation is usually even more difficult as the propagation is farther away from uniform. To summarize, in a practical system, we need to work with spatially correlated channels, and to evaluate the effectiveness of the MIMO signaling techniques, we need to consider the effects of spatial correlation.

We note that our discussion above is simplistic in the sense that the reasons listed for the spatial channel correlation in a practical MIMO communication system are not limited to the specific scattering environments at the transmitter and receiver sides. For instance, spatial correlations may also be the results of antenna coupling at the receiver and transmitter sides, and the structure of the specific electromagnetic propagation. As an example of the latter, in the case where there is a large separation between the transmitter and the receiver, and when the propagation is through "a narrow pipe" in space, the resulting channel model is a "pinhole" channel (see Gesbert et al. (2002)).

In this section, our objective is to characterize the channel capacity and specific code performance in the presence of spatial correlation.

10.2.1 Measurements and Modeling of Spatial Correlation

Extensive measurements in different environments (urban versus rural, many scatterers versus only a few, etc.), for different frequency ranges, types of antennas, antenna element separations, transmitter and receiver heights, and so on, are needed to gain insights on the actual characterization of MIMO wireless channels. Although some results are reported in the literature for various scenarios, they are not very comprehensive.

When taking many measurements is not practical, it is also possible to rely on propagation models, for instance, using certain scattering models both at the transmitter and the receiver sides, to gain some understanding of the channel properties, such as the correlation structure of the channel coefficients for different sub-channels. Several attempts in this direction are reported in the literature as referenced at the end of the chapter.

After developing a certain model using a theoretical approach, usually it is desirable to verify its correctness using measurements, or when experiments are not easy to perform, using ray-tracing techniques by assuming a certain geometry for the transmitter and receiver sides, and the propagation environment. The ray-tracing approach is also taken in the literature to develop models for MIMO wireless channels.

Experiments, modeling or ray-tracing methods are beyond our scope. Instead, we are interested in working with somewhat simplified channel models developed based on these techniques, characterizing spatial correlation structures for MIMO wireless channels, and examining their effects on the channel capacity and on the space-time coded system performance. We start with descriptions of several useful models in the next subsection.

10.2.2 Spatial Channel Correlation Models

Consider the basic MIMO flat Rayleigh fading channel model used throughout this book. Up to now, we have assumed that the fading coefficients of different sub-channels (between different pairs of transmit and receive antennas) are independent. That is, the channel matrix $\boldsymbol{H}$ is an $N_t \times N_r$ matrix whose coefficients are i.i.d. zero mean Gaussian with variance $1/2$ per dimension. We now extend this channel model to take the spatial channel correlations into account. Mathematically, the spatially correlated MIMO channel model we adopt is given by

$$\boldsymbol{Y} = \sqrt{\rho}\boldsymbol{X}\boldsymbol{H}_{corr} + \boldsymbol{N}, \tag{10.22}$$

where $\boldsymbol{H}_{corr}$ is the channel matrix including the correlations, and all the other variables are as defined before.

Semi-correlated MIMO Channels

Assume that the receiver is placed in an environment with rich scattering and there is sufficient separation between the antenna elements. Then, there will not be any correlation at the receiver. However, if the transmitter is not placed in a rich scattering environment, it is conceivable that there will be transmit-side correlation. This semi-correlated channel model can be described in terms of the channel matrix

$$\boldsymbol{H}_{corr} = \boldsymbol{A}\boldsymbol{H}, \tag{10.23}$$

where $\boldsymbol{A}$ is a deterministic matrix that describes the correlation at the transmitter, and $\boldsymbol{H}$ is a complex Gaussian matrix with i.i.d. entries. We assume that trace$\{\boldsymbol{A}^H\boldsymbol{A}\} = N_t$ to make sure that when we consider channels with and without transmit correlation, our comparisons are fair.

We assume that the transmitter does not know the channel matrix. However, it is interesting to note that for transmit-side correlation, if the correlation structure (long-term average) is known at the transmitter (i.e., the matrix $\boldsymbol{A}$) then it can exploit this knowledge, i.e., perform waterfilling to improve the channel capacity.

In a similar fashion, it may be possible to encounter receive-side correlation, but no transmit-side correlation. An example could be the uplink transmission in a cellular communication system. In this case, the channel matrix is modeled as

$$\boldsymbol{H}_{corr} = \boldsymbol{H}\boldsymbol{B}, \tag{10.24}$$

where $\boldsymbol{B}$ is a deterministic matrix inducing receive correlation. In a similar fashion to the transmit correlation case, we assume that trace$\{\boldsymbol{B}^H\boldsymbol{B}\} = N_r$ to make the comparisons fair.

Fully Correlated MIMO Channels

When spatial correlation exists both at the transmitter and the receiver sides, a useful model for the channel matrix is given by

$$\boldsymbol{H}_{corr} = \boldsymbol{A}\boldsymbol{H}\boldsymbol{B}, \tag{10.25}$$

where clearly we have combined the two semi-correlated channel models described previously.

Pinhole Channel

We note that the semi-correlated and fully correlated MIMO channel models described above are simplistic in the sense that they assume that the separation of the correlation structures is possible, and that multipath structures are the only effects inducing correlation. This is not true in a general MIMO system. For instance, in certain cases it is shown through detailed theoretical models, and verified by ray-tracing techniques and even experiments that the propagation mechanism is also important. Even when there is neither transmit-side correlation nor receive-side correlation, it is still possible to have correlated channels, for instance, due to the existence of only a narrow pipe between the transmitter and the receiver. A simple channel model obtained in such a case is what is called a "pinhole" channel, and is described by the channel matrix

$$\boldsymbol{H}_{corr} = \boldsymbol{h}_t \boldsymbol{h}_r^H, \tag{10.26}$$

where $\boldsymbol{h}_t$ and $\boldsymbol{h}_r$ are $N_t \times 1$ and $N_r \times 1$ random vectors denoting the fading coefficients for the transmitter and the receiver sides, respectively. In this case, the rank of the channel matrix is one regardless of the channel realization. Such a channel does not provide a significant capacity increase compared with the single-input single-output channels (compared to the i.i.d. fading case), but it can provide diversity of order $\min\{N_t, N_r\}$ (Gesbert et al. (2002)).

10.2.3 Channel Capacity with Spatial Correlation

We now have some idea of how the channel correlation at the transmitter side and/or the receiver side, and correlation due to the propagation environment can be modeled. Let us now consider the corresponding channel capacity calculations to see how much degradation in the MIMO system performance we should expect due to these nonidealities.

Transmit- and/or Receive-Side Correlation

Let us start with the full correlation model. Since the effective channel matrix is of the form $\boldsymbol{H}_{corr} = \boldsymbol{A}\boldsymbol{H}\boldsymbol{B}$, assuming that the transmitter does not perform "waterfilling" (based on the transmit correlation matrix $\boldsymbol{A}$), i.e., it splits its power equally among transmit antennas, the channel capacity as a function of the random matrix $\boldsymbol{H}$ is given by

$$C(\boldsymbol{H}) = \log\det\left(\boldsymbol{I}_{N_r} + \frac{\rho}{N_t}\boldsymbol{B}^H\boldsymbol{H}^H\boldsymbol{A}^H\boldsymbol{A}\boldsymbol{H}\boldsymbol{B}\right), \tag{10.27}$$

which is clearly a random variable. If the channel is non-ergodic, then we can compute the outage probability for a given transmission rate R by

$$P_{out} = P[C(\boldsymbol{H}) < R)], \tag{10.28}$$

which can be easily evaluated using Monte Carlo techniques, that is, by generating a large number of realizations of the channel, calculating the corresponding instantaneous channel capacities, and estimating the fraction of trials that do not support the given transmission rate.

If the channel is ergodic, we can simply calculate the ergodic capacity by taking the expectation of the above expression with respect to the distribution of the channel. That is, the ergodic channel capacity is given by

$$C = E_{\boldsymbol{H}}\left[\log\det\left(\boldsymbol{I}_{N_r} + \frac{\rho}{N_t}\boldsymbol{B}^H\boldsymbol{H}^H\boldsymbol{A}^H\boldsymbol{A}\boldsymbol{H}\boldsymbol{B}\right)\right], \tag{10.29}$$

where the expectation is over $\boldsymbol{H}$. We note that the expectation can easily be evaluated using Monte Carlo techniques as well.

We emphasize that the above discussion assumes that the transmitter does not perform transmit waterfilling. If we assume that it exploits the knowledge of the transmit correlation, and distributes the total power accordingly, the capacity will be improved. However, we do not go into the details of how this can be accomplished.

If we only have transmit correlation. In this case, $\boldsymbol{B} = \boldsymbol{I}_{N_r}$, and we obtain

$$C_{tx\ only}(\boldsymbol{H}) = \log\det\left(\boldsymbol{I}_{N_r} + \frac{\rho}{N_t}\boldsymbol{H}^H\boldsymbol{A}^H\boldsymbol{A}\boldsymbol{H}\right), \tag{10.30}$$

$$= \log\det\left(\boldsymbol{I}_{N_r} + \frac{\rho}{N_t}\boldsymbol{H}^H\boldsymbol{\Gamma}_A\boldsymbol{H}\right), \tag{10.31}$$

where $\boldsymbol{\Gamma}_A = \boldsymbol{A}^H\boldsymbol{A}$.

Considering receive correlation only, in a similar manner, we can write

$$C_{rx\ only}(\boldsymbol{H}) = \log\det\left(\boldsymbol{I}_{N_r} + \frac{\rho}{N_t}\boldsymbol{B}^H\boldsymbol{H}^H\boldsymbol{H}\boldsymbol{B}\right), \tag{10.32}$$

$$= \log\det\left(\boldsymbol{I}_{N_t} + \frac{\rho}{N_t}\boldsymbol{H}\boldsymbol{\Gamma}_B\boldsymbol{H}^H\right), \tag{10.33}$$

where $\boldsymbol{\Gamma}_B = \boldsymbol{B}\boldsymbol{B}^H$ and the second equality is obtained by noting that $\det(\boldsymbol{I} + \boldsymbol{E}\boldsymbol{F}) = \det(\boldsymbol{I} + \boldsymbol{F}\boldsymbol{E})$ (for any two matrices $\boldsymbol{E}$ and $\boldsymbol{F}$). We notice that for one-sided correlation, i.e., either transmit or receive correlation (not both), the expressions for the instantaneous channel capacity have the same form. Smith et al. (2003) obtain closed-form expressions for the ergodic channel capacity for this channel model demonstrating that one can evaluate the ergodic capacity for these special cases analytically.

Finally, we note that similar results can be obtained using specific signal constellations by computing achievable information rates (as described in more detail in Chapter 3).

Pinhole Channel

For the case of a pinhole channel, the instantaneous capacity is given by

$$C_{pinhole} = \log\det\left(\boldsymbol{I}_{N_r} + \frac{\rho}{N_t}\boldsymbol{h}_r\boldsymbol{h}_t^H\boldsymbol{h}_t\boldsymbol{h}_r^H\right), \tag{10.34}$$

which can be easily written as

$$C_{pinhole} = \log\left(1 + \frac{\rho}{N_t}\|\boldsymbol{h}_t\|^2\|\boldsymbol{h}_r\|^2\right), \tag{10.35}$$

where we note that both $\|\boldsymbol{h}_t\|^2$ and $\|\boldsymbol{h}_r\|^2$ are chi-squared distributed independent random variables. This expression can be used for the outage capacity or ergodic capacity calculations in a straightforward manner.

As a side note, we notice that the capacity growth as the numbers of transmit and receive antennas are increased is not very dramatic, that is, we only expect a logarithmic increase as opposed to the linear increase obtained with uncorrelated MIMO channels. This is because, by a simple use of the law of large numbers, for large values of N_t and N_r, we have $\frac{1}{N_t}\|\boldsymbol{h}_t\|^2\|\boldsymbol{h}_r\|^2 \approx N_r$, and thus $C_{pinhole} \approx \log(1 + \frac{\rho}{N_r})$. In fact, this logarithmic increase is due to the signal-to-noise ratio definition. If we normalize the total signal-to-noise ratio with respect to receive antennas as well, the capacity will approach a constant value.

Examples

In this section, we present an example to assess the potential loss in capacity when there is channel correlation. Consider a two transmit antenna system for an ergodic Rayleigh fading channel. Assume that the transmit correlation is described by

$$\boldsymbol{\Gamma}_A = \begin{bmatrix} 1 & r \\ r & 1 \end{bmatrix}. \tag{10.36}$$

We consider two cases for the number of receive antennas, $N_r = 2$ and $N_r = 1$. When $N_r = 2$, the receive-side correlation is taken as $\boldsymbol{\Gamma}_B = \boldsymbol{\Gamma}_A$. Clearly, when there is only one receive antenna, spatial correlation is induced only by the transmit side.

We show the ergodic channel capacity with no transmit waterfilling for several cases in Figure 10.5. For the $N_r = 2$ case, r is taken as 0, 0.5, 0.8 and 0.95. For $N_r = 1$, r is varied as 0, 0.8 and 0.95. We observe that with channel correlation, the system performance degrades. However, the degradation is not very severe, and large capacity gains offered by MIMO systems are still available. For example, when the correlation coefficient between the two fading gains $r = 0.95$ (for the system with $N_r = 1$), the capacity loss is less than 2 dB for a transmission rate of 5 bits per channel use. Such a small loss even at high level of correlation is very encouraging for practical applications.

As another example, we consider the ergodic channel capacity for the pinhole channel as a function of the number of antennas $N_t = N_r = n$ for several signal-to-noise ratios in Figure 10.6. Also shown is the capacity of the ideal case (i.e., full rank channel with i.i.d. coefficients). We observe that while the capacity increases linearly for the ideal case (no correlation), for the pinhole channel it is only logarithmic in the number of antennas. This shows that for this case using MIMO transmission does not increase the transmission rates significantly.

10.2.4 Space-Time Code Performance with Spatial Correlation

Let us now consider the performance of specific space-time codes over correlated MIMO fading channels. Although a theoretical exposition can be done in a straightforward manner, we will not go into these details here. Instead we will give several examples of codes using simulations in an effort to assess possible losses.

As an example, let us consider $N_t = 2$ and $N_r = 1, 2$ systems employing the Alamouti scheme in Figure 10.7. Transmit and receive correlations are modeled to be the same as the ones selected in the previous subsection with $r = 0.3,\ 0.5,\ 0.7$ (also shown are the results

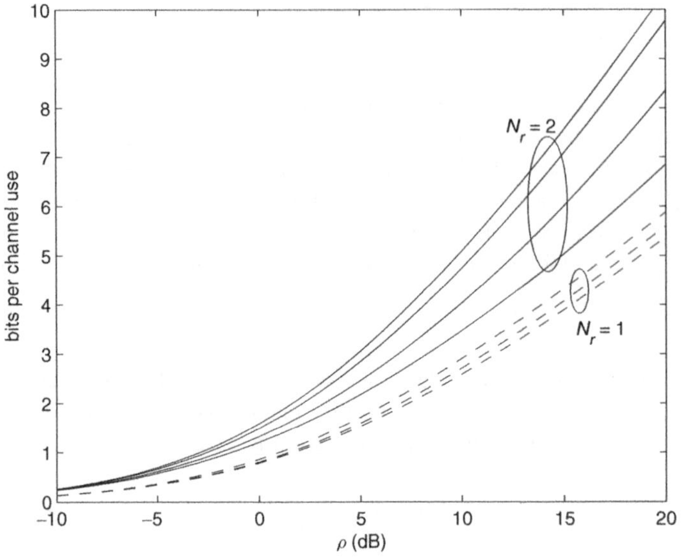

Figure 10.5 Ergodic capacity for a two transmit antenna system with spatial channel correlation. For $N_r = 1$, $r = 0, 0.8, 0.95$, and for $N_r = 2$, $r = 0,\ 0.5,\ 0.8,\ 0.95$ are used.

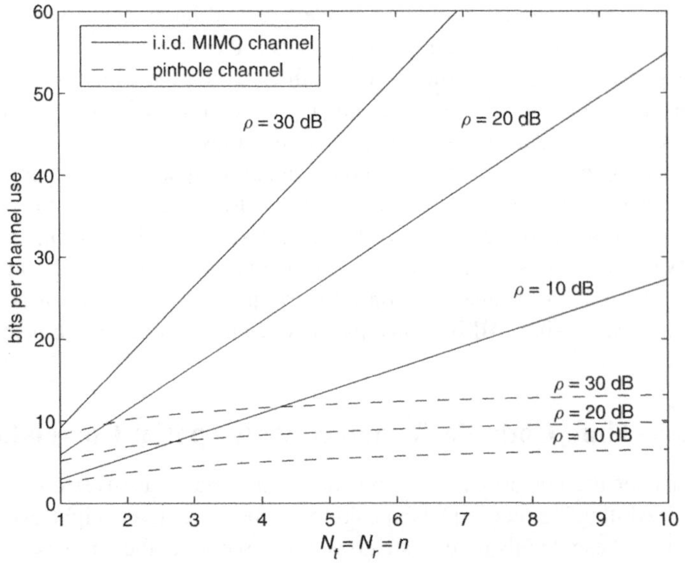

Figure 10.6 Ergodic capacity of pinhole channel as a function of number of antennas ($N_t = N_r = n$).

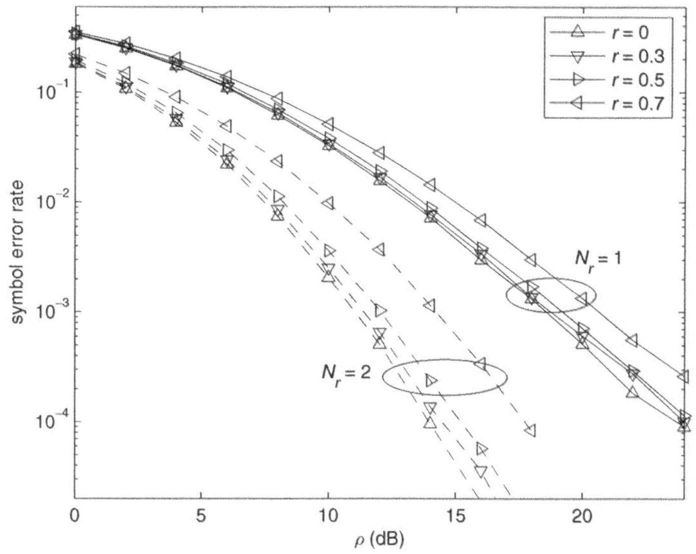

Figure 10.7 Symbol error rate of the Alamouti scheme with channel correlation (for one and two receive antennas).

with no spatial correlation, i.e., $r = 0$). We observe that the presence of channel correlation deteriorates the system performance, but not by much, and the diversity orders remain the same. For the single receive antenna case, the performance loss is less than the two-antenna case. This is because for $N_r = 1$ there is only transmit correlation, whereas when $N_r = 2$, both transmit and receive correlations exist. These observations are consistent with the capacity results of the previous subsection.

In Figures 10.8 and 10.9, we show the frame error rates of space-time trellis codes with channel correlation with one and two receive antennas, respectively. The codes used are developed in Tarokh et al. (1998), and described in Chapter 5. The same channel correlation model as in the previous example is used, but with $r = 0.5$ and $r = 0.8$. For both four-state and eight-state space-time trellis codes used, we observe that the error rates degrade slightly with channel correlation compared with the case with no spatial correlation.

To summarize, although the presence of channel correlation results in a performance loss for MIMO systems, even for relatively high correlation levels, the loss is not too much, and the use of MIMO systems over wireless channels remains beneficial. These observations are rather general in the sense that they also seem to hold for other MIMO communication systems, including concatenated coding schemes as reported in the literature.

10.3 Temporal Channel Correlation

In our study of MIMO systems, we concentrated on several channel models with different levels of time correlation, namely, quasi-static channels, block fading channels, and fully

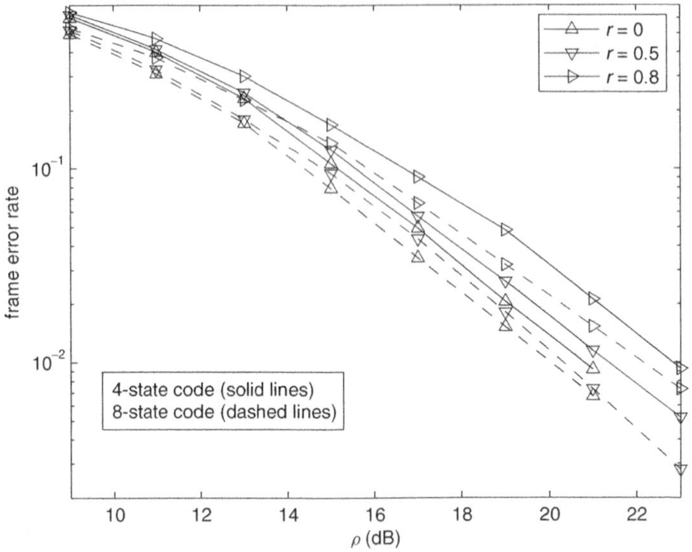

Figure 10.8 Space-time trellis code performance with channel correlation (two transmit and one receive antennas).

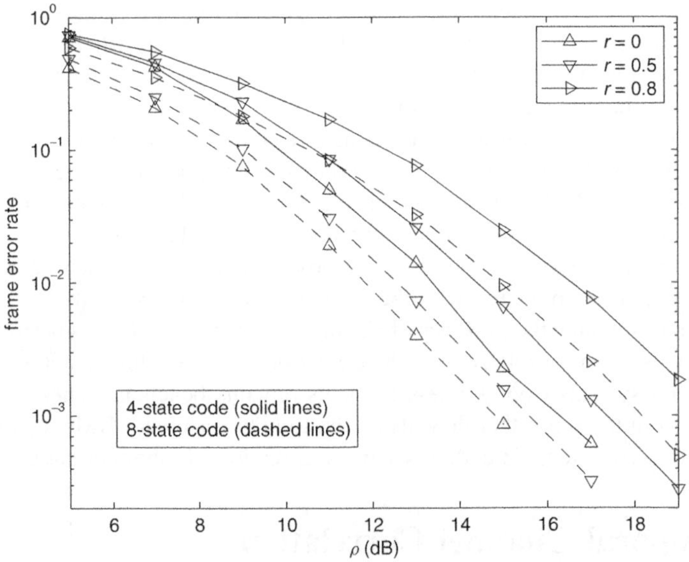

Figure 10.9 Space-time trellis code performance with channel correlation (two transmit and two receive antennas).

interleaved fading channel models. Quasi-static fading channels are observed in extremely slow fading environments, i.e., very low Doppler, and there is no change in the channel coefficients over a codeword. In reality, even if the change is very slow, we will observe a time-varying channel, in such a way that the fading coefficients for neighboring symbols are highly correlated (but, not identical). For example, time correlations are well modeled using a Bessel function for the case of a uniform scattering environment. Therefore, a channel model taking into account the time variations over a frame of data, such as a filtered Gaussian channel model or Jakes' model, could be more suitable (Stuber (2001)).

Continuous time variations over an entire frame may be good or bad. They may be good since time diversity will be available in addition to the spatial diversity, hence if properly handled, a better performance can be obtained. However, there may be some disadvantages in terms of CSI estimation, proper code design, or even proper receiver design. For instance, if we use a simple space-time block code, e.g., an Alamouti code, we need the fading coefficients to remain identical for several consecutive symbols so that the linear processing can decouple the detection of different symbols for optimal processing. The changes in the channel coefficients in time, even when they are small, may deteriorate the detector performance. This problem is tackled in Vielmon et al. (2004), and suitable detector structures are developed which closely approach the ideal performance (when the channel remains constant over consecutive symbols) with a reasonable complexity.

The problem of designing robust space-time codes that perform well when the channel variations are taken into account is addressed in several papers, see for example, Damen et al. (2001); Siwamogsatham and Fitz (2002); Siwamogsatham et al. (2002). Here we do not go into the details of the modifications in the earlier code design criteria that would be required.

10.4 MIMO Communication System Design Issues

We would like to point out that the problems we singled out in this chapter are not the only practical issues pertaining to MIMO communications. For example, the selection of a specific coding/decoding approach is a major issue. Different alternatives should be considered, and the scheme which offers the desired characteristics with the specific practical limitations should be adopted. For instance, if the fading channel is relatively fast varying (in time), trying to estimate the channel coefficients would require a large overhead, and may not be desirable. In such cases, differential encoding, or unitary signaling schemes could be more suitable. Another example of a trade-off we always need to consider is complexity versus performance, for instance, in the receiver design. Our objective in this section is to briefly describe some of the issues generally encountered in system design before closing the chapter.

We know that if high spectral efficiencies are desired, then spatial multiplexing schemes are suitable. However, for these schemes, possible diversity gains are very limited. If diversity gains are needed (to improve the error rate performance of the system), then (block, trellis or turbo based) space-time coding schemes should be employed, but these in general do not provide high spectral efficiencies (i.e., they do not achieve a significant fraction of the channel capacity). In general, there is a trade-off between the diversity offered by a system and its multiplexing gain as also briefly described in Chapter 6 (see Zheng and Tse (2003)). If we would like to obtain a reasonable multiplexing gain while maintaining a relatively high diversity gain, schemes that combine spatial multiplexing ideas with space-time coding, such as the multi-layered and threaded space-time coding schemes described in Chapter 6 should be used. Turbo-based schemes can also be adopted to offer this kind of trade-off. We further note that if there is CSI or statistical channel knowledge at the transmitter, this information can also be exploited to make a selection on the mode of operation (e.g., by selecting between space-time codes versus spatial multiplexing schemes).

Another important trade-off in the system design is the choice between the single-carrier versus multi-carrier (e.g., OFDM based) MIMO communications for frequency selective fading channels. For instance, if the channel ISI is very large, OFDM may be desirable, however this may cause some practical problems including frequency offset, loss of orthogonality of the subcarriers due to Doppler, etc.

Finally, we would like to comment on some hardware complexity issues. Implementation of MIMO communication systems, in general, would require multiple radio frequency (RF) chains to be employed both at the transmitter and the receiver. At the transmitter side, each RF chain generates the signal to be transmitted on a particular antenna. At the receiver side, each of them works with the signal picked up by a particular receive antenna to bring it to a form suitable for further processing (by the receiver). All these signals generated and picked up by the RF chains lie in the same frequency band, thus their isolation is absolutely essential for the MIMO transmitters and receivers to work properly. This may be very costly to accomplish, in particular, with the space constraints. Therefore, reduced complexity solutions to alleviate this problem should be produced. Performing antenna selection at the transmitter and/or at the receiver is one way of handling this problem. The main idea is to employ a reduced number of RF chains (maybe even a single RF chain) coupled with multiple transmit and receive antennas, in an effort to realize the benefits of MIMO communications. Since the antenna selection problem is studied thoroughly in Chapter 11, we do not present any details here.

10.5 Chapter Summary and Further Reading

In this chapter, we have concentrated on several practical issues encountered in MIMO communication systems. We started with a brief discussion of channel estimation techniques using pilot tones, and evaluation of system performance. We then concentrated on the problem of spatial channel correlations that will be observed in practice. We investigated their effects on the channel capacity, and specific space-time code performance. The results show that MIMO communications still remain very promising in achieving very high transmission rates over wireless links with such imperfections. We devoted the last part of the chapter to a brief discussion of channel variations in time that will be observed over fading links, and to various trade-offs that exist in MIMO systems.

Mismatch analysis where practical channel estimation is performed for space-time trellis codes has been studied in Tarokh et al. (1997). Channel estimation has been studied from an information-theoretic perspective in Hassibi and Hochwald (2003). Cozzo and Hughes (2003); Ronen et al. (2007) consider joint channel estimation and data detection. Turbo based MIMO system extensions for combined array processing and space-time coding are given in Stefanov and Duman (2000).

Ray-tracing methods for MIMO channel characterization are considered by Burr (2002); German et al. (2001); Tila et al. (2003). Channel characterization for practical wireless

MIMO systems are considered in Laurila et al. (2002) via experiments and in Gesbert et al. (2002); Molisch (2004) via modeling. Other papers studying MIMO communications channels via measurements and modeling include Baum et al. (2000); Chizhik et al. (2003); Kermoal et al. (2000); Stridh et al. (2000); Swindlehurst et al. (2001). The effects of channel correlations and other impairments on MIMO channel capacity are studied in Chuah et al. (2002); Ivrlac et al. (2003); Lozano et al. (2003); Moustakas et al. (2003); Shin and Lee (2003); Shiu et al. (2000); Smith et al. (2003). In addition, various papers report results on the effects of the spatial and/or temporal correlations on the performance of space-time codes, see, for example, El Gamal (2002b); Fozunbal et al. (2004); Hedayat et al. (2005b); Jorswieck and Sezgin (2004); Uysal and Georghiades (2004); Wang et al. (2004a); Younkins et al. (2004).

Problems

10.1 Derive the least squares MIMO channel estimator given in Equation (10.9).

10.2 Derive the linear minimum mean squared channel estimator given in Equation (10.20).

10.3 Reproduce Figure 10.2 using Matlab.

10.4 Reproduce Figure 10.4 using Matlab.

10.5 Reproduce Figure 10.5 using Matlab.

10.6 Reproduce Figure 10.7 using Matlab.

11

Antenna Selection for MIMO Systems

We have seen in previous chapters that employing multiple antennas results in a drastic increase in the channel capacity and leads to much more reliable wireless transmission, all relative to a single-input single output system. One of the drawbacks of using multiple antennas, however, is the complexity that arises from using a separate RF chain for every employed antenna. An RF chain normally comprises a low noise amplifier, frequency up- and down-converters, analog-to-digital and digital-to-analog converters, and several filters. This clearly results in a significant increase in the implementation cost. *Antenna selection* has emerged recently as a means to alleviate this complexity, while exploiting the advantages provided by the transmit and receive antennas. The idea behind antenna selection is to use only a subset of all available antennas in a MIMO system. The implication of this selection is that the number of RF chains required is reduced to as few as the number of selected antennas, thereby the deployment of MIMO systems would become less expensive and more feasible. Antenna selection can be performed at the transmitter, at the receiver, or at both ends simultaneously. Different antenna selection criteria can be considered, including maximization of the channel capacity and maximization of the received signal-to-noise ratio.

In this chapter, we study, in some depth, various aspects of antenna selection in MIMO systems. We first study the impact of antenna selection on channel capacity. We then consider antenna selection for STTCs and STBCs for various selection criteria and channel models. We also consider antenna selection with some nonidealities, including imperfect channel estimation, multipath fading and presence of spatial correlation. A brief summary of the chapter and some suggestions for further reading on antenna selection are given at the end of the chapter.

11.1 Capacity-based Antenna Selection

In this section, we study the impact of antenna selection on MIMO channel capacity which was treated in detail in Chapter 3. We begin with the case when optimal antenna selection is performed, and later consider suboptimal algorithms.

Coding for MIMO Communication Systems Tolga M. Duman and Ali Ghrayeb

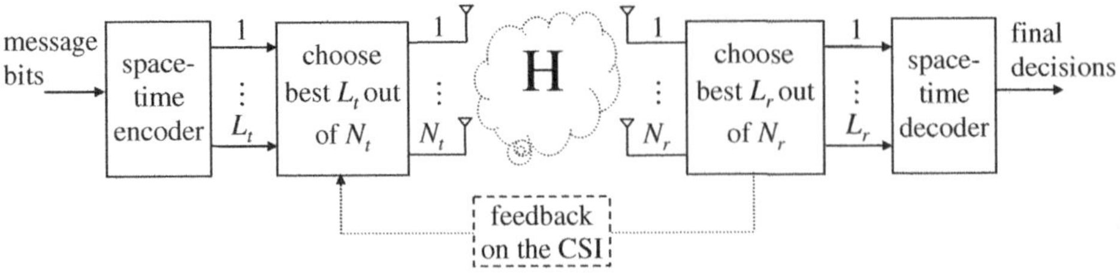

Figure 11.1 Generic MIMO system model with antenna selection.

11.1.1 System Model

Consider a MIMO system that employs N_t antennas at the transmitter side and N_r antennas at the receiver side (see Figure 11.1). When there is no antenna selection, $L_t = N_t$ and $L_r = N_r$, where L_t and L_r are the number of selected transmit and receive antennas, respectively. As shown in the figure, the incoming data is encoded by a space-time encoder. The output of the encoder is then fed into a serial-to-parallel converter that converts the input stream into parallel substreams. The resulting substreams are transmitted from the available transmit antennas simultaneously. When there is transmit antenna selection, as shown in the figure, some feedback on the CSI has to be sent back to the transmitter. Blocks that involve modulation, demodulation, etc., have been suppressed from the figure due to their irrelevance in the presentation. Therefore all analysis is done in baseband.

Let us denote the signal received by antenna j at time k by $y_j(k)$, which can be expressed as

$$y_j(k) = \sqrt{\rho}\sum_{i=1}^{N_t} x_i(k)h_{i,j} + n_j(k), \tag{11.1}$$

where $h_{i,j}$ denotes the fading coefficient between the ith transmit and jth receive antennas, and $n_j(k)$ is the complex Gaussian noise sample at time k corresponding to receive antenna j. The fading coefficients and the AWGN samples are modeled as independent and $\mathcal{CN}(0, 1)$ distributed. For notational convenience, we drop the time index k. With this, the signals received by all receive antennas can be expressed in a compact form as

$$\boldsymbol{y} = \sqrt{\rho}\boldsymbol{x}\boldsymbol{H} + \boldsymbol{n}, \tag{11.2}$$

where $\boldsymbol{H}$ is an $N_t \times N_r$ matrix whose (i, j)th entry is $h_{i,j}$, $\boldsymbol{x}$ is a $1 \times N_t$ vector that represents the signals transmitted simultaneously from the available N_t antennas, and $\boldsymbol{n}$ is a $1 \times N_r$ vector that represents noise samples. The signal constellation is normalized so that $E\left[\boldsymbol{x}\boldsymbol{x}^H\right] \leq 1$. Thus, ρ represents the average signal-to-noise ratio per receive antenna.

It is assumed, unless otherwise stated, that the subchannels fade independently, and the CSI is known perfectly at the receiver when receive antenna selection is involved, and the indices of the selected transmit antennas are available at the transmitter when transmit antenna selection is involved. Clearly, this is done by some limited feedback from the receiver to the transmitter. More practical assumptions will be dealt with later in this chapter. The channel is assumed to be flat quasi-static fading (or slow fading), where the fading coefficients are assumed to be constant over an entire frame and change independently from one frame to another.

11.1.2 Optimal Selection

When the CSI is not available at the transmitter, the transmitted power has to be distributed equally among the transmit antennas to maximize the channel capacity. Thus, the instantaneous capacity of a MIMO system using all available transmit and receive antennas is given by (see Chapter 3)

$$C = \log\det\left(\boldsymbol{I}_{N_r} + \frac{\rho}{N_t}\boldsymbol{H}^H\boldsymbol{H}\right). \tag{11.3}$$

Let us assume that antenna selection is performed at the receiver. As such, when the receiver selects the best L_r antennas that maximize the capacity, the resulting channel capacity can be written as

$$C_{\text{select}} = \arg\max_{\mathcal{S}(\tilde{\boldsymbol{H}})}\left(\log\det\left(\boldsymbol{I}_{L_r} + \frac{\rho}{N_t}\tilde{\boldsymbol{H}}^H\tilde{\boldsymbol{H}}\right)\right), \tag{11.4}$$

where $\tilde{\boldsymbol{H}}$ is formed by deleting $N_r - L_r$ columns from $\boldsymbol{H}$, and $\mathcal{S}\left(\tilde{\boldsymbol{H}}\right)$ represents the set of all such matrices.

Note that obtaining $\tilde{\boldsymbol{H}}$ that maximizes (11.4) requires perfect knowledge of $\boldsymbol{H}$. This, in turn, suggests that N_r RF chains have to be employed to be able to measure $\boldsymbol{H}$, which defeats the purpose of antenna selection. As an alternative, especially when the channel is slowly changing, one can still obtain an estimate of $\boldsymbol{H}$ while using only L_r RF chains, which can be accomplished through training sequences. In particular, the available L_r RF chains are cycled through the available N_r antennas during training. Once the training process is finished, the best L_r antennas are selected. The amount of overhead will be negligible particularly when the channel is slowly fading.

Driven by the fact that obtaining an exact analytical solution for (11.4) is difficult, solutions based on upper bounds were derived by Molisch et al. (2005). These upper bounds were based on the upper bound derived by Foschini and Gans (1998) for the full complexity system when $N_r \leq N_t$. The latter bound is given by

$$C \leq \sum_{i=1}^{N_t} \log\left(1 + \frac{\rho}{N_t}\gamma_i\right), \tag{11.5}$$

where γ_i represents the squared norm of the ith row of $\boldsymbol{H}$. The random variables γ_i are independent chi-squared distributed with $2N_r$ degrees of freedom. The equality in (11.5) corresponds to a hypothetical situation when each of the transmitted signals is received by a separate set of N_r receive antennas, i.e., the signal components are received without any interference from the other components.

By interchanging the roles of the transmitter and receiver, the expression in (11.5) can be used to derive an upper bound on C_{select} given by (11.4), which is

$$C_{\text{select}} \leq \sum_{i=1}^{L_r} \log\left(1 + \frac{\rho}{N_t}\gamma_i\right), \tag{11.6}$$

where γ_i are now ordered chi-squared distributed random variables with $2N_t$ degrees of freedom, i.e., $\gamma_1 > \gamma_2 > \cdots > \gamma_{N_r}$. This upper bound is quite tight for $L_r \leq N_t$, but it is loose for $L_r > N_t$. A tighter upper bound for the latter is derived by Molisch et al. (2005).

11.1.3 Simplified (Suboptimal) Selection

While antenna selection preserves most of the capacity promised by MIMO systems, the computational complexity arising from this selection is quite large because there are $\binom{N_r}{L_r}$ possibilities to search over. To reduce this complexity, several suboptimal selection algorithms have been introduced by Molisch et al. (2005), Gharavi-Alkhansari and Gershman (2004), and Gorokhov et al. (2003). Two algorithms are proposed by Molisch et al. (2005), one based on minimizing the correlation between the rows of $\boldsymbol{H}$ while maximizing their powers, and the other based on minimizing the mutual information between the receive antenna outputs. The first algorithm assumes no knowledge of the signal-to-noise ratio, whereas the latter does. The complexity of both algorithms is proportional to N_r^2 as opposed to $\binom{N_r}{L_r}$ for optimal selection, and little capacity reduction is incurred by using them.

The suboptimal algorithm presented by Gorokhov et al. (2003) achieves a near-optimal selection of the receive antennas. The algorithm starts with the full set of available antennas, and then removes one antenna in each step. The antenna removed is the one that contributes the least to the channel capacity at that step. The algorithm stops when the desired number of selected antennas is reached. In contrast, the suboptimal algorithm presented by Gharavi-Alkhansari and Gershman (2004) begins with an empty set of antennas and adds one antenna per step, where the added antenna is the one that contributes the most to the channel capacity. This algorithm also achieves a near-optimal selection, and it is faster than the one proposed by Gorokhov et al. (2003).

11.1.4 Examples

In Figure 11.2, we plot the cumulative distribution function (c.d.f.) of the capacity with receive antenna selection for $N_t = 3$, $N_r = 5$, and $L_r = 1, 2, 3, 4$. Optimal antenna selection is performed. The signal-to-noise ratio used in this plot is 20 dB with 10% outage capacity. As shown in the figure, a capacity of around 19 bits per channel use is achieved when $N_r = L_r = 5$, i.e., no selection, whereas the capacity decreases to about 18 bits per channel use and about 17 bits per channel use when $L_r = 4$ and 3, respectively. This is certainly not a big drop given the associated complexity reduction. The significant drop in capacity when $L_r = 1$ and 2 is also observed from the figure. This is due to the fact that having N_t independent parallel channels is no longer possible when $L_r < N_t$.

The above observation is confirmed in Figure 11.3 in which we plot the c.d.f. of the capacity for certain cases with and without antenna selection. In all cases, the number of transmit antennas is $N_t = 3$. We observe from the figure that the capacity of the case $N_r = L_r = 1$ is inferior to the capacity for $N_r = 5, L_r = 1$ by only 1.5 bits per channel use. This is due to the fact that $L_r < N_t$, and thus the capacity is dictated by L_r. In contrast, the capacity of the case $N_r = L_r = 3$ is inferior to the capacity of the case $N_r = 5, L_r = 3$ by about 3 bits per channel use. This strongly suggests that to retain most of the capacity

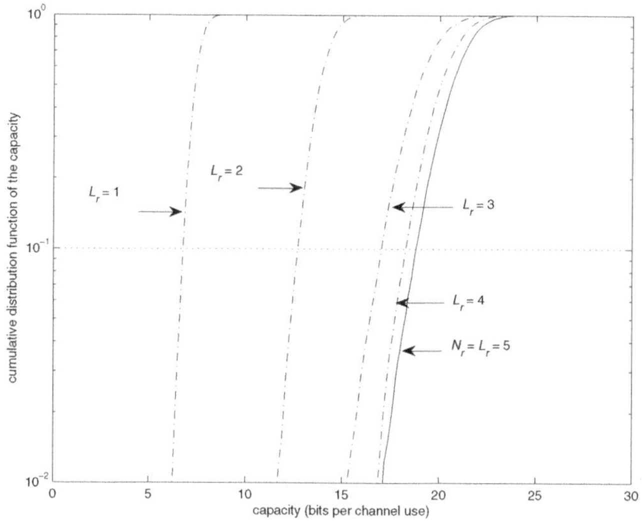

Figure 11.2 The cumulative distribution function of the capacity for a MIMO channel with $N_t = 3$, $N_r = 5$, and $L_r = 1, 2, 3, 4$ at a signal-to-noise ratio of 20 dB. Exhaustive search is performed to find the best L_r antennas.

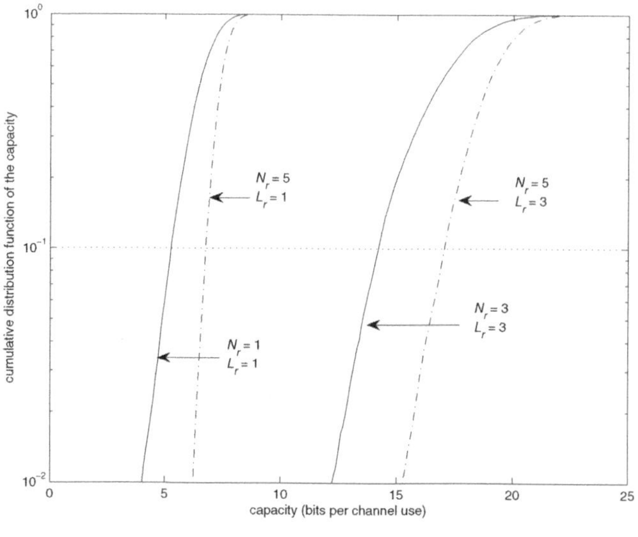

Figure 11.3 The cumulative distribution function of the capacity for a MIMO channel with $N_t = 3$, $N_r = 1, 3, 5$, and $L_r = 1, 3$ at a signal-to-noise ratio of 20 dB. Exhaustive search, where applicable, is performed to find the best L_r antennas.

of the MIMO channel, the number of selected antennas on one end should be at least as many as the number of available antennas on the other end.

11.2 Energy-based Antenna Selection

Another selection criterion would be to employ the subset of antennas that maximize the received signal-to-noise ratio, that is, to perform energy-based selection. This criterion is attractive because it is pragmatic as compared with the one that maximizes the channel capacity. However, maximizing the signal-to-noise ratio does not normally lead to maximizing the channel capacity. The only exception is when the number of selected antennas is $L_r = 1$. In this case, $\tilde{\boldsymbol{H}}$ is of size $N_t \times 1$, and thus, as per (11.4), maximizing the received signal-to-noise ratio results in maximizing the channel capacity. For the other cases, some capacity loss is incurred when the signal-to-noise ratio based selection criterion is used. This is attributed to the fact that the capacity is affected much more by the phase shifts between the antenna elements than by the instantaneous signal-to-noise ratio, as pointed out by Driessen and Foschini (1999).

Another way of explaining why maximizing the received signal-to-noise ratio is suboptimal is as follows. Consider, for example, the selection of two antennas from the available N_r antennas where the selected antennas are those that have the largest gains. But the specific channel gains from all the transmit antennas to these two antennas might be very similar. As a result, although the signal-to-noise ratio is maximized, the resulting capacity will be small (as effectively the channel will not be of high rank). On the other hand, when the two antennas are selected based on maximizing the capacity, the optimal selection algorithm will likely pick one of these two antennas along with another one from the remaining set, resulting in a high rank channel, hence better capacity.

We plot in Figure 11.4 the c.d.f. of the capacity of an $(N_t, N_r) = (3, 5)$ MIMO system with receive antenna selection. The purpose of this figure is to demonstrate that performing antenna selection based on maximizing the received signal-to-noise ratio is suboptimal. We observe from the figure that when $L_r = 4$, at 10^{-2} outage capacity, the energy-based selection criterion degrades the capacity by about 0.5 bits per channel use, whereas this degradation increases to about 1.8 bits per channel use when $L_r = 3$. It is interesting to note that the capacity degradation reduces to around 1.2 bits per channel use when $L_r = 2$. These results are in line with the above discussion, that is, the maximum degradation is expected when the uncertainty is maximum and in this case it is when $L_r = 3$. As for the case with $L_r = 1$, the capacities for both selection criteria are identical.

For the rest of the chapter, we will be employing the energy-based selection criterion. In addition, we will consider both STTCs and STBCs, as well as several different fading channel models. Since the impact of antenna selection on the performance depends on the underlying space-time code and channel model, we treat these cases separately.

Antenna selection has been limited, to a large extent, to receive antenna selection since transmit antenna selection is not appealing in general as it requires feeding an estimate of the CSI or the indices of the selected transmit antennas back to the transmitter, which

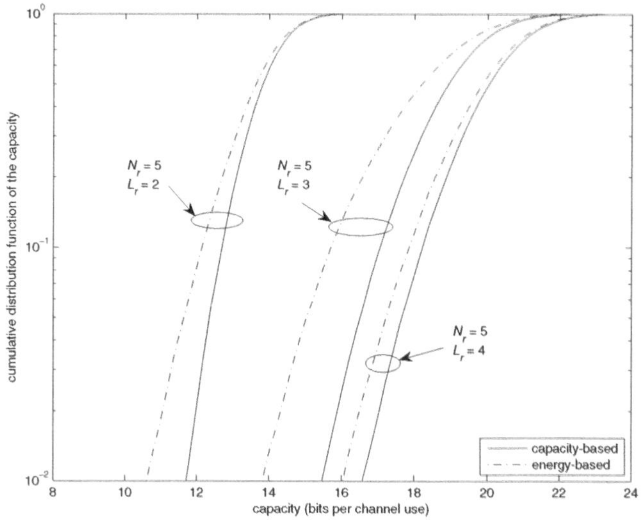

Figure 11.4 The cumulative distribution function of the capacity for an $N_t = 3$, $N_r = 5$ MIMO channel at a signal-to-noise ratio of 20 dB with receive antenna selection where selection is performed using the energy-based and capacity-based selection criteria.

would be feasible only if the channel changes relatively slowly. Therefore, more emphasis in this chapter will be on receive antenna selection.

11.3 Antenna Selection for Space-Time Trellis Codes

11.3.1 Quasi-Static Fading Channels

The pairwise error probability for STTCs over quasi-static fading channels was derived in Chapter 5. It was shown that a maximum diversity of $N_t N_r$ can be achieved when the codeword difference matrix $\boldsymbol{A} = \boldsymbol{X}_1 - \boldsymbol{X}_2$ is of column rank N_t, i.e., $\boldsymbol{A}^H \boldsymbol{A}$ is full rank. We now examine the pairwise error probability with receive antenna selection.

Receive Antenna Selection

When the receiver selects the L_r antennas that maximize the received instantaneous signal-to-noise ratio, assuming ML decoding and that the CSI is perfectly known at the receiver, the conditional pairwise error probability can be expressed as (see Chapter 5)

$$P(\boldsymbol{X}_1 \rightarrow \boldsymbol{X}_2|\boldsymbol{H}) \leq \exp\left(-\,\|\boldsymbol{D}_s\|^2 \frac{\rho}{4}\right) \tag{11.7}$$

where

$$\|\boldsymbol{D}_s\|^2 \triangleq \sum_{j=1}^{L_r} \left\|(\boldsymbol{X}_1 - \boldsymbol{X}_2)\boldsymbol{h}_j\right\|^2 \text{ such that } \gamma_1 \geq \cdots \geq \gamma_{N_r} \tag{11.8}$$

where $\boldsymbol{X}_1$ and $\boldsymbol{X}_2$ are the transmitted and erroneously decoded codewords, respectively, $\boldsymbol{h}_j$ representing the jth column of $\boldsymbol{H}$, and

$$\gamma_j = \left\|\boldsymbol{h}_j\right\|^2 = \boldsymbol{h}_j^H \boldsymbol{h}_j, \text{ for } j = 1, 2, \ldots, N_r. \tag{11.9}$$

To simplify the analysis, let us assume that if $\gamma_i \geq \gamma_j$ then

$$\|(\boldsymbol{X}_1 - \boldsymbol{X}_2)\boldsymbol{h}_i\|^2 \geq \left\|(\boldsymbol{X}_1 - \boldsymbol{X}_2)\boldsymbol{h}_j\right\|^2$$

for the worst-case codeword pairs. Clearly, this assumption is not necessarily true for all codeword pairs. However, it is a reasonable assumption for the codeword pairs that dominate the performance at high signal-to-noise ratios. As such, when L_r antennas are selected at the receiver, the resulting squared distance can be expressed as

$$\|\boldsymbol{D}_s\|^2 = \sum_{j=1}^{L_r} \left\|(\boldsymbol{X}_1 - \boldsymbol{X}_2)\boldsymbol{h}_j\right\|^2. \tag{11.10}$$

It is implied in (11.10) that maximizing the received signal-to-noise ratio results in maximizing the squared Euclidean distance, unlike the case in (11.8).

Computing an upper bound on the average pairwise error probability requires averaging (11.7) with respect to the random variables $\left\|(\boldsymbol{X}_1 - \boldsymbol{X}_2)\boldsymbol{h}_j\right\|^2$ for $j = 1, 2, \ldots, N_r$. This has been done by Bahceci et al. (2003), but it is mathematically tedious as it requires using order statistics to find the marginal p.d.f.s of the random variables $\left\|(\boldsymbol{X}_1 - \boldsymbol{X}_2)\boldsymbol{h}_j\right\|^2$ and then finding the probability density function of the random variable $\|\boldsymbol{D}_s\|^2$. As an alternative, by knowing that the sum of the largest L_r out of N_r nonnegative numbers is always greater than or equal to the average of these N_r numbers multiplied by L_r, we conclude that (Ghrayeb and Duman (2003))

$$\exp\left(-\|\boldsymbol{D}_s\|^2 \frac{\rho}{4}\right) \leq \exp\left(-\frac{L_r}{N_r}\|\boldsymbol{D}\|^2 \frac{\rho}{4}\right). \tag{11.11}$$

By averaging the right-hand expression in (11.11) with respect to the random variables $\left\|(\boldsymbol{X}_1 - \boldsymbol{X}_2)\boldsymbol{h}_j\right\|^2$ for $j = 1, 2, \ldots, N_r$, the pairwise error probability can be upper bounded as

$$P(\boldsymbol{X}_1 \to \boldsymbol{X}_2) \leq \left(\prod_{i=1}^{N_t} \lambda_i\right)^{-N_r} \left(\frac{L_r}{N_r} \cdot \frac{\rho}{4}\right)^{-N_t N_r}, \tag{11.12}$$

where λ_i for $i = 1, 2, \ldots, N_t$ are the eigenvalues of $\boldsymbol{A}^H \boldsymbol{A}$.

Equation (11.12) then suggests that using all the receive antennas after scaling the signal-to-noise ratio of each by a factor of L_r/N_r results in the same upper bound as if all receive antennas were used without scaling, i.e., the probability of error curves in both cases will asymptotically have the same slope. As for the coding gain, we observe that the only effect scaling has on the performance is that the signal-to-noise ratio is reduced by a factor of L_r/N_r, resulting in a degradation of $10\log_{10}(N_r/L_r)$ dB at high signal-to-noise ratios. Consequently, when selecting the best L_r antennas, the diversity order is maintained and the loss in coding gain is upper bounded by $10\log_{10}(N_r/L_r)$ dB.

We remark that the above result holds only for full-rank STTCs, i.e., when matrix $\boldsymbol{A}$ is of rank N_t. When the underlying STTC is rank-deficient, the spatial diversity degrades substantially with antenna selection as pointed out by Bahceci et al. (2003). Furthermore, the above analysis is based on the approximation that maximizing the received instantaneous signal-to-noise ratio results in maximizing the squared Euclidean distance. Similar conclusions are obtained by Bahceci et al. (2003) without resorting to any approximations.

Transmit Antenna Selection

Since most STTCs are designed for two transmit antennas, when transmit antenna selection is considered, it makes sense to select only two out of the available N_t antennas. One selection criterion would be the one based on selecting the two antennas that correspond to the largest square norms of the rows of $\boldsymbol{H}$. That is, the pair of antennas with indices (i_1, i_2) are selected such that $\|\boldsymbol{h}_{i_1}\|^2$ and $\|\boldsymbol{h}_{i_2}\|^2$ are the largest where $1 \leq i_1, i_2 \leq N_t$.

The results obtained for receive antenna selection hold for transmit antenna selection as well, as was shown by Chen et al. (2002). That is, the diversity order is maintained with antenna selection provided that the underlying STTC is full rank. Otherwise the diversity degrades. For example, for $N_r = 1$ and when the best two transmit antennas are selected, the pairwise error probability is approximated at high signal-to-noise ratios as

$$P(\boldsymbol{X}_1 \rightarrow \boldsymbol{X}_2) \leq \lambda_{\min}^{-N_t} \left(\frac{N_t!}{2^{N_t}}\right) \left(\frac{\rho}{4}\right)^{-N_t}, \qquad (11.13)$$

where $\lambda_{\min}$ is the smallest eigenvalue of the matrix $\boldsymbol{A}^H\boldsymbol{A}$, where $\boldsymbol{A}$ is as defined above. When $N_t = 2$, the pairwise error probability is approximated at high signal-to-noise ratios as

$$P(\boldsymbol{X}_1 \rightarrow \boldsymbol{X}_2) \leq \lambda_{\min}^{-2N_t} \left(\frac{N_t^2\,(2N_t - 2)!}{2^{3N_t-2}}\right) \left(\frac{\rho}{4}\right)^{-2N_t}. \qquad (11.14)$$

Both expressions suggest that the diversity is maintained with transmit antenna selection. For the pairwise error probability for an arbitrary number of receiver antennas, the reader may refer to Chen et al. (2002).

11.3.2 Block Fading Channels

In this section, we analyze the performance of STTCs with receive antenna selection over block fading channels. Let N denote the length of the data frame transmitted from each antenna, T denote the length of the faded block (i.e., coherence time of the channel), and L denote the number of blocks in a frame, i.e., $L = N/T$. Note that when $T = N$ the channel becomes quasi-static fading, whereas when $T = 1$ the channel becomes fast fading (fully-interleaved). Since the fading coefficients change for every T symbols within a frame, we incorporate this change by expressing $\boldsymbol{H}$ as $\boldsymbol{H}^l$ for $l = 1, 2, \ldots, L$. Similarly, we can partition $\boldsymbol{X}_1$ and $\boldsymbol{X}_2$ according to the number of blocks in a frame as

$$\boldsymbol{X}_1 = \begin{bmatrix} \boldsymbol{X}_1^1 \\ \boldsymbol{X}_1^2 \\ \vdots \\ \boldsymbol{X}_1^L \end{bmatrix}, \tag{11.15}$$

and

$$\boldsymbol{X}_2 = \begin{bmatrix} \boldsymbol{X}_2^1 \\ \boldsymbol{X}_2^2 \\ \vdots \\ \boldsymbol{X}_2^L \end{bmatrix}, \tag{11.16}$$

respectively. Note that each of the submatrices $\boldsymbol{X}_1^l$ and $\boldsymbol{X}_2^l$ is of size $L \times N_t$. Therefore, the channel matrix corresponding to $\boldsymbol{X}_1^l$ is $\boldsymbol{H}^l$.

Assuming ML decoding, the conditional pairwise error probability is then given by

$$P\left(\boldsymbol{X}_1 \to \boldsymbol{X}_2 | \boldsymbol{H}\right) \le \exp\left(-\left\|\boldsymbol{D}\right\|^2 \frac{\rho}{4}\right), \tag{11.17}$$

where

$$\left\|\boldsymbol{D}\right\|^2 = \sum_{l=1}^{L}\sum_{j=1}^{N_r} \left\| \left(\boldsymbol{X}_1^l - \boldsymbol{X}_2^l\right) \boldsymbol{h}_j^l \right\|^2, \tag{11.18}$$

where $\boldsymbol{h}_j^l$ is the jth column of $\boldsymbol{H}^l$. Define $\boldsymbol{A}^l \triangleq \boldsymbol{X}_1^l - \boldsymbol{X}_2^l$. Following steps similar to those that led to Equation (5.22), one can easily show that averaging (11.17) with respect to $\boldsymbol{H}$ yields

$$P\left(\boldsymbol{X}_1 \to \boldsymbol{X}_2\right) \le \prod_{l=1}^{L} \left(\prod_{i=1}^{r\left(\boldsymbol{A}^l\right)} \left(\frac{1}{1 + \lambda_i^l \frac{\rho}{4}} \right)^{N_r} \right), \tag{11.19}$$

where $r\left(\boldsymbol{A}^l\right)$ is the rank of $\boldsymbol{A}^l$, λ_i^l for $i = 1, 2, \ldots, \left(\boldsymbol{A}^l\right)$ are the eigenvalues of $\boldsymbol{A}^l$.

To simplify the above expression further, let Ψ denote the set of block indices over which an error event extends. The implication of this is that the matrices over which an error event extends all have non-zero rank. As such, when the signal-to-noise ratio is sufficiently large, (11.19) can be expressed as

$$P\left(\boldsymbol{X}_1 \to \boldsymbol{X}_2\right) \le \left(\prod_{l\in\Psi} \left(\prod_{i=1}^{r\left(\boldsymbol{A}^l\right)} \lambda_i^l \right)^{-N_r} \right) \left(\frac{\rho}{4}\right)^{-N_r \sum_{l\in\Psi} r\left(\boldsymbol{A}^l\right)}, \tag{11.20}$$

which is an upper bound on the pairwise error probability for the full-complexity system over block fading. Note that setting $L = 1$ or $L = N$ converts (11.20) to the pairwise error probability upper bounds for the full-complexity systems over slow and fast fading channels, respectively.

It is important to note that in the above analysis it is assumed that $T \ge N_t$, that is, the coherence time of the channel is greater than the number of transmit antennas. It is easy to show that for the case $T < N_t$, assuming the block matrices are rank deficient, which is normally the case, an upper bound on the pairwise error probability is the same as (11.20).

The rank of A_l depends on the location of the error events. For any error event that starts in a block and ends in a different block, some of the corresponding block matrices are likely to be rank deficient. Since error events occur randomly, there is no guarantee that the block matrices are all full rank. Therefore, it makes sense to assume the worst-case scenario, that is, some of the block matrices are rank deficient because they are the ones that dominate the performance at high signal-to-noise ratios. The analysis that follows is based on this assumption.

When receive antenna selection is considered, the receiver selects the best L_r antennas that maximize the received instantaneous signal-to-noise ratio. Antenna selection in this case is performed every T symbol intervals, which is the coherence time of the channel. To examine the feasibility of performing antenna selection from a practical point of view, let us consider, for example, a wideband CDMA system with $f_0 = 2$ GHz. The coherence time at $v = 10$ km/h with Doppler frequency $f_m = 18.51$ Hz is $T = 9.7$ ms. On the other extreme, when $v = 150$ km/h and Doppler frequency $f_m = 277.77$ Hz, $T = 0.60$ ms. Since the chip rate for this system is 3.84 Mcps, it is clear that, even at high mobility, the fading remains the same over many consecutive symbols, which makes antenna selection for block fading a practical solution.

Define $\gamma_j^l \triangleq \|\boldsymbol{h}_j^l\|^2$. Assume, without loss of generality, that $\gamma_1^l \geq \gamma_2^l \geq \cdots \geq \gamma_{N_r}^l$. As such, when the best L_r receive antennas are selected, the conditional pairwise error probability can be expressed as

$$P\left(\boldsymbol{X}_1 \rightarrow \boldsymbol{X}_2 \,\middle|\, \gamma_1^l, \gamma_2^l, \ldots, \gamma_{L_r}^l,\ l \in \{1, 2, \ldots, L\}\right) \leq \exp\left(-\|\boldsymbol{D}_s\|^2 \frac{\rho}{4}\right), \tag{11.21}$$

where

$$\|\boldsymbol{D}_s\|^2 = \sum_{l=1}^{L} \sum_{j=1}^{L_r} \left\|\left(\boldsymbol{X}_1^l - \boldsymbol{X}_2^l\right) \boldsymbol{h}_j^l\right\|^2. \tag{11.22}$$

Averaging (11.21) with respect to $\gamma_1^l, \gamma_2^l, \ldots, \gamma_{L_r}^l$ for $l \in \{1, 2, \ldots, L\}$ yields an upper bound on the pairwise error probability, which can be shown to be (Sanei (2006))

$$P\left(\boldsymbol{X}_1 \rightarrow \boldsymbol{X}_2\right) \leq (\text{constant}) \cdot \left(\frac{\rho}{4}\right)^{-L_r \sum\limits_{l \in \Psi} r\left(\boldsymbol{A}^l\right)}. \tag{11.23}$$

Expression (11.23) suggests that the diversity order of the system is $L_r \sum_{l\in\Psi} r\left(\boldsymbol{A}^l\right)$. By comparing this diversity order with that of the full complexity system (see (11.20)), it is clear that employing antenna selection deteriorates the diversity order and reduces it from $N_r \sum_{l\in\Psi} r\left(\boldsymbol{A}^l\right)$ to $L_r \sum_{l\in\Psi} r\left(\boldsymbol{A}^l\right)$.

However, considering the fact that (11.23) gives an upper bound on the pairwise error probability, we also need to find a lower bound on it to conclude that the diversity order indeed deteriorates as suggested by (11.23). This can be accomplished by using Craig's formula for $Q(x)$, as shown by Sanei et al. (2006a). This lower bound can be shown to be

$$P\left(\boldsymbol{X}_1 \rightarrow \boldsymbol{X}_2\right) \geq (\text{constant}) \cdot \left(\frac{\rho}{4}\right)^{-L_r \sum\limits_{l \in \Psi} r\left(\boldsymbol{A}^l\right)}. \tag{11.24}$$

The diversity order suggested by (11.24) confirms the diversity order given by (11.23).

11.3.3 Fast Fading Channels

As mentioned previously, fast fading channels essentially model fully interleaved block fading channels where the depth of interleaving is much longer than the coherence time of the channel. Therefore, performing antenna selection for this channel may be feasible since the selection is done at a rate lower than the symbol rate.

When the channel is fully interleaved, we have $N = L$ and $T = 1$. Consequently, we can express $\boldsymbol{X}_1$ and $\boldsymbol{X}_2$ as

$$\boldsymbol{X}_1 = \begin{bmatrix} \boldsymbol{X}_1^1 \\ \boldsymbol{X}_1^2 \\ \vdots \\ \boldsymbol{X}_1^N \end{bmatrix}, \quad (11.25)$$

and

$$\boldsymbol{X}_2 = \begin{bmatrix} \boldsymbol{X}_2^1 \\ \boldsymbol{X}_2^2 \\ \vdots \\ \boldsymbol{X}_2^N \end{bmatrix}, \quad (11.26)$$

respectively. Note now that $\boldsymbol{X}_1^l$ and $\boldsymbol{X}_2^l$ are vectors of size $1 \times N_t$. The implication of this change from block fading is that the matrix $\left(\boldsymbol{X}_1^l - \boldsymbol{X}_2^l\right)^H \left(\boldsymbol{X}_1^l - \boldsymbol{X}_2^l\right)$ will have rank *zero* or *one*, which is always less than $N_t \geq 2$. That is, the codeword difference matrices are always rank deficient.

In light of the above discussion, one can easily adapt the analysis for the block fading case to the fast fading case. In particular, one can show that the pairwise error probability for the full complexity system is given by

$$P\left(\boldsymbol{X}_1 \to \boldsymbol{X}_2\right) \leq \left(\prod_{l \in \Delta} \left\|\boldsymbol{X}_1^l - \boldsymbol{X}_2^l\right\|^2\right)^{-N_r} \left(\frac{\rho}{4}\right)^{-\mu N_r}, \quad (11.27)$$

where Δ in this case denotes the set of time indices $1 \leq l \leq N$ such that $\left|\boldsymbol{X}_1^l - \boldsymbol{X}_2^l\right| \neq 0$, and $\mu = |\Delta|$. The expression in (11.27) suggests that the diversity order of the full complexity system is μN_r. We remark that μ represents the minimum symbol-wise Hamming distance of the code, which is directly related to the shortest error event length in the code trellis.

When the receiver selects the L_r antennas that maximize the instantaneous received signal-to-noise ratio, the pairwise error probability can be shown to be

$$P\left(\boldsymbol{X}_1 \to \boldsymbol{X}_2\right) \leq \left(\frac{N_r! f(N_t, N_r, L_r)}{(N_r - L_r)! L_r! \left(N_t!\right)^{N_r - L_r}}\right)^L \cdot \left(\prod_{l \in \Delta} \left\|\boldsymbol{X}_1^l - \boldsymbol{X}_2^l\right\|^2\right)^{-L_r} \left(\frac{\rho}{4}\right)^{-\mu L_r}, \quad (11.28)$$

where $f(N_t, N_r, L_r)$ is a scalar that is a function of N_t, N_r and L_r. By comparing (11.28) and (11.27), it is clear that the diversity order resulting from selecting L_r antennas is reduced from μN_r to μL_r. Similar to the block fading case, since the inequality in (11.28) gives an upper bound on the pairwise error probability, we still need to find a lower bound on it to conclude that the diversity order is reduced to μL_r. However, the derivation of a lower bound on (11.28) also follows the same steps used to derive the lower bound for the block fading case. Therefore, we do not repeat it here; we only give the final expression, which is (Sanei et al. (2006b))

$$P\left(X_1 \to X_2\right) \geq (\text{constant}) \cdot \left(\prod_{l\in\Delta} \left\|X_1^l - X_2^l\right\|^2\right)^{-L_r} \left(\frac{\rho}{4}\right)^{-\mu L_r}. \quad (11.29)$$

Expressions (11.28) and (11.29) clearly suggest that the diversity order resulting from selecting L_r antennas is reduced from μN_r to μL_r. In Table 11.1, we give the closed-form expressions for the coefficients $f(N_t, N_r, L_r)$ for specific values of N_t.

To summarize, we list in Table 11.2 the diversity order achieved with antenna selection over various fading channel models. In the slow fading case, it is assumed that the STTC is full rank.

11.3.4 Examples

We now present several examples to demonstrate the performance of STTCs with antenna selection. In Figure 11.5, we plot the frame error rate against ρ in dB for various cases with $N = 130$. We examine the cases $N_t = 2$, $N_r = 3$ with $L_r = 1, 2, 3$. We observe from the figure that the full-complexity scheme, $L_r = 3$, is superior to all other schemes, as expected. We also observe that the cases with $N_r = 3$ and $L_r = 1, 2, 3$ have the same diversity order. Moreover, the losses in coding gain for $L_r = 2$ and $L_r = 1$ are about 0.9 dB

Table 11.1 Values of the constant $f(N_t, N_r, L_r)$ for specific values of N_t.

	$L_r = 1$	$L_r > 1$
$N_t = 2$	$[2(N_r - 1)]!$	$[2(N_r - L_r)]!$
$N_t = 3$	$[3N_r - 2]!$	$[3(N_r - L_r) + 1]!$
$N_t = 4$	$\frac{1}{2}[4N_r - 2]!$	$\frac{1}{2}[4(N_r - L_r) + 2]!$

Table 11.2 Diversity order of STTCs with antenna selection for various Rayleigh fading channel models

		Diversity order	
	Block length	$L_r = N_r$	$1 \leq L_r < N_r$
Fast fading	1	μN_r	μL_r
Block fading	$1 < T < N$	$N_r \sum_{l\in\Psi} r\left(A^l\right)$	$L_r \sum_{l\in\Psi} r\left(A^l\right)$
Slow fading	N	$N_t N_r$	$N_t N_r$

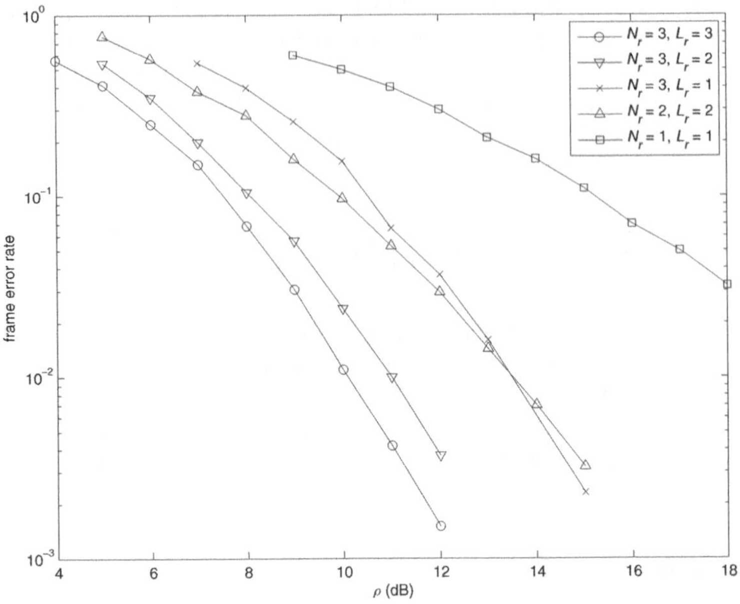

Figure 11.5 Frame error rate performance of the four-state, QPSK STTC presented in Chapter 5 over quasi-static fading for $N_t = 2$, $N_r = 1, 2, 3$ with receive antenna selection where $L_r = 1, 2, 3$.

and 3.4 dB at a frame error rate of 10^{-2}, respectively, relative to the case $N_r = 3$, $L_r = 3$. These losses are significantly smaller than the approximate upper bounds derived earlier (i.e., $10 \log_{10}(3/2) = 1.76$ and $10 \log_{10}(3) = 4.77$ dB). Clearly, these results agree with the analytical results presented above.

We plot in Figure 11.6 the frame error rate against ρ in dB for various cases of antenna selection for the QPSK case. We examine the following cases: (i) $N_r = 3$ with $L_r = 1, 2, 3$; (ii) $N_r = 2$ with $L_r = 1, 2$; and (iii) $N_r = 1$ with $L_r = 1$. We observe from the figure that the slopes of the performance curves corresponding to a specific L_r, for all N_r, are the same, suggesting that their diversity order is the same, which agrees with the analytical results derived above. In other words, the diversity order appears to be linearly proportional to L and it does not depend on N_r. On the other hand, for a specific L_r, as N_r increases, an additional coding gain can be achieved, as expected. However, this additional coding gain becomes smaller as N_r increases. For instance, at a frame error rate of 10^{-3}, the case $N_r = 2, L_r = 1$ achieves a gain of about 3 dB over the case $N_r = 1, L_r = 1$, whereas only an additional 0.5 dB is achieved in the case with $N_r = 3, L_r = 1$.

In Figure 11.7, we plot the frame error rate of the four-state STTC code used above for various transmit antenna selection scenarios. In particular, the cases $N_t = 2, 3, 6$ and $N_r = 1$ with $L_t = 2$ are considered over quasi-static fading with QPSK modulation. As shown in the figure, full diversity order is maintained with antenna selection.

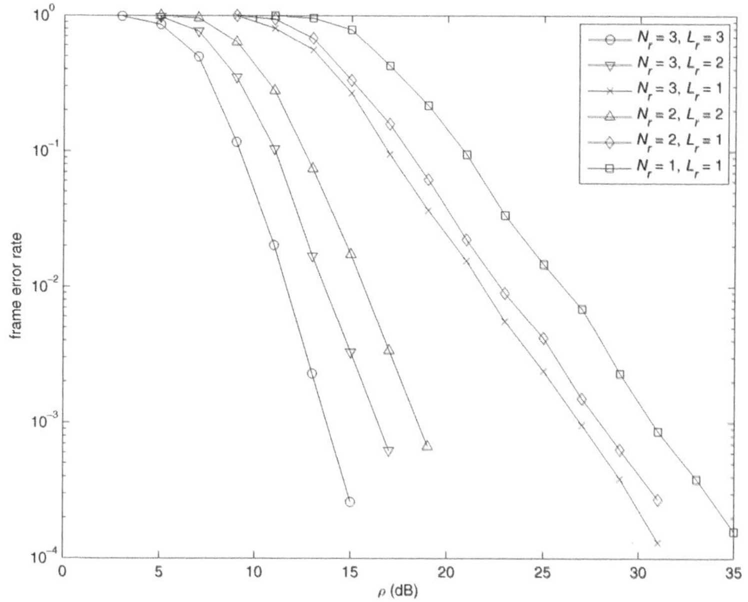

Figure 11.6 Frame error rate performance of the STTC considered in Figure 11.5 over fast fading for $N_t = 2$, $N_r = 1, 2, 3$ with receive antenna selection where $L_r = 1, 2, 3$.

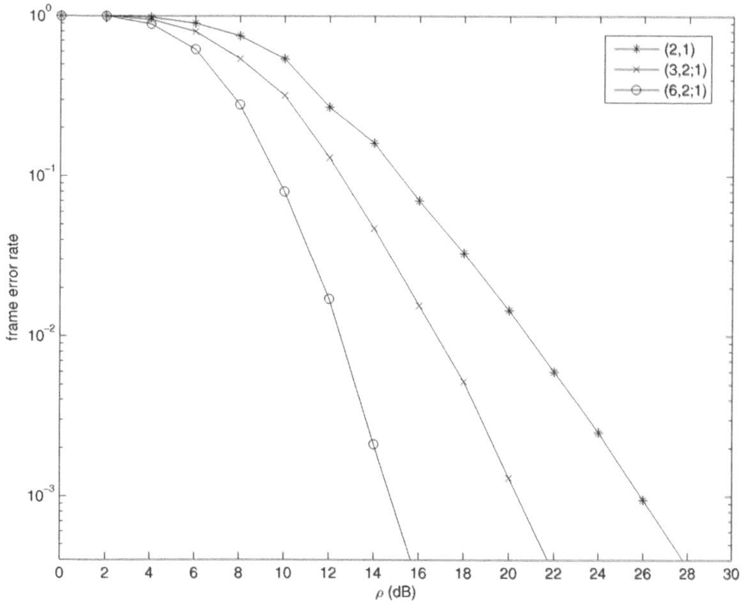

Figure 11.7 Frame error rate performance for the four-state, QPSK STTC in quasi-static fading for $N_t = 2, 4, 6$, $N_r = 1$ with $L_t = 2$.

11.4 Antenna Selection for Space-Time Block Codes

In this section, we study antenna selection for orthogonal space-time block coded MIMO systems. Our emphasis is on these codes because they are easy to design and they outperform other STBCs for the same number of transmit and receive antennas. We first consider the case when the MIMO system employs an STBC only. The case when the STBC is combined with an outer channel code is treated in the following section.

11.4.1 Receive Antenna Selection

We have seen in Chapter 4 that the bit error rate for space-time block coded MIMO systems with BPSK modulation conditioned on the channel gains is given as (see (4.50))

$$P_b(\boldsymbol{H}) = Q\left(\sqrt{\frac{2\rho}{N_t}\sum_{j=1}^{N_r}\gamma_j}\right) \tag{11.30}$$

where γ_j is defined by (11.9). Assume that $\gamma_1 \geq \gamma_2 \geq \cdots \geq \gamma_{N_r}$. Consequently, when the receiver selects the L_r antennas that have the largest signal-to-noise ratios, the conditional bit error rate in (11.30) is expressed as

$$P_b(\boldsymbol{H}) = Q\left(\sqrt{\frac{2\rho}{N_t}\sum_{j=1}^{L_r}\gamma_j}\right). \tag{11.31}$$

Using a similar argument that we have used for the STTC case, that is, the largest L_r out of N_r nonnegative numbers is always greater than or equal to the average of these N_r numbers multiplied by L_r, the expression in (11.31) can be upper bounded as (Zeng and Ghrayeb (2004b))

$$P_b(\boldsymbol{H}) \leq Q\left(\sqrt{\frac{2\rho}{N_t}\frac{L_r}{N_r}\sum_{j=1}^{N_r}\gamma_j}\right). \tag{11.32}$$

By comparing the right-hand side of the expression in (11.32) and that of the full complexity system, we observe that both are the same except that the signal-to-noise ratio in the former expression is scaled by L_r/N_r. Therefore, the average bit error rate performance, with antenna selection, can be approximated at high signal-to-noise ratio as

$$P_b \lesssim \binom{2N_tN_r - 1}{N_tN_r}\left(\frac{L_r}{N_r}\frac{4\rho}{N_t}\right)^{-N_tN_r}, \tag{11.33}$$

which suggests that the antenna selection system has the same diversity as that of the full-complexity system for any $L_r \leq N_r$ and the corresponding reduction in signal-to-noise ratio is upper bounded by $10\log_{10}(N_r/L_r)$ dB.

Note that the results for this case are the same as those for STTCs over quasi-static fading channels. However, the results for orthogonal STBCs hold for all fading channel models as long as the fading remains constant over N_t consecutive symbols to allow for decoupling of symbols at the receiver. This is a major difference in the analysis for orthogonal STBCs versus that of STTCs with antenna selection.

The upper bound in (11.33) is rather loose for large N_r and small L_r, which renders it not useful in predicting the exact signal-to-noise ratio degradation due to antenna selection. Motivated by the fact that schemes with small L_r are of interest from a practical point of view, we derive in this section an exact expression for the bit error rate, specifically for the Alamouti scheme when $L_r = 1$.

When the best receive antenna is selected, the conditional bit error rate is expressed as

$$P_b(\boldsymbol{H}) = Q\left(\sqrt{\frac{2\rho}{N_t}\gamma_1}\right). \tag{11.34}$$

Now define $Z \triangleq \gamma_1$. The p.d.f. of Z can be shown to be

$$\begin{aligned} p_Z(z) &= N_r F(z)^{N_r-1} f(z), \\ &= N_r \sum_{j=0}^{N_r-1} \sum_{i=0}^{j} A_{j,i} z^{i+1} e^{-(j+1)z}, \end{aligned} \tag{11.35}$$

where $f(z)$ is the p.d.f. of the elements γ_j without ordering, $F(z)$ is the corresponding c.d.f., and $A_{j,i} \triangleq (-1)^j \binom{N_r-1}{j}\binom{j}{i}$. The average bit error rate is obtained by averaging (11.34) with respect to Z, which yields (Zeng and Ghrayeb (2004b))

$$\begin{aligned} P_b &= \int_0^\infty Q\left(\sqrt{\frac{2\rho}{N_t}z}\right) p_Z(z)dz, \\ &= N_r \sum_{j=0}^{N_r-1} \sum_{i=0}^{j} A_{j,i} \int_0^\infty Q\left(\sqrt{\frac{2\rho}{N_t}z}\right) z^{i+1} e^{-(j+1)z} dz. \end{aligned} \tag{11.36}$$

The integrand in the above expression can be evaluated using integration by parts as

$$\begin{aligned} &\int_0^\infty Q\left(\sqrt{\frac{2\rho}{N_t}z}\right) z^{i+1} e^{-(j+1)z} dz \\ &= \frac{(i+1)!}{2\,(j+1)^{i+2}} - \frac{(i+1)!}{2}\sqrt{\frac{2\rho}{\pi N_t}} \sum_{l=0}^{i+1} \frac{\Gamma(l+1/2)}{l!\,(j+1)^{i+2-l}\left(1+j+\frac{2\rho}{N_t}\right)^{l+1/2}}. \end{aligned} \tag{11.37}$$

We finally remark that receive antenna selection for orthogonal STBCs is mathematically similar to the receive antenna selection for multiple antenna systems over Nakagami fading channels with a single transmit antenna, which has been treated by Alouini and Simon (1999). Both problems become equivalent when the Nakagami fading parameter $m = N_t$.

11.4.2 Transmit Antenna Selection

When antenna selection is performed at the transmit side, the transmitter selects the L_t antennas whose corresponding $\|\boldsymbol{h}_i\|^2$ are the largest, where $\boldsymbol{h}_i$ denotes the ith row of $\boldsymbol{H}$. It is assumed here that the indices of antennas to be selected are known at the transmitter via some feedback from the receiver. Obviously such antenna selection would be effective only when the coherence time of the channel is relatively long.

Expressions for the bit error rate with transmit antenna selection were derived by Chen et al. (2003) for any number of transmit and receive antennas with $L_t = 2$, which is essentially the Alamouti scheme. The results for these cases are in line with the receive antenna selection results. That is, the diversity order is maintained with antenna selection as if all the available antennas are used. Several approximate expressions were also given for specific number of receive antennas, including $N_r = 1$ and 2, which are

$$P_b \simeq \frac{(2N_t - 1)!}{2^{2N_t-1}(N_t - 1)!}\left(\frac{\rho}{N_t}\right)^{-N_t}, \tag{11.38}$$

and

$$P_b \simeq \frac{N_t\,(4N_t - 1)!}{2^{5N_t-2}(2N_t - 1)(2N_t - 1)!}\left(\frac{\rho}{N_t}\right)^{-2N_t}, \tag{11.39}$$

respectively.

11.4.3 Examples

We demonstrate here the performance of STBCs with antenna selection via several examples. In Figure 11.8, we plot the bit error rate performance for receive antenna selection for several cases and compare these simulation results against their upper bounds given by (11.33). In particular, we consider the cases with $N_t = 2$, $N_r = 3$ and $L_r = 1, 2, 3$. It is clear from the figure that the diversity order is maintained with antenna selection. Moreover, the losses in signal-to-noise ratio due to antenna selection for $L_r = 2$ and $L_r = 1$ are about $10\log_{10}(3/2) = 1.76$ and $10\log_{10}(3) = 4.77$ dB, all at a bit error rate of 10^{-5}. We point out here that the gap between this bound and simulations gets smaller as L_r approaches N_r. This is simply because the approximation used in the analysis becomes more and more accurate as L_r increases.

In Figure 11.9, we plot the bit error rate performance for the Alamouti scheme, i.e., $N_t = 2$, with receive antenna selection for the cases with $N_r = 2$, $L_r = 1, 2$ and $N_r = 3$, $L_r = 1, 3$. We also plot on the same figure the exact analysis results for these cases based on the expression given by (11.36). It is clear from the figure that the simulation results match perfectly with their theoretical counterparts, as expected.

In Figure 11.10, we plot the frame error performance for various transmit antenna selection cases. In particular, we consider the cases with $N_t = 2, 4, 6$, $N_r = 1$, and $L_t = 2$. Therefore, in all cases, the transmitter always selects the best two antennas where the Alamouti scheme is used. We plot on the same figure the approximate bit error rate performance results given by (11.38). It is shown in the figure that the diversity order is maintained

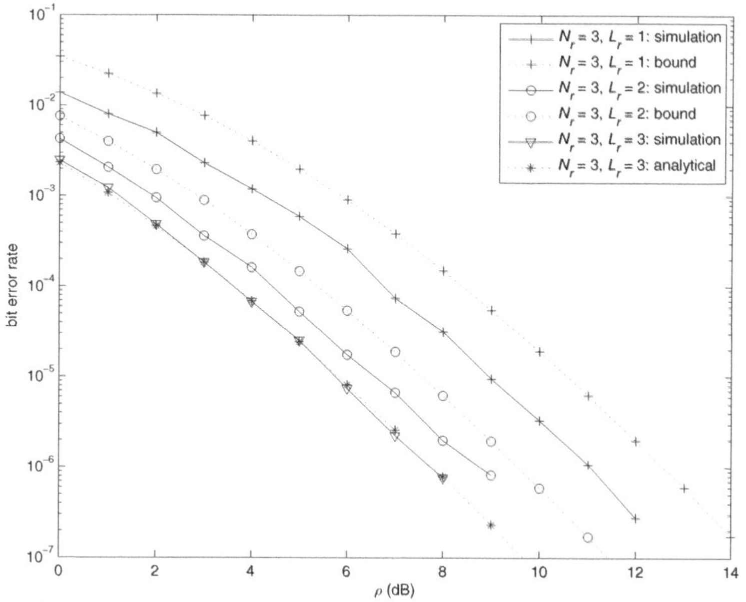

Figure 11.8 Bit error rate performance for the Alamouti scheme with receive antenna selection for the cases with $N_r = 3$ and $L_r = 1, 2, 3$.

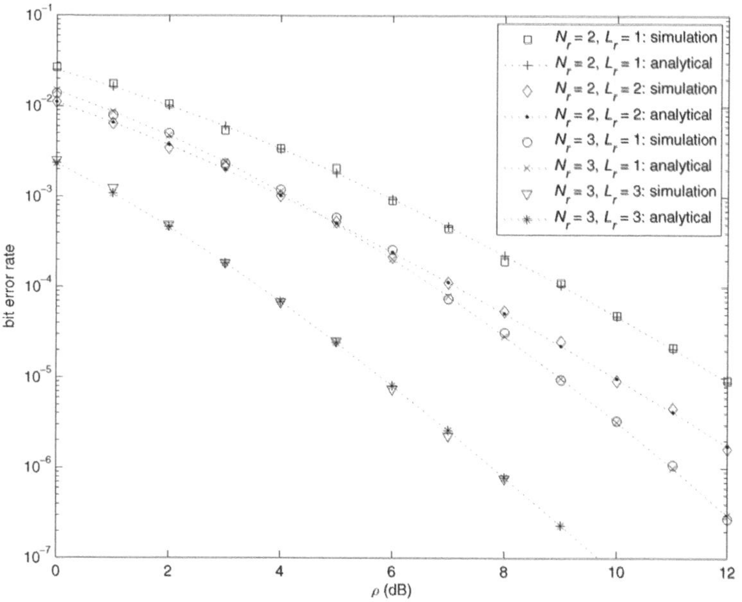

Figure 11.9 Bit error rate performance for the Almouti scheme with receive antenna selection for the cases with $N_r = 2, 3$ and $L_r = 1, 2, 3$ along with their exact theoretical results given by (11.36).

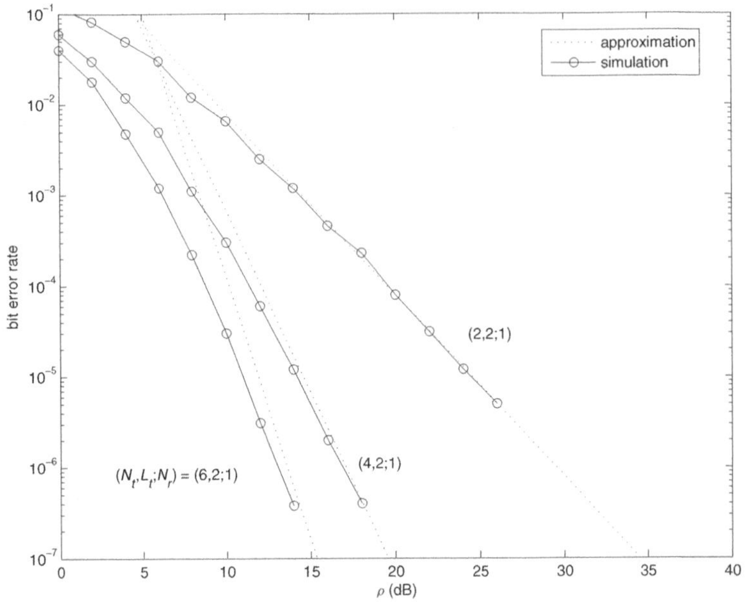

Figure 11.10 Bit error rate performance of the Alamouti scheme with transmit antenna selection for the cases $N_t = 2, 4, 6$, $N_r = 1$ and $L_t = 2$.

with transmit antenna selection. The good match between the simulated performance and the corresponding approximation is also clear from the figure.

11.5 Antenna Selection for Combined Channel Coding and Orthogonal STBCs

In this section, we consider receive antenna selection for the case when space-time block coding is combined with channel coding. As for transmit antenna selection, the results reported in the previous section for the single STBC case can be easily adapted to this concatenation scheme. We derive an upper bound on the bit error rate for any number of transmit and receive antennas. We also derive a tight upper bound on the bit error rate of the Alamouti scheme when $L_r = 1$.

11.5.1 Performance Analysis

Expressions for the pairwise error probability of this concatenation scheme were derived in Section 7.5.3 for fully interleaved and quasi-static fading channels. When receive antenna selection is considered where the receiver selects the best L_r antennas, using the argument that the largest L_r out of N_r nonnegative numbers is always greater than or equal to the average of these N_r numbers multiplied by L_r, the pairwise error probability can be easily found approximately upper bounded as (Zeng and Ghrayeb (2006))

$$P(\boldsymbol{x}_1 \to \boldsymbol{x}_2) \lesssim \binom{2N_tN_rd(\boldsymbol{x}_1,\boldsymbol{x}_2)-1}{N_tN_rd(\boldsymbol{x}_1,\boldsymbol{x}_2)} \left(\prod_{k=1}^{d(\boldsymbol{x}_1,\boldsymbol{x}_2)} \left(|x_1(k)-x_2(k)|^2\right)^{-N_tN_r} \right) \cdot \left(\frac{L_r}{N_r}\frac{\rho}{N_t}\right)^{-N_tN_rd(\boldsymbol{x}_1,\boldsymbol{x}_2)}, \tag{11.40}$$

and

$$P(\boldsymbol{x}_1 \to \boldsymbol{x}_2) \lesssim \binom{2N_tN_r-1}{N_tN_r} \left(d^2(\boldsymbol{x}_1,\boldsymbol{x}_2)\frac{L_r}{N_r}\frac{\rho}{N_t}\right)^{-N_tN_r}, \tag{11.41}$$

for the fully interleaved and quasi-static fading channels, respectively. These upper bounds clearly suggest that, similar to the previous cases, the diversity order is maintained with antenna selection and the loss in signal-to-noise ratio due to antenna selection is upper bounded by $10\log_{10}(N_r/L_r)$ dB. Note that $d(\boldsymbol{x}_1,\boldsymbol{x}_2)$ in (11.40) denotes the minimum (Hamming or symbol-wise Hamming) distance of the code, whereas it denotes the squared Euclidean distance in (11.41). Expressions for the bit error rate can be easily obtained similar to those obtained in Section 7.5.3.

We now derive a tighter upper bound on the bit error rate performance when the receiver selects the best antenna, i.e., $L_r = 1$. In this case, the conditional pairwise error probability for fully interleaved channels is expressed as (Zeng and Ghrayeb (2006))

$$P(\boldsymbol{x}_1 \to \boldsymbol{x}_2|\mathbf{Z}) = Q\left(\sqrt{\frac{\rho}{2N_t}\prod_{k=1}^{d(\boldsymbol{x}_1,\boldsymbol{x}_2)}\left(|x_1(k)-x_2(k)|^2\right)Z_k}\right), \tag{11.42}$$

where $Z_k \triangleq \gamma_{k,1}$ and $\mathbf{Z} \triangleq \{Z_k : k = 1, 2, \ldots, d(\boldsymbol{x}_1,\boldsymbol{x}_2)\}$. Note that Z_k represents the largest received signal-to-noise ratio (among the N_r received signal-to-noise ratios) at time k. The p.d.f. of Z_k is given by (11.35). The pairwise error probability is then obtained by averaging (11.43) with respect to $\mathbf{Z}$, which can be shown to be

$$P(\boldsymbol{x}_1 \to \boldsymbol{x}_2) = \frac{1}{\pi}\int_0^{\pi/2} \prod_{k=1}^{d(\boldsymbol{x}_1,\boldsymbol{x}_2)} N_r \sum_{j=0}^{N_r-1}\sum_{i=0}^{j} A_{j,i}(i+1)!\left(1+j+\frac{\delta_k^2}{\sin^2\theta}\right)^{-(i+2)} d\theta, \tag{11.43}$$

where $A_{j,i}$ is defined above. As for the bit error rate, we upper bound it using (7.64), with $D(\theta)$ replaced by

$$D(\theta) = \left(N_r \sum_{j=0}^{N_r-1}\sum_{i=0}^{j} A_{j,i}(i+1)!\left(1+j+\frac{\delta_k^2}{\sin^2\theta}\right)^{-(i+2)}\right). \tag{11.44}$$

11.5.2 Examples

In this section we present several examples to demonstrate the performance of antenna selection for combined channel coding and space-time block coding. In all cases, it is assumed that the channel fading coefficients remain fixed over a block of N_t consecutive symbols and change independently from one block to another. We consider two outer codes, a convolutional code and a TCM code. For the convolutional coding, we use a rate 1/2 code with generator polynomials $(7, 5)_{octal}$, and BPSK modulation. For the TCM code, we use a rate 2/3, 4-state, 8-PSK TCM scheme presented in Wilson and Leung (1987). The effective length for this code is $d_{\min} = 2$. We plot the bit error rate performance against ρ in dB for all the simulation and numerical results. For the numerical results of the convolutional coding case, only the first seven terms in (7.66) are considered. Therefore, the curves labeled with *bound* in the figures are actually *truncated* bounds.

In Figure 11.11, we plot the bit error rate performance for the cases with $N_t = 2$, $N_r = 3$, and $L_r = 1, 2, 3$ along with their general upper bounds derived above (for the convolutional coded case). It is clear from the figure that all curves have the same slope at high signal-to-noise ratios, suggesting that they have the same diversity order as that of the full complexity system, which agrees with the analytical results derived above. Moreover, the losses in performance due to antenna selection for $L_r = 2$ and $L_r = 1$ are about 0.8 dB and 3 dB at bit error rate 10^{-5}, respectively, relative to the case $L_r = 3$, i.e., no selection.

We repeat the above experiment for the TCM case. The simulation results along with their respective general upper bounds are plotted in Figure 11.12. It is clear from the figure that all the curves have the same slope at high signal-to-noise ratios, which, again, suggests that the diversity order is maintained with antenna selection. Moreover, the losses in signal-to-noise ratio due to antenna selection for $L_r = 2$ and $L_r = 1$ are about 0.8 dB and 3 dB at a bit error rate of 10^{-5}, respectively, relative to the case $L_r = 3$ (i.e., no selection.).

In Figure 11.13, we plot the bit error rate performance along with the corresponding bounds for the TCM scheme for the cases with $N_t = 2$, $N_r = 3$, $L_r = 1, 3$. We observe from the figure the *tight bound* virtually overlaps with the corresponding simulation results.

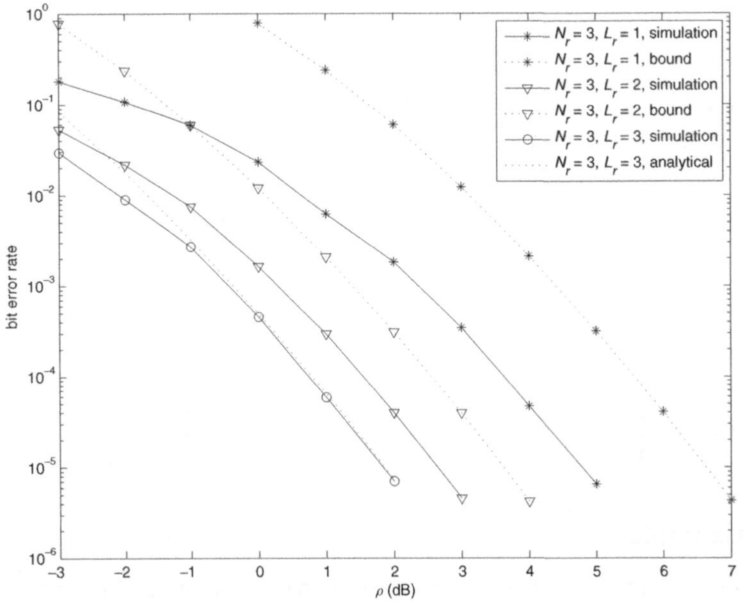

Figure 11.11 Bit error rate performance comparison between various antenna selection scenarios along with their upper bounds (for the convolutional code).

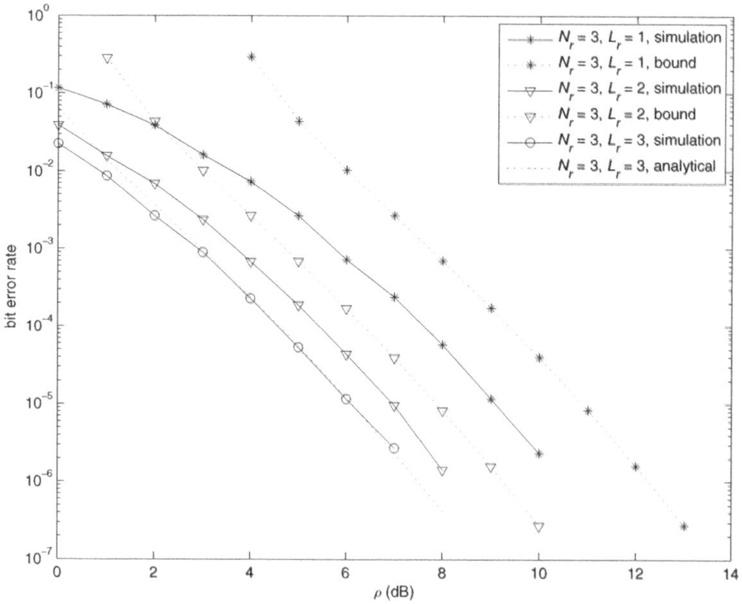

Figure 11.12 Bit error rate performance comparison between various antenna selection scenarios along with their upper bounds (for the TCM case).

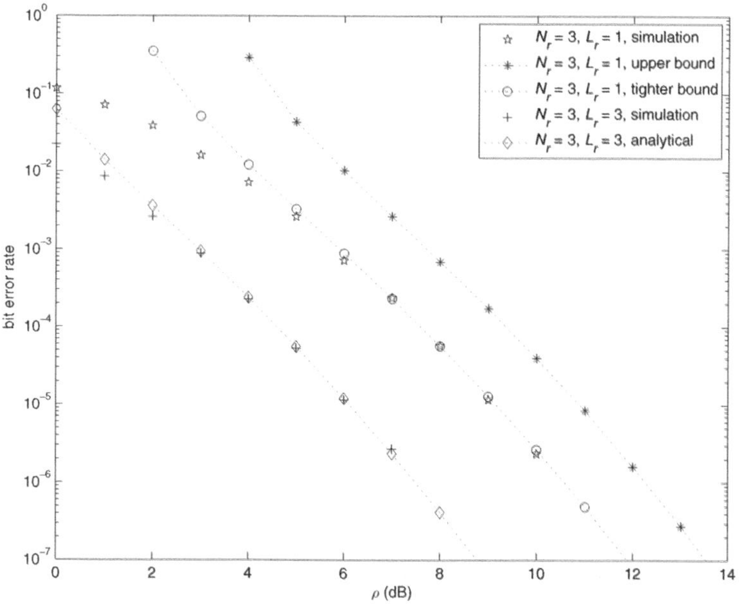

Figure 11.13 Bit error rate performance comparison between simulations and the upper bounds for the cases with $N_t = 2$, $N_r = 3$, $L_r = 1, 3$ (for the TCM case).

11.6 Antenna Selection for Frequency Selective Channels

We have seen in Section 9.3 that a MIMO frequency selective channel can be viewed as a MIMO flat fading channel after properly expressing the transmitted space-time codeword and the channel coefficient matrix as given by (9.14). As such, the results we have presented for the flat fading channel readily apply to the frequency selective fading channels as well (Gucluoglu et al. (2004)). The maximum diversity that can be achieved in this case is $N_t N_r L$, where L represents the number of multipath components.

In Figure 11.14, we plot the frame error rate for a coded MIMO frequency selective fading channel where $N_t = N_r = 2$. The system is coded using two convolutional codes, one with generator $(5, 7)_{octal}$ and another with generator $(4, 2)_{octal}$. The number of multipaths is $L = 2$ with uniform power profile. The antenna selection algorithm employs the receive antenna that maximizes the instantaneous received signal-to-noise ratio, that is, the antenna with the larger $\sum_{i=1}^{N_t} \sum_{l=1}^{L} \left|h_{i,j}^{(l)}\right|^2$ for $j = 1, 2$ is selected, where $h_{i,j}^{(l)}$ denotes the fading coefficient between the ith transmit and jth receive antennas for the lth path.

For this particular case, the $(5, 7)_{octal}$ code achieves the full rank whereas the $(4, 2)_{octal}$ code does not. Consequently, the diversity achieved by the former code is the maximum which is $2 \cdot 2 \cdot 2 = 8$, but it degrades to $3 \cdot 2 = 6$ for the latter one. This is clear from the curves presented in Figure 11.14.

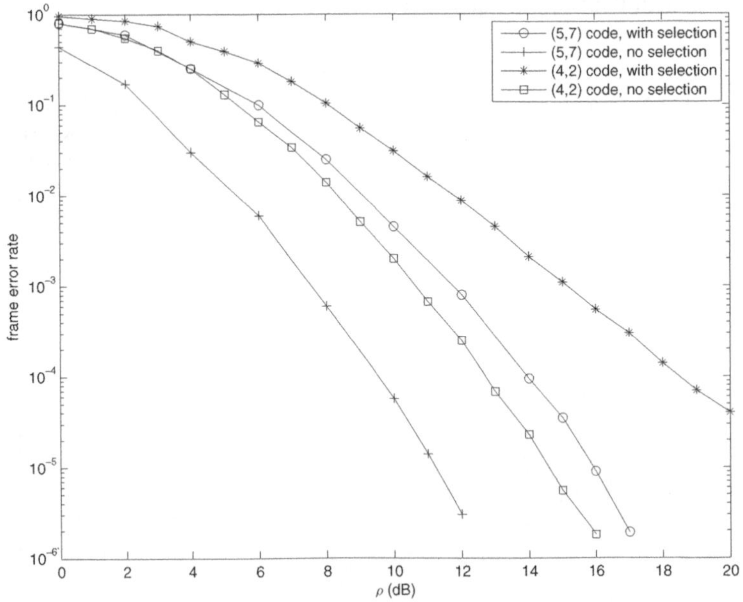

Figure 11.14 Frame error rate performance for a coded system with receive antenna selection over frequency selective fading.

11.7 Antenna Selection with Nonidealities

So far, we have assumed ideal conditions in assessing the impact of antenna selection on the performance of MIMO systems. In this section, we take some of the MIMO system nonidealities, discussed in Chapter 10, into consideration. Specifically, we consider the presence of spatial correlation and channel estimation error on the performance with antenna selection.

11.7.1 Impact of Spatial Correlation

As detailed in Chapter 10, it may be difficult to achieve spatially uncorrelated channels in practice, particularly when the wireless device is relatively small where it is not possible to keep enough distance between adjacent antennas or when scattering is not sufficiently rich. A widely accepted model for the correlated Rayleigh fading channel is the one given by (10.25). The impact of the presence of spatial correlation on the system performance is examined by Zeng and Ghrayeb (2004a) where it is shown that the bit error rate performance degrades, relative to independent fading, by

$$10\log_{10}\left(\left[\prod_{i=1}^{r(\boldsymbol{A})}\lambda_{\boldsymbol{A}}^{i}\right]^{-\frac{1}{r(\boldsymbol{A})}}\left[\prod_{i=1}^{r(\boldsymbol{B})}\lambda_{\boldsymbol{B}}^{i}\right]^{-\frac{1}{r(\boldsymbol{B})}}\right)\ \text{dB}, \tag{11.45}$$

where $r(\boldsymbol{A})$ and $r(\boldsymbol{B})$ are the ranks of the transmit correlation matrix $\boldsymbol{A}$ and the receive correlation matrix $\boldsymbol{B}$, respectively. The corresponding eigenvalues are $\lambda_{\boldsymbol{A}}^{i}$ for $i = 1, 2, \ldots, r(\boldsymbol{A})$ and $\lambda_{\boldsymbol{B}}^{i}$ for $i = 1, 2, \ldots, r(\boldsymbol{B})$.

When the receiver selects the best L_r antennas that maximize the received signal-to-noise ratio, using a similar argument to that used in the analysis of STTCs with antenna selection, the bit error rate can be approximated at high signal-to-noise ratio by (Zeng and Ghrayeb (2004a))

$$P_b \approx \binom{2r(\boldsymbol{A})r(\boldsymbol{B})-1}{r(\boldsymbol{A})r(\boldsymbol{B})}\left[\prod_{i=1}^{r(\boldsymbol{A})}\lambda_{\boldsymbol{A}}^{i}\right]^{-r(\boldsymbol{B})}\left[\prod_{i=1}^{r(\boldsymbol{B})}\lambda_{\boldsymbol{B}}^{i}\right]^{-r(\boldsymbol{A})}\left(\frac{L_r}{N_r}\frac{4\rho}{N_t}\right)^{-r(\boldsymbol{A})r(\boldsymbol{B})}, \tag{11.46}$$

which suggests that the diversity order with antenna selection is the same as that of the full complexity system, whereas the loss in coding gain is upper bounded by $10\log_{10}(N_r/L_r)$ dB.

Similar to the independent fading case, when the orthogonal STBC is concatenated with an outer channel code and assuming ideal interleaving, the pairwise error probability for the antenna selection system can be approximated at high signal-to-noise ratios to be

$$P(\boldsymbol{x}_1 \to \boldsymbol{x}_2) \approx \binom{2r(\boldsymbol{A})r(\boldsymbol{B})d(\boldsymbol{x}_1,\boldsymbol{x}_2)-1}{r(\boldsymbol{A})r(\boldsymbol{B})d(\boldsymbol{x}_1,\boldsymbol{x}_2)}\left(\prod_{k\in\varphi}\left(|x_1(k)-x_2(k)|^2\right)^{-r(\boldsymbol{A})r(\boldsymbol{B})}\right)$$

$$\cdot \left[\prod_{i=1}^{r(A)} \lambda_A^i \right]^{-r(B)d(x_1,x_2)} \left[\prod_{i=1}^{r(B)} \lambda_B^i \right]^{-r(A)d(x_1,x_2)} \left(\frac{L_r}{N_r} \frac{\rho}{N_t} \right)^{-r(A)r(B)d(x_1,x_2)} \quad (11.47)$$

where $d(x_1, x_2)$ is the distance between the transmitted codeword, x_1, and the erroneously decoded codeword, x_2. Therefore, for combined convolutional coding and orthogonal STBC coding, the diversity in the joint transmit-receive correlation case is $r(A)\, r(B)\, d_{\min}(x_1, x_2)$.

We remark that receive antenna selection for STTCs over spatially correlated quasi-static fading channels is considered by Bahceci et al. (2006). The results are in line with the ones above. That is, provided that the underlying STTC is full rank, the diversity order is dictated by the ranks of the transmit and receive correlation matrices. Otherwise, the diversity degrades with antenna selection, similar to the independent fading case.

11.7.2 Example

In this example, we consider antenna selection for a MIMO system employing the Alamouti scheme with $N_t = 2$ and $N_r = 3$. The correlation matrices considered are

$$A = \begin{bmatrix} 1.0 & 0.6 \\ 0.6 & 1.0 \end{bmatrix} \text{ and } B = \begin{bmatrix} 1.0 & 0.6 & 0.4 \\ 0.6 & 1.0 & 0.45 \\ 0.4 & 0.45 & 1.0 \end{bmatrix}.$$

The eigenvalues of A are $\lambda_A^1 = 1.6$, $\lambda_A^2 = 0.4$, whereas the eigenvalues of B are $\lambda_B^1 = 1.9720$, $\lambda_B^2 = 0.6321$, $\lambda_B^3 = 0.3959$. Obviously these matrices are full rank, and thus the maximum diversity order is achieved with antenna selection. In all cases, it is assumed that the channel fading coefficients remain fixed over a block of N_t consecutive symbols and change independently from one block to another. It is also assumed, where applicable, that antenna selection is based on maximizing the signal-to-noise ratio. For the concatenated coding scheme, we consider a rate $1/2$ convolutional code with generator $(7, 5)_{octal}$ and BPSK modulation. To achieve the maximal time diversity, an interleaver is used which swaps the odd indexed convolutional coded symbols between two consecutive convolutional coded frames before they are fed into the orthogonal STBC encoder. Thus, independent fading is guaranteed among coded symbols within a frame.

The bit error rate performance for this concatenation scheme is plotted in Figure 11.15. We also plot in the same figure the performance of the full complexity system over independent fading, which we used as a baseline. We observe from the figure that all the curves have the same slope, suggesting that the diversity order is maintained with antenna selection.

11.7.3 Impact of Channel Estimation Error

It was been established in Section 10.1 that using noisy CSI estimates in place of perfect CSI has no impact on the diversity order of MIMO systems. The only consequence for having imperfect CSI is a degradation in the signal-to-noise ratio. In the context of antenna selection, a MIMO system with imperfect CSI can be viewed as a MIMO system with perfect CSI operating at a lower signal-to-noise ratio. Therefore, the results obtained for

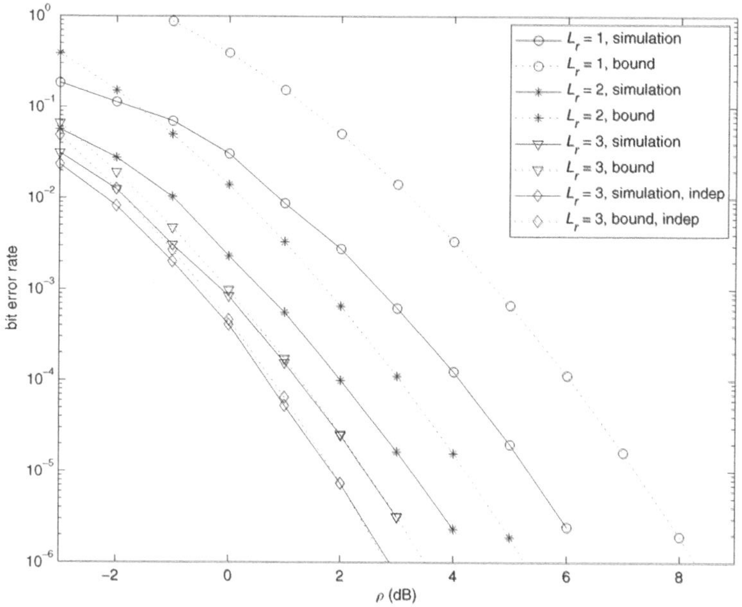

Figure 11.15 Frame error rate performance with receive antenna selection over spatially correlated Rayleigh fading.

the perfect CSI cases apply in a straightforward manner to the imperfect CSI cases. This argument has been confirmed by Ma and Tepedelenlioglu (2007) and Li and Beaulieu (2006).

11.8 Chapter Summary and Further Reading

In this chapter we addressed the problem of antenna selection for MIMO systems. We considered the impact of antenna selection on the channel capacity, and we showed that only a small loss in capacity is incurred. We also looked at the impact of antenna selection on the performance of STTCs and orthogonal STBCs over various Rayleigh fading channels. A pragmatic selection criterion based on maximizing the instantaneous received signal-to-noise ratio was used. It was shown that the spatial diversity is retained with antenna selection for STTCs over quasi-static fading channels. However, this diversity is lost when the underlying STTC is rank deficient or the channel is block or fast fading. It should be noted that when the transmitter is equipped with only one antenna, the spatial diversity is always maintained regardless of the underlying channel model or space-time coding scheme simply because the problem of antenna selection become equivalent to selection combining with one transmit antenna. On the other hand, it was shown that the spatial diversity is retained for orthogonal STBCs with antenna selection independent of the underlying Rayleigh fading channel model. This is attributed to the difference in the structures of both classes of codes. A concatenation scheme comprising an outer channel code, such as a TCM or convolutional code, and an orthogonal STBC was analyzed with antenna selection, it was shown that the spatial diversity is preserved with antenna selection for this case as well.

The MIMO system performance with antenna selection with some nonidealities was also considered. It was shown that the presence of spatial correlation at both ends of the communication link does not impact the overall spatial diversity only if the transmit and receive correlation matrices are full rank. Otherwise, the resulting diversity order becomes the product of the ranks of these correlation matrices. Another negative impact of the presence of spatial correlation is a loss in signal-to-noise ratio as compared with the case of independent fading. We also considered the case when the channel is modeled as frequency selective fading. We showed that a frequency selective channel with L paths can be viewed as a flat fading channel with $N_t L$ virtual transmit antennas. As such, the results obtained for flat fading apply in a straightforward manner. The impact of imperfect CSI on the system performance with antenna selection was also considered. It was shown that imperfect CSI does not degrade the diversity order with antenna selection. Other nonidealities in conjunction with antenna selection are considered by Sudarshan et al. (2004, 2006).

There are other issues pertaining to antenna selection that we have not considered in this chapter. In the following we provide some references that the reader may find useful. Joint transmit and receive antenna selection is considered by Thoen et al. (2001), Zhou and Dai (2006) and Cai and Giannakis (2004). Transmit antenna selection for layered space-time codes is considered by Mun (2006), Zhang et al. (2006) and Blum and Winters (2002). Receive antenna selection for layered space-time codes is considered by Lin et al. (2006) and Gorokhov et al. (2003). Multimode antenna selection for spatial multiplexing systems is considered by Heath and Love (2005). Antenna selection for spatial multiplexing with linear receiver is considered by Heath et al. (2001). Sanayei and Nosratinia (2007) consider antenna selection for keyhole channels. The outage probability with antenna selection over independent and correlated Rayleigh fading is analyzed by Shen and Ghrayeb (2005, 2006). Antenna selection for unitary space-time codes is considered by Ma and Tepedelenlioglu (2005).

Problems

11.1 Reproduce Figure 11.2.

11.2 Show that the capacity-based and energy-based receive antenna selection criteria are equivalent for any N_r and L_r when the number of transmit antennas is $N_t = 1$.

11.3 Starting with (11.11), prove (11.12).

11.4 Derive Equation (11.21).

11.5 Show that the codeword difference matrix in fast fading will always have a rank of one or less, i.e., will always be rank deficient.

11.6 The results reported in Figure 11.6 are for fast fading. Regenerate this figure using the same parameters, including the same STTC, but for block fading channels with $T = 2$ and 50. Comment on the diversity order for these cases.

11.7 Reproduce Figure 11.9.

11.8 Derive Equations (11.40) and (11.41).

11.9 Regenerate Figure 11.15.

Bibliography

Abou-Faycal IC, Trott MD and Shamai S 2001 The capacity of discrete-time memoryless Rayleigh fading channels. *IEEE Transactions on Information Theory* **47**(4), 1290–1301.

Agrawal D, Tarokh V, Naguib A and Seshadri N 1998 Space-time coded OFDM for high data-rate wireless communication over wideband channels *Proceedings of IEEE Vehicular Technology Spring Conference (VTC–Spring)*, vol. 3, pp. 2232–2236, Ottawa, Canada.

Akay E and Ayanoglu E 2006 Achieving full frequency and space diversity in wireless systems via BICM, OFDM, STBC and Viterbi decoding. *IEEE Transactions on Communications* **54**(12), 2164–2172.

Aktas D and Fitz MP 2003 Distance spectrum analysis of space-time trellis coded modulations in quasi-static Rayleigh-fading channels. *IEEE Transactions on Information Theory* **49**, 3335–3344.

Aktas E and Mitra U 2003 Blind equalization for an application of unitary space-time modulation in ISI channels. *IEEE Transactions on Signal Processing* **51**(11), 2931–2942.

Al-Dhahir N 2001 FIR channel-shortening equalizers for MIMO ISI channels. *IEEE Transactions on Communications* **49**(2), 213–218.

Al-Dhahir N and Sayed AH 2000 The finite-length multi-input multi-output MMSE-DFE. *IEEE Transactions on Signal Processing* **48**(10), 2921–2936.

Al-Dhahir N, Naguib AF and Calderbank AR 2001 Finite-length MIMO decision feedback equalization for space-time block-coded signals over multipath-fading channels. *IEEE Transactions on Vehicular Technology* **50**(4), 1176–1182.

Alamouti SM 1998 A simple transmit diversity technique for wireless communications. *IEEE Journal on Selected Areas in Communications* **16**(8), 1451–1458.

Alexander PD, Grant AJ and Reed MC 1998 Iterative detection in code-division multiple-access with error control coding. *European Transactions on Telecommunications* **9**, 419–425.

Alon N and Luby M 1996 A linear time erasure-resilient code with nearly optimal recovery. *IEEE Transactions on Information Theory* **42**(6), 1732–1736.

Alouini MS and Simon MK 1999 Performance of coherent receivers with hybrid SC/MRC over Nakagami-m fading channels. *IEEE Transactions on Vehicular Technology* **48**(4), 1155–1164.

AlRustamani A and Vojcic BR 2002 A new approach to greedy multiuser detection. *IEEE Transactions on Communications* **50**(8), 1326–1336.

AlRustamani A, Vojcic BR and Stefanov A 2002 Greedy detection. *Journal of VLSI Signal Processing Systems* **30**(1–3), 179–195.

Ariyavisitakul SL 2000 Turbo space-time processing to improve wireless channel capacity. *IEEE Transactions on Communications* **48**(8), 1347–1359.

Arnold D and Loeliger HA 2001 On the information rate of binary-input channels with memory *Proceedings of IEEE International Conference on Communications (ICC)*, pp. 2692–2695, Helsinki.

Arnold DM, Loeliger HA, Vontobel PO, Kavcic A and Zeng W 2006 Simulation-based computation of information rates for channels with memory. *IEEE Transactions on Information Theory* **52**(8), 3498–3508.

Bahceci I and Duman TM 2002 Combined turbo coding and unitary space-time modulation. *IEEE Transactions on Communications* **50**(8), 1244–1249.

Bahceci I and Duman TM 2004 Trellis-coded unitary space-time modulation. *IEEE Transactions on Wireless Communications* **3**(6), 2005–2012.

Bahceci I, Altunbasak Y and Duman TM 2006 Space-time coding over correlated fading channels with antenna selection. *IEEE Transactions on Wireless Communications* **5**(1), 34–39.

Bahceci I, Duman TM and Altunbasak Y 2003 Antenna selection for multiple-antenna transmission systems: performance analysis and code construction. *IEEE Transactions on Information Theory* **49**(10), 2669–2681.

Bahl L, Cocke J, Jelinek F and Raviv J 1974 Optimal decoding of linear codes for minimizing symbol error rate. *IEEE Transactions on Information Theory* **20**(2), 284–287.

Barbulescu AS and Pietrobon SS 1995 Terminating the trellis of turbo-codes in the same state. *Electronics Letters* **31**(1), 22–23.

Baro S, Bauch G and Hansmann A 2000 Improved codes for space-time trellis coded modulation. *IEEE Communications Letters* **4**(1), 20–22.

Bauch G 1999 Concatenation of space-time block codes and turbo-TCM *Proceedings of IEEE International Conference on Communications (ICC)*, pp. 1202–1206.

Bauch G 2003 Space-time block codes versus space-frequency block codes. *Proceedings of IEEE Vehicular Technology Spring Conference (VTC–Spring)* pp. 567–571.

Bauch G and Al-Dhahir N 2002 Reduced-complexity space-time turbo-equalization for frequency-selective MIMO channels. *IEEE Transactions on Wireless Communications* **1**(4), 819–828.

Baum DS, Gore D, Nabar R, Panchanathan S, Hari KVS, Erceg V and Paulraj AJ 2000 Measurement and characterization of broadband MIMO fixed wireless channels at 2.5 GHz *Proceedings of IEEE International Conference on Personal Wireless Communications (ICPWC)*, pp. 203–206.

Benedetto S and Montorsi G 1997 Performance of continuous and blockwise decoded turbo codes. *IEEE Communications Letters* **1**(3), 77–79.

Benedetto S, Divsalar D, Montorsi G and Pollara F 1998 Serial concatenation of interleaved codes: performance analysis, design, and iterative decoding. *IEEE Transactions on Information Theory* **44**(3), 909–926.

Berrou C, Glavieux A and Thitimajshima P 1993 Near Shannon limit error-correcting coding and decoding: Turbo-codes *Proceedings of IEEE International Conference on Communications (ICC)*, pp. 1064–1070, Geneva, Switzerland.

Biglieri E 2005 *Coding for Wireless Channels*. Springer.

Biglieri E, Calderbank R, Constantinides A, Goldsmith A, Paulraj A and Poor HV 2007 *MIMO Wireless Communications*. Cambridge University Press, New York.

Biglieri E, Divsalar D, McLane PJ and Simon MK 1991 *Introduction to Trellis-Coded Modulation with Applications*. Macmillan, New York.

Biglieri E, Proakis J and Shamai S 1998 Fading channels: information–theoretic and communications aspects. *IEEE Transactions on Information Theory* **44**(6), 2619–2692.

Biguesh M and Gershman AB 2004 MIMO channel estimation: optimal training and tradeoffs between estimation techniques *Proceedings of IEEE International Conference on Communications (ICC)*, vol. 5, pp. 2658–2662.

Blackert WJ, Hall EK and Wilson SG 1995 Turbo code termination and interleaver conditions. *Electronics Letters* **31**(24), 2082–2084.

Blum RS 2002 Some analytical tools for the design of space-time convolutional codes. *IEEE Transactions on Communications* **50**(10), 1593–1599.

Blum RS and Winters JH 2002 On optimum MIMO with antenna selection. *IEEE Communications Letters* **6**(8), 322–324.

Blum RS, Li YG, Winters JH and Yan Q 2001 Improved space-time coding for MIMO-OFDM wireless communication. *IEEE Transactions on Communications* **49**(11), 1873–1878.

Boariu A and Ionescu DM 2003 A class of nonorthogonal rate-one space-time block codes with controlled interference. *IEEE Transactions on Wireless Communications* **2**(2), 270–276.

Boelcskei H 2006 MIMO-OFDM wireless systems: basics, perspectives and challenges. *IEEE Wireless Communications* **13**(4), 31–37.

Boelcskei H, Borgmann M and Paulraj AJ 2003 Impact of the propagation enviroment on the performance of space-frequency coded MIMO-OFDM. *IEEE Journal on Selected Areas in Communications* **21**(3), 427–439.

Boelcskei H, Gesbert D and Paulraj AJ 2002 On the capacity of OFDM-based spatial multiplexing systems. *IEEE Transactions on Communications* **50**(2), 225–234.

Boubaker N, Letaief KB and Murch RD 2001 A layered space-time coded wideband OFDM architechture for dispersive wireless links. *Proceedings of IEEE Symposium on Computers and Communications* pp. 518–523.

Boutros J, Gresset N, Brunel L and Fossorier M 2003 Soft-input soft-output lattice sphere decoder for linear channels *Proceedings of IEEE Global Communications Conference (GLOBECOM)*, vol. 3, pp. 1583–1587, San Francisco, CA.

Burr A 2002 Evaluation of capacity of indoor wireless MIMO channel using ray tracing *Proceedings of International Zurich Seminar on Broadband Communications, Access, Transmission, Networking*, pp. 28.1–28.6, Zurich, Switzerland.

Cai X and Giannakis GB 2004 Performance analysis of combined transmit selection diversity and receive generalized selection combining in Rayleigh fading channels. *IEEE Transactions on Wireless Communications* **3**(6), 1980–1983.

Cai X and Giannakis GB 2006 Differential space-time modulation with eigen-beamforming for correlated MIMO fading channels. *IEEE Transactions on Signal Processing* **54**(4), 1279–1288.

Caire G, Taricco G and Biglieri E 1998 Bit-interleaved coded modulation. *IEEE Transactions on Information Theory* **44**(3), 927–946.

Caire G and Colavolpe G 2003 On low complexity space-time coding for quasi-static channels. *IEEE Transactions on Information Theory* **49**(6), 1400–1416.

Catreux S, Erceg V, Gesbert D and Heath Jr RW 2002 Adaptive modulation and MIMO coding for broadband wireless data networks. *IEEE Communications Magazine* **40**(6), 108–115.

Chang R and Hancock J 1966 On receiver structures for channels having memory. *IEEE Transactions on Information Theory* **12**(4), 463–468.

Chen X, Zhou K and Aravena JL 2006 A new family of unitary space-time codes with a fast parallel sphere decoder algorithm. *IEEE Transactions on Information Theory* **52**(1), 115–140.

Chen Z, Vucetic B and Yuan J 2003 Space-time trellis codes with transmit antenna selection. *Electronics Letters* **39**(11), 854–855.

Chen Z, Vucetic BS, Yuan J and Lo KL 2002 Space-time trellis codes for 4-PSK with three and four transmit antennas in quasi-static flat fading channels. *IEEE Communications Letters* **6**(2), 67–69.

Cheung SK and Schober R 2006 Differential spatial multiplexing. *IEEE Transactions on Wireless Communications* **5**(8), 2127–2135.

Chizhik D, Ling J, Wolniansky PW, Valenzuela RA, Costa N and Huber K 2003 Multiple-input-multiple-output measurements and modeling in Manhattan. *IEEE Journal on Selected Areas in Communications* **21**(3), 321–331.

Chuah CN, Tse DNC, Kahn JM and Valenzuela RA 2002 Capacity scaling in MIMO wireless systems under correlated fading. *IEEE Transactions on Information Theory* **48**(3), 637–650.

Cover TM and Thomas JA 2006 *Elements of Information Theory*. 2nd edition Wiley, Hoboken, NJ.

Cozzo C and Hughes BL 2003 Joint channel estimation and data detection in space-time communications. *IEEE Transactions on Communications* **51**(8), 1266–1270.

Cui D and Haimovich AM 2001 Performance of parallel concatenated space-time codes. *IEEE Communications Letters* **5**(6), 236–238.

Dalton LA and Georghiades CN 2005 A full-rate, full-diversity four-antenna quasi-orthogonal space-time block code. *IEEE Transactions on Wireless Communications* **4**(2), 363–366.

Damen MO, Abdi A and Kaveh M 2001 On the effect of correlated fading on several space-time coding and detection schemes *Proceedings of IEEE Vehicular Technology Fall Conference (VTC–Fall)*, vol. 1, pp. 13–16, Atlantic City, NJ.

Damen MO and Beaulieu NC 2003 On diagonal algebraic space-time block codes. *IEEE Transactions on Communications* **51**(6), 911–919.

Damen MO, Chkeif A and Belfiore JC 2000 Lattice code decoder for space-time codes. *IEEE Communications Letters* **4**(5), 161–163.

Damen MO, Tewfik A and Belfiore JC 2002 A construction of a space-time code based on number theory. *IEEE Transactions on Information Theory* **48**(3), 753–760.

Deng RH and Costello Jr DJ 1989 High rate concatenated coding systems using bandwidth efficient trellis inner codes. *IEEE Transactions on Communications* **37**(5), 420–427.

Diggavi SN, Al-Dhahir N, Stamoulis A and Calderbank AR 2002 Differential space-time coding for frequency-selective channels. *IEEE Communications Letters* **6**(6), 253–255.

Divsalar D and Pollara F 1995 Multiple turbo-codes for deep-space communications. *TDA Progress Report 42-120*, Jet Propulsion Laboratory, California Institute of Technology, Pasadena, CA.

Divsalar D and Pollara F 2000 Hybrid concatenated codes and iterative decoding. US Patent 6023783.

Divsalar D and Simon MK 1987 Trellis coded modulation for 4800–9600 bits/s transmission over a fading mobile satellite channel. *IEEE Journal on Selected Areas in Communications* **5**(2), 162–175.

Divsalar D and Simon MK 1988a The design of trellis coded MPSK for fading channels: performance criteria. *IEEE Transactions on Communications* **36**(9), 1004–1012.

Divsalar D and Simon MK 1988b The design of trellis coded MPSK for fading channels: set partitioning for optimum code design. *IEEE Transactions on Communications* **36**(9), 1013–1021.

Driessen PF and Foschini GJ 1999 On the capacity formula for multiple input-multiple output wireless channels: a geometric interpretation. *IEEE Transactions on Communications* **47**(2), 173–176.

Duffin RJ and Schaeffer AC 1952 A class of nonharmonic Fourier series. *Trans. Amer. Math. Soc.* **72**, 341–366.

Duman TM 1998 *Turbo Codes and Turbo Coded Modulation Systems: Analysis and Performance Bounds*. Ph.D. Dissertation, Electrical and Computer Engineering Department, Northeastern University, Boston, MA.

Duman TM 2002 Interleavers for serial and parallel concatenated (turbo) codes, *Wiley Encyclopedia of Telecommunications* (J. G. Proakis, ed.) 1st edn. Wiley and Sons.

Duman TM and Salehi M 1998 New performance bounds for turbo codes. *IEEE Transactions on Communications* **46**(6), 717–723.

El Gamal H 2002a On the design of layered space-time systems for autocoding. *IEEE Transactions on Communications* **50**(9), 1451–1461.

El Gamal H 2002b On the robustness of space-time coding. *IEEE Transactions on Signal Processing* **50**(10), 2417-2428.

El Gamal H and Geraniotis E 2000 Iterative multiuser detection for coded CDMA signals in AWGN and fading channels. *IEEE Journal on Selected Areas in Communications* **18**(1), 30–41.

El Gamal H and Hammons Jr AR 2001 A new approach to layered space-time coding and signal processing. *IEEE Transactions on Information Theory* **47**(6), 2321–2334.

El Gamal H, Hammons Jr. AR, Liu Y, Fitz MP and Takeshita OY 2003 On the design of space-time and space-frequency codes for MIMO frequency-selective fading channels. *IEEE Transactions on Information Theory* **49**(9), 2277–2292.

Femenias G and Furio I 2000 A new, simple, and exact union bound for reference-based predetection and postdetection diversity TCM-MPSK systems in Rayleigh fading. *IEEE Transactions on Vehicular Technology* **49**(2), 540–549.

Fincke U and Pohst M 1985 Improved methods for calculating vectors of short lengths in a lattice, including a complexity analysis. *Math. Comput.* **44**, 463–471.

Forney, Jr. GD 1966 *Concatenated Codes* 1st edn. MIT Press, Cambridge, MA.

Forney, Jr. GD 1971 Burst correcting codes for the classic bursty channel. *IEEE Transactions on Communications Technology* **19**(5), 772–781.

Forney, Jr. GD 1989 A bounded-distance decoding algorithm for the Leech lattice with generalizations. *IEEE Transactions on Information Theory* **35**(4), 906–909.

Foschini GJ 1996 Layered space-time architecture for wireless communication in a fading environment when using multi-element antennas. *Bell Laboratories Technical Journal* **1**(2), 41–59.

Foschini GJ and Gans M 1998 On limits of wireless communications in a fading environment when using multiple antennas. *Wireless Personal Communications* **6**(3), 311–335.

Fozunbal M, McLaughlin SW and Schafer RW 2005 On space-time-frequency coding over MIMO-OFDM systems. *IEEE Transactions on Wireless Communications* **4**(1), 320–331.

Fozunbal M, McLaughlin SW, Schafer RW and Landsberg JM 2004 On space-time coding in the presence of spatio-temporal correlation. *IEEE Transactions on Information Theory* **50**(9), 1910–1926.

Gallager R 1962 Low-density parity-check codes. *IRE Transactions on Information Theory* **8**(1), 21–28.

Gallager RG 1968 *Information Theory and Reliable Communication*. John Wiley and Sons, New York.

Ganesan G and Stoica P 2001 Space-time block codes: a maximum SNR approach. *IEEE Transactions on Information Theory* **47**(4), 1650–1656.

German G, Spencer Q, Swindlehurst L and Valenzuela R 2001 Wireless indoor channel modeling: statistical agreement of raytracing simulations and channel sounding measurements *Proceedings of International Conference on Acoustic, Speech and Signal Processing*, pp. 2501–2504.

Gershman AB and Sidiropoulos ND 2005 *Space-Time Processing for MIMO Communications (Edited)*. Wiley, Hoboken, NJ.

Gesbert D, Bolcskei H, Gore DA and Paulraj AJ 2002 Outdoor MIMO wireless channels: Models and performance prediction. *IEEE Transactions on Communications* **50**(12), 1926–1934.

Gharavi-Alkhansari M and Gershman AB 2004 Fast antenna subset selection in MIMO systems. *IEEE Transactions on Signal Processing* **52**(2), 339–347.

Ghrayeb A 2003 On the SOVA for extremely high code rates over partial response channels. *Journal of Communications and Networks* **5**(1), 1–6.

Ghrayeb A and Duman TM 2003 Performance analysis of MIMO systems with antenna selection over quasi-static fading channels. *IEEE Transactions on Vehicular Technology* **52**(2), 281–288.

Ghrayeb A and Ryan WE 1999 Performance of high rate turbo codes employing soft-output Viterbi algorithm (SOVA) *Proceedings of the 33rd Asilomar Conference on Signals, Systems, and Computers*, pp. 1665–1669.

Ghrayeb A and Ryan WE 2001 Concatenated code system design for storage channels. *IEEE Journal on Selected Areas in Communications* **19**(4), 709–718.

Giannakis GB, Liu Z, Ma X and Zhou S 2007 *Space-Time Coding for Broadband Wireless Communications*. Wiley-Interscience, Hoboken, NJ.

Goldsmith A 2005 *Wireless Communications*. Cambridge University Press.

Gong Y and Letaief KB 2002 Concatenated space-time block coding with trellis coded modulation in fading channels. *IEEE Transactions on Wireless Communications* **1**(4), 580–590.

Gong Y and Letaief KB 2003 An efficient space-frequency coded OFDM system for broadband wireless communications. *IEEE Transactions on Communications* **51**(12), 2019–2029.

Gore D, Sandhu S and Paulraj A 2001 Delay diversity code for frequency selective channels. *Electronics Letters* **37**(20), 1230–1231.

Gorokhov A, Gore DA and Paulraj AJ 2003 Receive antenna selection for MIMO spatial multiplexing: theory and algorithms. *IEEE Transactions on Signal Processing* **51**(11), 2796–2807.

Gucluoglu T and Duman TM 2005 Soft input soft output stack equalization for MIMO frequency selective fading channels *Proceedings of IEEE International Conference on Communications (ICC)*, pp. 510–514, Seoul, S. Korea.

Gucluoglu T, Duman TM and Ghrayeb A 2004 Antenna selection for space time coding over frequency-selective fading channels *Proceedings of International Conference on Acoustic, Speech and Signal Processing*, pp. iv–709 – iv–712, Montreal, Canada.

Gulati V and Narayanan KR 2003 Concatenated codes for fading channels based on recursive space-time trellis codes. *IEEE Transactions on Wireless Communications* **2**(1), 118–128.

Hagenauer J 1997 The turbo principle: Tutorial introduction and the state of the art *Proc. International Symposium on Turbo Codes and Related Topics*, pp. 1–11, Brest, France.

Hagenauer J and Hoeher P 1989 Concatenated Viterbi decoding *Proceedings of the 4th Joint Swedish–Soviet International Workshop on Information Theory*, pp. 29–33, Gotland, Sweden.

Hagenauer J, Offer E and Papke L 1996 Iterative decoding of binary block and convolutional codes. *IEEE Transactions on Information Theory* **42**(2), 429–445.

Hammons AR and El Gamal H 2000 On the theory of space-time codes for PSK modulation. *IEEE Transactions on Information Theory* **46**(2), 524–542.

Hassibi B 2000 An efficient square-root algorithm for BLAST *Proceedings of International Conference on Acoustic, Speech and Signal Processing*, vol. 2, pp. 737–740, Istanbul, Turkey.

Hassibi B and Hochwald BM 2002 High-rate codes that are linear in space and time. *IEEE Transactions on Information Theory* **48**(7), 1804–1824.

Hassibi B and Hochwald BM 2003 How much training is needed in multiple-antenna wireless links?. *IEEE Transactions on Information Theory* **49**(4), 951–963.

Hassibi B and Vikalo H 2005 On the sphere-decoding algorithm I. Expected complexity. *IEEE Transactions on Signal Processing* **53**(8), 2806–2818.

Heath Jr. RW and Love DJ 2005 Multimode antenna selection for spatial multiplexing systems with linear receivers. *IEEE Transactions on Signal Processing* **53**(8), 3042–3056.

Heath Jr. RW and Paulraj AJ 2002 Linear dispersion codes for MIMO systems based on frame theory. *IEEE Transactions on Signal Processing* **50**(10), 2429–2441.

Heath Jr. RW, Sandhu S and Paulraj A 2001 Antenna selection for spatial multiplexing systems with linear receivers. *IEEE Communications Letters* **5**(4), 142–144.

Hedayat A, Nosratinia A and Al-Dhahir N 2005a Linear equalizers for flat Rayleigh MIMO channels *Proceedings of International Conference on Acoustic, Speech and Signal Processing*, vol. 3, pp. 445–448, Philadelphia, PA.

Hedayat A, Shah H and Nosratinia A 2005b Analysis of space-time coding in correlated fading channels. *IEEE Transactions on Wireless Communications* **4**(6), 2882–2891.

Hirt W and Massey JL 1988 Capacity of the discrete-time Gaussian channel with intersymbol interference. *IEEE Transactions on Information Theory* **34**(3), 380–388.

Hochwald BM and Marzetta TL 2000 Unitary space-time modulation for multiple-antenna communications in Rayleigh flat fading. *IEEE Transactions on Information Theory* **46**(2), 543–564.

Hochwald BM and Sweldens W 2000 Differential unitary space-time modulation. *IEEE Transactions on Communications* **48**(12), 2041–2052.

Hochwald BM, Marzetta TL and Hassibi B 2001 Space-time autocoding. *IEEE Transactions on Information Theory* **47**(7), 2761–2781.

Hochwald BM, Marzetta TL, Richardson TJ, Sweldens W and Urbanke R 2000 Systematic design of unitary space-time constellations. *IEEE Transactions on Information Theory* **46**(6), 1962–1973.

Horn RA and Johnson CR 1985 *Matrix Analysis*. Cambridge University Press.

Hou J, Siegel PH and Milstein LB 2005 Design of multi-input multi-output systems based on low-density parity-check codes. *IEEE Transactions on Communications* **53**(4), 601–611.

Hu J and Duman TM 2007 Graph-based detector for BLAST architecture *Proceedings of IEEE International Conference on Communications (ICC)*, Glasgow, Scotland.

Huang CX and Ghrayeb A 2006 A simple remedy for the exaggerated extrinsic information produced by the SOVA algorithm. *IEEE Transactions on Wireless Communications* **5**(5), 996–1002.

Huang Y, Zhang J and Djuric PM 2005 Bayesian detection for BLAST. *IEEE Transactions on Signal Processing* **53**(3), 1086–1096.

Hughes B 2000 Differential space-time modulation. *IEEE Transactions on Information Theory* **46**(7), 2567–2578.

IEEE 802.16e Part 16 2004 : Air interface for fixed boroadband wireless access systems.

IEEE 802.16e/D12: Part 16 2006 : Air interface for fixed and mobile boroadband wireless access systems.

IEEE P802.11n/D1.0 2006 : Part 11: Wireless LAN medium access control (MAC) and physival layer (PHY) specifications: Enhancements for higher throughputs.

Ivrlac MT, Utschick W and Nossek JA 2003 Fading correlation in wireless MIMO communication systems. *IEEE Journal on Selected Areas in Communications* **21**(5), 819–828.

Jafar SA and Goldsmith A 2004 Transmitter optimization and optimality of beamforming for multiple antenna systems. *IEEE Transactions on Wireless Communications* **3**(4), 1165–1175.

Jafarkhani H 2001 A quasi-orthogonal space-time block code. *IEEE Transactions on Communications* **49**(1), 1–4.

Jafarkhani H 2005 *Space-Time Coding: Theory and Practice*. Cambridge University Press.

Jafarkhani H and Seshadri N 2003 Super-orthogonal space-time trellis codes. *IEEE Transactions on Information Theory* **49**(4), 937–950.

Jamali SH and Le-Ngoc T 1994 *Coded Modulation Techniques for Fading Channels*. Kluwer Academic Publishers, Boston, MA.

Jankiraman M 2004 *Space-Time Codes and MIMO Systems*. Artech House.

Joerssen O and Meyr M 1994 Terminating the trellis of turbo codes. *Electronics Letters* **30**(16), 1285–1286.

Jorswieck EA and Sezgin A 2004 Impact of spatial correlation on the performance of orthogonal space-time block codes. *IEEE Communications Letters* **8**(1), 21–23.

Jung T and Cheun K 2003 Design of concatenated space-time block codes using signal space diversity and the Alamouti scheme. *IEEE Communications Letters* **7**(7), 329–331.

Kadous T 2003 Ordered H-BLAST for MIMO/OFDM systems over multipath channels. *Proceedings of IEEE Symposium of Computers and Communications* pp. 481–485.

Kaiser T, Bourdoux A, Boche H, Fonollosa JR, Andersen JB and Utschick W 2004 *Smart Antennas – State of the Art*. Hindawi Publishing Corp., New York.

Kammoun I and Belfiore JC 2003 A new family of Grassmann space-time codes for non-coherent MIMO systems. *IEEE Communications Letters* **7**(11), 528–530.

Kay S 1993 *Fundamentals of Statistical Signal Processing: Estimation Theory* vol. 1. Prentice Hall.

Kaynak MN, Duman TM and Kurtas EM 2004 *Turbo Codes*. Coding and Signal Processing for Magnetic Recording Systems, pp. 35.1–35.21, Editors: Bane Vasic and Erozan M. Kurtas, CRC Press.

Kaynak MN, Duman TM and Kurtas EM 2005 Belief propagation over MIMO frequency selective fading channels *Joint Conference on Autonomic and Autonomous Systems and International Conference on Networking and Services*, p. 45 (6 pages).

Kaynak MN, Duman TM and Kurtas EM 2006 Belief propagation over SISO/MIMO frequency selective fading channels (accepted). *IEEE Transactions on Wireless Communications*.

Kermoal JP, Schumacher L, Mogensen PE and Pedersen KI 2000 Experimental investigation of correlation properties of MIMO radio channels for indoor picocell scenarios *Proceedings of IEEE Vehicular Technology Conference (VTC)*, pp. 14–21.

Kiran T and Rajan BS 2005 STBC-schemes with nonvanishing determinant for certain number of transmit antennas. *IEEE Transactions on Information Theory* **51**(8), 2984–2992.

Lampe LHJ and Schober R 2002 Bit-interleaved coded differential space-time modulation. *IEEE Transactions on Communications* **50**(9), 1429–1439.

Larsson EG and Stoica P 2003 *Space-Time Block Coding for Wireless Communications*. Cambridge University Press.

Laurila J, Kalliola K, Toeltsch M, Hugl K, Vainikainen P and Bonek E 2002 Wide-band 3-D characterization of mobile radio channels in urban environment. *IEEE Transactions on Antennas and Propagation* **50**(2), 233–243.

Lee H, Lee B and Lee I 2006 Iterative detection and decoding with an improved V-BLAST for MIMO-OFDM systems. *IEEE Journal on Selected Areas in Communications* **24**(3), 504–513.

Lee KF and Williams DB 2000 A space-frequency transmitter diversity technique for OFDM systems. *Proceedings of IEEE Global Communications Conference (GLOBECOM)* pp. 1473–1477.

Lehne PH and Pettersen M 1999 An overview of smart antenna technology for mobile communication systems. *IEEE Communication Surveys and Tutorials* **2**(4), 2–13.

Li G, Fair IJ and Krzymien WA 2006 Low-density parity-check codes for space-time wireless transmission. *IEEE Transactions on Wireless Communications* **5**(2), 312–322.

Li H 2005 Differential space-time modulation over frequency-selective channels. *IEEE Transactions on Signal Processing* **53**(6), 2228–2242.

Li W and Beaulieu NC 2006 Effects of channel-estimation errors on receiver selection-combining schemes for alamouti MIMO systems with BPSK. *IEEE Transactions on Communications* **54**(1), 169–178.

Li Y and Xia XG 2006 Iterative demodulation/decoding methods based on Gaussian approximations for lattice based space-time coded systems. *IEEE Transactions on Wireless Communications* **5**(8), 1976–1983.

Li Y, Guo X and Wang X 2005a Design of recursive convolutional space-time codes with an arbitrary number of transmit antennas. *IEEE Communications Letters* **9**(7), 637–639.

Li Y, Vucetic B, Zhang Q and Huang Y 2005b Assembled space-time turbo trellis codes. *IEEE Transactions on Vehicular Technology* **54**(5), 1768–1772.

Li Z and Hu G 2003 Space-time block codes based coordinate symmetric orthogonal design. *Electronics Letters* **39**(8), 670–671.

Liang XB 2003a A high rate orthogonal space-time block code. *IEEE Communications Letters* **7**(5), 222–223.

Liang XB 2003b Orthogonal designs with maximal rates. *IEEE Transactions on Information Theory* **49**(10), 2468–2503.

Liang XB and Xia XG 2002 Unitary signal constellations for differential space-time modulation with two transmit antennas: parametric codes, optimal designs, and bounds. *IEEE Transactions on Information Theory* **48**(8), 2291–2322.

Liberti Jr. JC and Rappaport TS 1999 *Smart Antennas for Wireless Communications:* IS-95 and Third Generation CDMA Applications. Prentice Hall PTR, Upper Saddle River, NJ.

Liew TH and Hanzo L 2002 Space-time codes and concatenated channel codes for wireless communications. *Proceedings of the IEEE* **90**(2), 187–219.

Lin S and Costello, Jr. DJ 2004 *Error Control Coding*. Pearson Prentice Hall, Upper Saddle River, NJ.

Lin X and Blum RS 2000 Improved space-time codes using serial concatenation. *IEEE Communications Letters* **4**(7), 221–223.

Lin Z, Premkumar AB and Madhukumar BS 2006 Least squares-based receive antenna selection for MIMO spatial multiplexing systems with linear receivers. *IEEE Communications Letters* **10**(4), 254–256.

Litva J and Lo TK 1996 *Digital Beamforming in Wireless Communications*. Artech House Inc., Norwood, MA.

Liu S, and Tian Z 2004 Near-optimum soft decision equalization for frequency selective MIMO channels. *IEEE Transactions on Signal Processing* **52**(3), 721–733.

Liu Y, Fitz MP and Takeshita OY 2002a A rank criterion for QAM space-time codes. *IEEE Transactions on Information Theory* **48**(12), 3062–3079.

Liu Y, Fitz MP and Takeshita OY 2001a Full-rate space-time turbo codes. *IEEE Journal on Selected Areas in Communications* **19**(5), 969–980.

Liu Z, Giannakis GB and Hughes BL 2001b Double differential space-time block coding for time-selective fading channels. *IEEE Transactions on Communications* **49**(9), 1529–1539.

Liu Z, Xin Y and Giannakis GB 2002b Space-time-frequency coded OFDM over frequency-selective fading channels. *IEEE Transactions on Signal Processing* **50**(10), 2465–2476.

Loyka S and Gagnon F 2004 Performance analysis of the V-BLAST algorithm: an analytical approach. *IEEE Transactions on Wireless Communications* **3**(4), 1326–1337.

Lozano A and Papadias C 2002 Layered space-time receivers for frequency-selective wireless channels. *IEEE Transactions on Communications* **50**(1), 65–73.

Lozano A, Tulino AM and Verdu S 2003 Multiple-antenna capacity in the low power regime. *IEEE Transactions on Information Theory* **49**(10), 2527–2544.

Lu B, Yue G and Wang X 2004 Performance analysis and design optimization of LDPC-coded MIMO OFDM systems. *IEEE Transactions on Signal Processing* **52**(2), 348–361.

Lu K, Fu S and Xia XG 2005 Closed-form design of complex orthogonal space-time block codes of rates (k+1)/(2k) for 2k-1 or 2k transmit antennas. *IEEE Transactions on Information Theory* **51**(12), 4340–4347.

Luo P and Leib H 2005 Full-rate full-diversity space-time code for four-transmit-antenna systems. *IEEE Transactions on Wireless Communications* **4**(5), 1974–1979.

Ma Q and Tepedelenlioglu C 2005 Antenna selection for unitary space-time modulation. *IEEE Transactions on Information Theory* **51**(10), 3620–3631.

Ma Q and Tepedelenlioglu C 2007 Antenna selection for space-time coded systems with imperfect channel estimation. *IEEE Transactions on Wireless Communications* **6**(2), 710–719.

Ma Q, Tepedelenlioglu C and Liu Z 2005 Differential space-time-frequency coded OFDM with maximum multipath diversity. *IEEE Transactions on Wireless Communications* **4**(5), 2232–2243.

Ma X and Giannakis GB 2003 Full-diversity full-rate complex-field space-time coding. *IEEE Transactions on Signal Processing* **51**(11), 2917–2930.

MacKay DJC 1999 Good error correcting codes based on very sparse matrices. *IEEE Transactions on Information Theory* **45**(2), 399–431.

MacKay DJC and Neal RM 1995 Good codes based on very sparse matrices, in C. Boyd, (Ed.), *Cryptography and Coding, 5th IMA Conference, 1995, (Lecture Notes in Computer Science 1025)*, pp. 100–111, Berlin, Germany.

Maddah-Ali MA and Khandani AK 2006 A new non-orthogonal space-time code with low decoding complexity. *IEEE Transactions on Wireless Communications* **5**(5), 1115–1121.

Malkamaki E and Leib H 1999 Coded diversity on block fading channels. *IEEE Transactions on Information Theory* **45**(2), 771–782.

Marzetta TL and Hochwald BM 1999 Capacity of a mobile multiple-antenna communication link in Rayleigh flat fading. *IEEE Transactions on Information Theory* **45**(1), 139–157.

Matache A and Wesel RD 2003 Universal trellis codes for diagonally layered space-time systems. *IEEE Transactions on Signal Processing* **51**(11), 2773–2783.

McAdam PL, Welch LR and Weber CL 1972 M.A.P. bit decoding of convolutional codes *Proceedings of the International Symposium on Information Theory*, Asilomar, CA.

McCloud ML, Brehler M and Varanasi MK 2002 Signal design and convolutional coding for non-coherent space-time communication on the block-Rayleigh-fading channel. *IEEE Transactions on Information Theory* **48**(5), 1186–1194.

McDonald V, Sullivan P, Duman TM, Roy S and Proakis J 2006 Comprehensive MIMO testing in the 2005 Makai experiment *Proceedings of the 8th European Conference in Underwater Acoustics*, pp. 855–862.

McEliece RJ, MacKay DJC and Jung-Fu C 1998 Turbo decoding as an instance of Pearl's 'belief-propagation' algorithm. *IEEE Journal on Selected Areas in Comuunications* **16**(2), 140–152.

Moher M 1998 An iterative multiuser decoder for near-capacity communications. *IEEE Transactions on Communications* **46**(7), 870–880.

Molisch AF 2001 *Wideband Wireless Digital Communications (Edited)*. Prentice Hall, Upper Saddle River, NJ.

Molisch AF 2004 A generic model for MIMO wireless propagation channels in macro- and microcells. *IEEE Transactions on Signal Processing* **52**(1), 61–71.

Molisch AF, Win MZ and Winters JH 2002 Space-time-frequency (STF) coding for MIMO-OFDM systems. *IEEE Communications Letters* **6**(9), 370–372.

Molisch AF, Win MZ, Choi YS and Winters JH 2005 Capacity of MIMO systems with antenna selection. *IEEE Transactions on Wireless Communications* **4**(4), 1759–1772.

Moustakas AL, Simon SH and Sengupta AM 2003 MIMO capacity through correlated channels in the presence of correlated interferers and noise: a (not so) large N analysis. *IEEE Transactions on Information Theory* **49**(10), 2545–2561.

Mun C 2006 Transmit-antenna selection for spatial multiplexing with ordered successive interference cancellation. *IEEE Transactions on Communications* **54**(3), 423–429.

Nguyen VK 2007 A differential space-time modulation scheme for correlated Rayleigh fading channels: Performance analysis and design. *IEEE Transactions on Signal Processing* **55**(1), 299–312.

Papoulis A and Pillai SU 2002 *Probability, Random Variables and Stochastic Processes*. McGraw-Hill, New York.

Paulraj A, Nabar R and Gore D 2003 *Introduction to Space-Time Wireless Communications* 1st edn. Cambridge University Press.

Paulraj AJ, Gore D, Nabar RU and Boelcskei H 2004 An overview of MIMO communications – a key to gigabit wireless. *Proceedings of IEEE* **92**(2), 198–218.

Pearl J 1988 *Probabilistic Reasoning in Intelligent Systems: Network of Plausible Inference*. Morgan Kaumann, San Francisco, CA.

Perez LC, Seghers J and Costello Jr. DJ 1996 A distance spectrum interpretation of turbo codes. *IEEE Transactions on Information Theory* **42**(6), 1698–1709.

Pfister HD, Soriaga JB and Siegel PH 2001 On the achievable information rates of finite-state isi channels *Proceedings of IEEE Global Communications Conference (GLOBECOM)*, pp. 2992–2996, San Antonio, TX.

Piechocki RJ, Fletcher PN, Nix AR, Canagarajah CN and McGeehan JP 2001 Performance evaluation of BLAST-OFDM enhanced Hiperlan/2 using simulated and measured channel data. *Electronics Letters* **37**(18), 1137–1139.

Proakis JG 2001 *Digital Communications*. McGraw-Hill, New York.

Rappaport TS 2001 *Wireless Communications: Principles and Practice* 2nd ed. Prentice Hall, Upper Saddle River, NJ.

Reddy H and Duman TM 2003 Space-time coded OFDM with low PAR *Proceedings of IEEE Global Communications Conference (GLOBECOM)*, vol. 2, pp. 799–803, San Francisco, CA.

Reed MC, Schlegel CB, Alexander PD and Asenstorfer JA 1997 Iterative multiuser detection for DS-CDMA with FEC *Proceedings of the International Symposium on Turbo Codes and Related Topics*, pp. 162–165.

Reial A and Wilson SG 2005 Multistage iterative decoding with complexity reduction for concatenated space-time codes. *IEEE Transactions on Communications* **53**(8), 1300–1309.

Rende D and Wong TF 2005 Bit-interleaved space-frequency coded modulation for OFDM systems. *IEEE Transactions on Wireless Communications* **4**(5), 2256–2266.

Robertson P, Villebrun E and Hoeher P 1995 A comparison of optimal and suboptimal MAP decoding algorithms operating in the log domain *Proceedings of the IEEE International Conference on Communications (ICC)*, vol. 2, pp. 1009–1013.

Ronen S, Bross SI, Shamai (Shitz) S and Duman TM 2007 Iterative channel estimation and decoding in turbo coded space-time systems. *European Transactions on Telecommunications*.

Roy S and Duman TM 2004 Soft input soft output Kalman filter for equalization of MIMO frequency selective fading channels *Proceedings of IEEE Vehicular Technology Fall Conference (VTC–Fall)*, vol. 3, pp. 1753–1757.

Roy S and Duman TM 2007 Soft input soft output Kalman equalizer for MIMO frequency selective fading channels. *IEEE Transactions on Wireless Communications* **6**(2), 506–514.

Roy S, Duman TM, Ghazikhanian L, McDonald V, Proakis J and Zeidler J 2004 Enhanced underwater acoustic communication performance using space-time coding and processing *Proceedings of IEEE Oceans*, vol. 1, pp. 26–33.

Roy S, Duman TM and McDonald VK 2006 Error rate improvement in underwater MIMO communications using sparse partial response equalization *Proceedings of IEEE Oceans*, pp. 1–6.

Roy S, Duman TM, McDonald V and Proakis JG 2007 High rate communication for underwater acoustic channels using multiple transmitters and space-time coding: receiver structures and experimental results. Accepted for publication in *IEEE Journal of Oceanic Engineering*.

Ryan WE 2003 Concatenated Codes and Iterative Decoding, *Wiley Encyclopedia of Telecommunications* (J. G. Proakis, ed.) 1st ed. Wiley and Sons.

Sanayei S and Nosratinia A 2007 Antenna selection in keyhole channels. *IEEE Transactions on Communications* **55**(3), 404–408.

Sanei A 2006 *Antenna selection for space-time trellis codes over Rayleigh fading channels*. Ph.D. Dissertation, Electrical and Computer Engineering Department, Concordia University, Montreal, Canada.

Sanei A, Ghrayeb A and Shayan Y 2006a Antenna selection for space-time trellis codes over block Rayleigh fading channels *Proceedings of IEEE Vehicular Technology Conference (VTC-Fall)*, pp. 1–5.

Sanei A, Ghrayeb A, Shayan Y and Duman TM 2006b On the diversity order of space-time trellis codes with receive antenna selection over fast fading channels. *IEEE Transactions on Wireless Communications* **5**(7), 1579–1585.

Sason I and Shamai S 2000 Improved upper bounds on the ML decoding error probability of parallel and serial concatenated turbo codes via their ensemble distance spectrum. *IEEE Transactions on Information Theory* **46**(1), 24–47.

Schlegel C and Costello, Jr. DJ 1989 Bandwidth efficient coding for fading channels: code construction and performance analysis. *IEEE Journal on Selected Areas in Communications* **7**(9), 1356–1368.

Schlegel C and Grant A 2003 Differential space-time turbo codes. *IEEE Transactions on Information Theory* **49**(9), 2298–2306.

Schober R and Lampe LHJ 2002 Noncoherent receivers for differential space-time modulation. *IEEE Transactions on Communications* **50**(5), 768–777.

Schober R, Gerstacker WH and Lampe LHJ 2004 Performance analysis and design of STBCs for frequency slective fading channels. *IEEE Transactions on Wireless Communications* **3**(3), 734–744.

Sellathurai M and Haykin S 2002 Turbo-BLAST for wireless communications: Theory and experiments. *IEEE Transactions on Signal Processing* **50**(10), 2538–2546.

Sethuraman BA, Rajan BS and Shashidhar V 2003 Full-diversity, high-rate space-time block codes from division algebras. *IEEE Transactions on Information Theory* **49**(10), 2596–2616.

Sezgin A and Jorswieck EA 2005 Capacity achieving high rate space-time block codes. *IEEE Communications Letters* **9**(5), 435–437.

Shannon CE 1948 A mathematical theory of communication. *Bell System Technical Journal* **27**, 379–423 and 623–656.

Shao L and Roy S 2005 Rate-one space-frequency block codes with maximum diversity for MIMO-OFDM. *IEEE Transactions on Wireless Communications* **4**(4), 1674–1687.

Sharma N and Papadias CB 2003 Improved quasi-orthogonal codes through constellation rotation. *IEEE Transactions on Communications* **51**(3), 332–335.

Sharma N and Papadias CB 2004 Full-rate full-diversity linear quasi-orthogonal space-time codes for any number of transmit antennas. *EURASIP J. Applied Signal Processing* **2004**(9), 1246–1256.

Shen H and Ghrayeb A 2005 Analysis of the outage probability for spatially correlated MIMO channels with receive antenna selection *Proceedings of IEEE Global Communications Conference (GLOBECOM)*, pp. 2460–2464.

Shen H and Ghrayeb A 2006 Analysis of the outage probability for MIMO systems with receive antenna selection. *IEEE Transactions on Vehicular Technology* **55**(4), 1435–1440.

Shin H and Lee JH 2002 Exact symbol error probability of orthogonal space-time block codes *Proceedings of IEEE Global Communications Conference (GLOBECOM)*, vol. 2, pp. 1197–1201.

Shin H and Lee JH 2003 Capacity of multiple-antenna fading channels: Spatial fading correlation, double scattering, and keyhole. *IEEE Transactions on Information Theory* **49**(10), 2636–2647.

Shiu DS, Foschini GJ, Gans MJ and Kahn JM 2000 Fading correlation and its effect on capacity of multielement antenna systems. *IEEE Transactions on Communications* **48**(3), 502–513.

Simon MK and Alouini MS 2004 *Digital Communication over Fading Channels* 2nd edn. Wiley, New York.

Siwamogsatham S and Fitz MP 2002 Robust space-time codes for correlated Rayleigh fading channels. *IEEE Transactions on Signal Processing* **50**(10), 2408–2416.

Siwamogsatham S and Fitz MP 2005 High-rate concatenated space-time block code M-TCM designs. *IEEE Transactions on Information Theory* **51**(12), 4173–4183.

Siwamogsatham S, Fitz MP and Grimm JH 2002 A new view performance analysis of transmit diversity schemes in correlated Rayleigh fading. *IEEE Transactions on Information Theory* **48**(4), 950–956.

Smith PJ, Roy S and Shafi M 2003 Capacity of MIMO systems with semicorrelated flat fading. *IEEE Transactions on Information Theory* **49**(10), 2781–2788.

So DKC and Cheng RS 2006 Layered maximum likelihood detection for MIMO systems in frequency selective fading channels. *IEEE Transactions on Wireless Communications* **5**(4), 752–762.

Sohn I 2003 Space frequency block coded turbo-BLAST detection for MIMO-OFDM systems. *Electronics Letters* **39**(21), 1557–1558.

Song HK, Kang SJ, Kim MJ and You YH 2004 Error performance analysis of STBC-OFDM systems with parameter imbalances. *IEEE Transaction on Broadcasting* **50**(1), 76–82.

Stefanov A 2001 *Space-Time Turbo Coding for High Spectral Efficiency Wireless Communications*. Ph.D. Dissertation, Electrical Engineering Department, Arizona State University, Tempe, AZ.

Stefanov A and Duman TM 2000 Turbo coded modulation as an alternative to space-time codes: The case of large number of antennas *Proceedings of SPIE, Digital Wireless Communications II*, pp. 54–61, Orlando, FL.

Stefanov A and Duman TM 2001a Performance analysis of space-time trellis codes over quasi-static fading channels *Proceedings of SPIE, Modeling and Design of Wireless Networks*, pp. 191–203, Denver, CO.

Stefanov A and Duman TM 2001b Performance bounds for space-time trellis codes *Proceedings of IEEE International Symposium on Information Theory (ISIT)*, p. 82, Washington DC.

Stefanov A and Duman TM 2001c Turbo coded modulation for systems with transmit and receive antenna diversity over block fading channels: System model, decoding approaches and practical considerations. *IEEE Journal on Selected Areas in Communications* **19**(5), 958–968.

Stefanov A and Duman TM 2002 Performance bounds for turbo coded multiple antenna systems *Proceedings of IEEE International Symposium on Information Theory (ISIT)*, p. 302, Lausanne, Switzerland.

Stefanov A and Duman TM 2003a Performance bounds for space-time trellis codes. *IEEE Transactions on Information Theory* **49**(9), 2134–2140.

Stefanov A and Duman TM 2003b Performance bounds for turbo coded multiple antenna systems. *IEEE Journal on Selected Areas in Communications* **21**(3), 374–381.

Stridh R, Ottersten B and Karlsson P 2000 MIMO channel capacity of a measured indoor radio channelat 5.8 GHz *Proc. Asilomar Conf. Signals, Systems and Computers*, **1**(1), pp. 733–737, Pacific Grove, CA.

Stuber G, Barry J, McLaughlin SW, Li Y, Ingram MA and Pratt TG 2004 Broadband MIMO-OFDM wireless communications. *Proceedings of IEEE* **92**(2), 271–294.

Stuber GL 2001 *Principles of Mobile Communication* 2nd edn. Springer, New York.

Su W and Xia XG 2003 Two generalized complex orthogonal space-time block codes of rates 7/11 and 3/5 for 5 and 6 transmit antennas. *IEEE Transactions on Information Theory* **49**(1), 313–316.

Su W and Xia XG 2004 Signal constellations for quasi-orthogonal space-time block codes with full diversity. *IEEE Transactions on Information Theory* **50**(10), 2331–2347.

Su W, Safar Z and Liu KJR 2005 Full-rate full-diversity space-frequency codes with optimum coding advantage. *IEEE Transactions on Information Theory* **51**(1), 229–249.

Sudarshan P, Mehta NB, Molisch AF and Zhang J 2004 Antenna selection with RF pre-processing: robustness to RF and selection non-idealities *Proceedings of the IEEE Radio and Wireless Conference (RAWCON)*, pp. 391–394, Atlanta, GA.

Sudarshan P, Mehta NB, Molisch AF and Zhang J 2006 Channel statistics-based RF pre-processing with antenna selection. *IEEE Transactions on Wireless Communications* **5**(12), 3501–3511.

Swindlehurst AL, German G, Wallace J and Jensen M 2001 Experimental measurements of capacity for MIMO indoor wireless channels *Proc. IEEE Third Workshop on Signal Processing Advances in Wireless Communications*, pp. 30–33.

Tan M, Latinovic Z and Bar-Ness Y 2005 STBC MIMO-OFDM peak-to-average power ratio reduction by cross-antenna rotation and inversion. *IEEE Communications Letters* **9**(7), 592–594.

Tanner RM 1981 A recursive approach to low complexity codes. *IEEE Transactions on Information Theory* **27**(5), 533–547.

Tao M and Cheng RS 2001 Improved design criteria and new trellis codes for space-time coded modulation in slow fading channels. *IEEE Communications Letters* **5**(7), 313–315.

Tao M and Cheng RS 2003 Trellis-coded differential unitary space-time modulation over flat fading channels. *IEEE Transactions on Communications* **51**(4), 587–596.

Tao M and Cheng RS 2004 Generalized layered space-time codes for high data rate wireless communications. *IEEE Transactions on Wireless Communications* **3**(4), 1067–1075.

Tarasak P and Bhargava VK 2004 Analysis and design criteria for trellis-coded modulation with differential space-time transmit diversity. *IEEE Transactions on Wireless Communications* **3**(5), 1374–1378.

Tarokh V and Jafarkhani H 2000 A differential detection scheme for transmit diversity. *IEEE Journal on Selected Areas in Communications* **18**(7), 1169–1174.

Tarokh V, Jafarkhani H and Calderbank AR 1999 Space-time block coding for wireless communications: performance results. *IEEE Journal on Selected Areas in Communications* **17**(3), 451–460.

Tarokh V, Naguib A, Seshadri N and Calderbank AR 1999 Combined array processing and space-time coding. *IEEE Transactions on Information Theory* **45**(4), 1121–1128.

Tarokh V, Naguib AF, Seshadri N and Calderbank AR 1997 Space-time codes for high data rate wireless communication: mismatch analysis *Proceedings of IEEE International Conference on Communications (ICC)*, vol. 1, pp. 309–313, Montreal, Canada.

Tarokh V, Seshadri N and Calderbank AR 1998 Space-time codes for high data rate wireless communication: Performance criterion and code construction. *IEEE Transactions on Information Theory* **44**(2), pp. 744–765.

Telatar IE 1999 Capacity of multi-antenna Gaussian channels. *European Transactions on Telecommunications* **10**(6), 585–595.

Tellambura C 1996 Evaluation of the exact union bound for trellis-coded modulation over fading channels. *IEEE Transactions on Communications* **44**(12), 1693–1699.

Thoen S, Van der Perre L, Gyselinckx B and Engels M 2001 Performance analysis of combined transmit-SC/receive-MRC. *IEEE Transactions on Communications* **49**(1), 5–8.

Tila F, Shepherd PR and Pennock SR 2003 Theoretical capacity evaluation of indoor micro- and macro-MIMO systems at 5 GHz using site specific ray tracing. *Electronics Letters* **39**(5), 471–472.

Tran LC, Seberry J, Wang Y, Wysocki BJ, Wysocki TA, Xia T and Zhao Y 2004 Two complex orthogonal space-time codes for eight transmit antennas. *Electronics Letters* **40**(1), 55–57.

Tse D and Viswanath P 2005 *Fundamentals of Wireless Communication*. Cambridge University Press, New York.

TSG-R1(04)0336 G 2004 : Double TxAA for MIMO.

Tuchler M, Singer A and Koetter R 2002 Minimum mean squared error equalization using a priori information. *IEEE Transactions on Signal Processing* **50**(3), 673–683.

Tujkovic D, Juntti M and Latva-Aho M 2001 Space-frequency-time turbo coded modulation. *IEEE Communications Letters* **5**(12), 480–482.

Ungerboeck G 1982 Channel coding with multilevel/phase signals. *IEEE Transactions on Information Theory* **28**(1), 55–67.

Ungerboeck G 1987a Trellis-coded modulation with redundant signal sets – part I: introduction. *IEEE Communications Magazine* **25**(2), 5–11.

Ungerboeck G 1987b Trellis-coded modulation with redundant signal sets – part II: state of art. *IEEE Communications Magazine* **25**(2), 12–21.

Uysal M and Georghiades CN 2000 Error performance analysis of space-time trellis codes over Rayleigh fading channels. *Journal of Communications and Networks (JCN)* **2**(4), 351–355.

Uysal M and Georghiades CN 2004 On the error performance analysis of space-time trellis codes. *IEEE Transactions on Wireless Communications* **3**(4), 1118–1123.

Venkataraman A, Reddy H and Duman TM 2006 Space-time coded OFDM with low PAPR. *Eurasip Journal on Applied Signal Processing* **2006**(1), 9 pages.

Verdu S 1987 Maximum likelihood sequence detection for intersymbol interference channels: a new upper bound on error probability. *IEEE Transactions on Information Theory* **33**(1), 62–68.

Vielmon A, Ye L and Barry JR 2004 Performance of Alamouti transmit diversity over time-varying Rayleigh-fading channels. *IEEE Transactions on Wireless Communications* **3**(5), 1369–1375.

Vikalo H and Hassibi B 2002 Maximum-likelihood sequence detection of multiple antenna systems over dispersive channels via sphere decoding. *EURASIP Journal on Applied Signal Processing* **2002**(5), 525–531.

Vikalo H, Hassibi B and Kailath T 2004 Iterative decoding for MIMO channels via modified sphere decoding. *IEEE Transactions on Wireless Communications* **3**(6), 2299–2311.

Viterbo E and Boutros J 1999 A universal lattice decoder for fading channels. *IEEE Transactions on Information Theory* **45**(5), 1639–1642.

Voulgarelis A, Joham M and Utschick W 2003 Space-time equalization based on V-BLAST and DFE for frequency-selective MIMO channels. *Proc. IEEE Int. Conf. Acoustics, Speech and Signal Processing, 4(4), Hong Kong* pp. 381–384.

Vucetic B and Yuan J 2003 *Space-Time Coding*. John Wiley and Sons.

Wang J, Simon MK, Fitz MP and Yao K 2004a On the performance of space-time codes over spatially correlated Rayleigh fading channels. *IEEE Transactions on Communications* **52**(6), 877–881.

Wang X and Poor HV 1999 Iterative (turbo) soft interference cancellation and decoding for coded CDMA. *IEEE Transactions on Communications* **47**(7), 1046–1061.

Wang Y, Omrani R, Chugg KM and Kumar PV 2004b Low-density parity check space-time codes: performance analysis and code construction. *Proceedings of IEEE International Symposium on Information Theory, Chicago*, p. 156.

Wang Z and Giannakis GB 2004 Outage mutual information of space-time MIMO systems. *IEEE Transactions on Information Theory* **50**(4), 657–662.

Wicker SB 1995 *Error Control Systems for Digital Communication and Storage*. Prentice Hall, NJ.

Wilson SG and Leung YS 1987 Trellis-coded phase modulation on Rayleigh fading channels *Proceedings of IEEE International Conference on Communications (ICC)*, pp. 2131–2135.

Winters JH 1998 Smart antennas for wireless systems. *IEEE Personal Communication* **5**(1), 23–27.

Wu Y, Lau VKN and Patzold M 2006 Constellation design for trellis-coded unitary spacetime modulation systems. *IEEE Transactions on Communications* **54**(11), 1948–1959.

Xia Y, Wang Z and Giannakis GB 2003 Space-time diversity systems based on linear constellation precoding. *IEEE Transactions on Wireless Communications* **2**(2), 294–309.

Xin Y and Giannakis GB 2002 High-rate space-time layered OFDM. *IEEE Communications Letters* **6**(5), 187–189.

Xu C and Kwak KS 2005 On decoding algorithm and performance of space-time block codes. *IEEE Transactions on Wireless Communications* **4**(3), 825–829.

Yan Q and Blum RS 2002 Improved space-time convolutional codes for quasi-static slow fading channels. *IEEE Transactions on Wireless Communications* **1**(4), 563–571.

Yang H 2005 A road to future broadband wireless access: MIMO-OFDM-based air interface. *IEEE Communications Magazine* **43**(1), 53–60.

Younkins LT, Su W and Liu KJR 2004 On the robustness of space-time coding for spatially and temporally correlated wireless channels *IEEE Wireless Commununication and Networking Conference*, **1**(1) pp. 587–592, Atlanta, GA.

Zehavi E 1992 8–PSK trellis codes for a Rayleigh channel. *IEEE Transactions on Communications* **40**(5), 873–884.

Zeng X and Ghrayeb A 2004a Antenna selection for space-time block codes over correlated Rayleigh fading channels. *IEEE Canadian Journal of Electrical and Computer Engineering* **29**(4), 219–226.

Zeng X and Ghrayeb A 2004b Performance bounds for space-time block codes with receive antenna selection. *IEEE Transactions on Information Theory* **50**(9), 2130–2137.

Zeng X and Ghrayeb A 2006 Performance bounds for combined channel coding and space-time block coding with receive antenna selection. *IEEE Transactions on Vehicular Technology* **55**(4), 1441–1446.

Zeng X and Ghrayeb A 2007 A blind carrier frequency offset estimation scheme for OFDM systems with constant modulus signaling *Proceedings of IEEE Workshop on Signal Processing Advances in Wireless Communications (SPAWC).*

Zhang H, Dai H, Zhou Q and Hughes BL 2006 On the diversity order of spatial multiplexing systems with transmit antenna selection: A geometrical approach. *IEEE Transactions on Information Theory* **52**(12), 5297–5311.

Zhang W, Xia XG and Ching PC 2007 High-rate full-diversity space-time-frequency codes for broadband MIMO block-fading channels. *IEEE Transactions on Communications* **55**(1), 25–34.

Zhang Z, Duman TM and Kurtas EM 2004 Achievable information rates and coding for MIMO systems over ISI channels and frequency-selective fading channels. *IEEE Transactions on Communications* **52**(10), 1698–1710.

Zhao W and Giannakis GB 2005 Sphere decoding algorithms with improved radius search. *IEEE Transactions on Communications* **53**(7), 1104–1109.

Zhao W, Leus G and Giannakis GB 2004 Orthogonal design of unitary constellations for uncoded and trellis-coded noncoherent space-time systems. *IEEE Transactions on Information Theory* **50**(6), 1319–1327.

Zheng L and Tse DNC 2002 Communication on the Grassmann manifold: A geometric approach to the noncoherent multiple-antenna channel. *IEEE Transactions on Information Theory* **48**(2), 359–383.

Zheng L and Tse DNC 2003 Diversity and multiplexing: A fundamental tradeoff in multiple antenna channels. *IEEE Transactions on Information Theory* **49**(5), 1073–1096.

Zhou Q and Dai H 2006 Joint antenna selection and link adaptation for MIMO systems. *IEEE Transactions on Vehicular Technology* **55**(1), 243–255.

Zhou S and Giannakis GB 2001 Space-time coding with maximum diversity gains over frequency-selective fading channels. *IEEE Signal Processing Letters* **8**(10), 269–272.

Zhou S and Giannakis GB 2003 Single-carrier space-time block coded transmission over frequency selective fading channels. *IEEE Transactions on Information Theory* **49**(1), 164–179.

Zhu H, Lei Z and Chin FPS 2004 An improved square-root algorithm for BLAST. *IEEE Signal Processing Letters* **11**(9), 772–775.

Zhu X and Murch RD 2003 Layered space-time equalization for wireless MIMO systems. *IEEE Transactions on Wireless Communications* **2**(6), 1189–1203.

Index

Coding for MIMO Communication Systems Tolga M. Duman and Ali Ghrayeb

Printed and bound by CPI Group (UK) Ltd, Croydon, CR0 4YY

05/04/2023

03208635-0001